T0137456

Lecture Notes in Networks and Systems

Volume 19

Series editor

Janusz Kacprzyk, Polish Academy of Sciences, Warsaw, Poland
e-mail: kacprzyk@ibspan.waw.pl

The series "Lecture Notes in Networks and Systems" publishes the latest developments in Networks and Systems—quickly, informally and with high quality. Original research reported in proceedings and post-proceedings represents the core of LNNS.

Volumes published in LNNS embrace all aspects and subfields of, as well as new challenges in, Networks and Systems.

The series contains proceedings and edited volumes in systems and networks, spanning the areas of Cyber-Physical Systems, Autonomous Systems, Sensor Networks, Control Systems, Energy Systems, Automotive Systems, Biological Systems, Vehicular Networking and Connected Vehicles, Aerospace Systems, Automation, Manufacturing, Smart Grids, Nonlinear Systems, Power Systems, Robotics, Social Systems, Economic Systems and other. Of particular value to both the contributors and the readership are the short publication timeframe and the world-wide distribution and exposure which enable both a wide and rapid dissemination of research output.

The series covers the theory, applications, and perspectives on the state of the art and future developments relevant to systems and networks, decision making, control, complex processes and related areas, as embedded in the fields of interdisciplinary and applied sciences, engineering, computer science, physics, economics, social, and life sciences, as well as the paradigms and methodologies behind them.

Advisory Board

Fernando Gomide, Department of Computer Engineering and Automation—DCA, School of Electrical and Computer Engineering—FEEC, University of Campinas—UNICAMP, São Paulo, Brazil
e-mail: gomide@dca.fee.unicamp.br
Okyay Kaynak, Department of Electrical and Electronic Engineering, Bogazici University, Istanbul, Turkey
e-mail: okyay.kaynak@boun.edu.tr
Derong Liu, Department of Electrical and Computer Engineering, University of Illinois at Chicago, Chicago, USA and
Institute of Automation, Chinese Academy of Sciences, Beijing, China
e-mail: derong@uic.edu
Witold Pedrycz, Department of Electrical and Computer Engineering, University of Alberta, Alberta, Canada and
Systems Research Institute, Polish Academy of Sciences, Warsaw, Poland
e-mail: wpedrycz@ualberta.ca
Marios M. Polycarpou, KIOS Research Center for Intelligent Systems and
Networks, Department of Electrical and Computer Engineering, University of Cyprus, Nicosia, Cyprus
e-mail: mpolycar@ucy.ac.cy
Imre J. Rudas, Óbuda University, Budapest Hungary
e-mail: rudas@uni-obuda.hu
Jun Wang, Department of Computer Science, City University of Hong Kong, Kowloon, Hong Kong
e-mail: jwang.cs@cityu.edu.hk

More information about this series at http://www.springer.com/series/15179

Yu-Chen Hu · Shailesh Tiwari
Krishn K. Mishra · Munesh C. Trivedi
Editors

Intelligent Communication and Computational Technologies

Proceedings of Internet of Things
for Technological Development,
IoT4TD 2017

 Springer

Editors
Yu-Chen Hu
Department of Computer Science
  and Information Management
Providence University
Taichung City
Taiwan

Shailesh Tiwari
CSED
ABES Engineering College
Ghaziabad, Uttar Pradesh
India

Krishn K. Mishra
Department of Computer Science
  and Engineering
Motilal Nehru National Institute
  of Technology Allahabad
Allahabad, Uttar Pradesh
India

Munesh C. Trivedi
Department of Computer Science
  and Engineering
ABES Engineering College
Ghaziabad
India

ISSN 2367-3370   ISSN 2367-3389 (electronic)
Lecture Notes in Networks and Systems
ISBN 978-981-13-5417-5   ISBN 978-981-10-5523-2 (eBook)
https://doi.org/10.1007/978-981-10-5523-2

This book was advertised with a copyright holder name of the "editor(s)/author(s)" in error, whereas the publisher holds the copyright.
© Springer Nature Singapore Pte Ltd. 2018
Softcover re-print of the Hardcover 1st edition 2018
This work is subject to copyright. All rights are reserved by the Publisher, whether the whole or part of the material is concerned, specifically the rights of translation, reprinting, reuse of illustrations, recitation, broadcasting, reproduction on microfilms or in any other physical way, and transmission or information storage and retrieval, electronic adaptation, computer software, or by similar or dissimilar methodology now known or hereafter developed.
The use of general descriptive names, registered names, trademarks, service marks, etc. in this publication does not imply, even in the absence of a specific statement, that such names are exempt from the relevant protective laws and regulations and therefore free for general use.
The publisher, the authors and the editors are safe to assume that the advice and information in this book are believed to be true and accurate at the date of publication. Neither the publisher nor the authors or the editors give a warranty, express or implied, with respect to the material contained herein or for any errors or omissions that may have been made. The publisher remains neutral with regard to jurisdictional claims in published maps and institutional affiliations.

Printed on acid-free paper

This Springer imprint is published by Springer Nature
The registered company is Springer Nature Singapore Pte Ltd.
The registered company address is: 152 Beach Road, #21-01/04 Gateway East, Singapore 189721, Singapore

Preface

The International Conference on Internet of Things for Technological Development (IoT4TD 2017) has been held at Gandhinagar, India, during 1–2 April, 2017. The IoT4TD 2017 has been organized and supported by the **Kadi Sarva Vishvavidyalaya, Gandhinagar, Gujarat, India**.

The IoT4TD 2017 is an international forum for researchers, developers, students, and academicians to explore cutting-edge ideas and results for the problems involved in the emerging area of Internet of Things to disseminate and share novel research solutions to real-life problems that fulfill the needs of heterogeneous applications and environments, as well to identify new issues and directions for future research and development. IoT4TD also provides an international delivery platform for the universities, colleges, engineering field experts, and professionals.

Nowadays, IoT is an increasingly growing area that advances with speedy pace. IoT believes in the rule that '*Anything can be connected, will be connected*'. Indeed there is a lot of thrust when we need to operate any device through Internet. Keeping this ideology in preference, **Kadi Sarva Vishvavidyalaya, Gandhinagar, Gujarat, India** has come up with an event—IoT4TD 2017. IoT4TD 2017 has a foreseen objective of enhancing the research activities at a large scale. Technical Program Committee and Advisory Board of IoT4TD 2017 include eminent academicians, researchers, and practitioners from abroad as well as from all over the nation.

Gandhinagar is the capital city of Gujarat that got its name from the Father of the Nation, India, Mahatma Gandhi. Gandhinagar city was established around 1960, located on the banks of the River Sabarmati, having the administrative center of Gandhinagar and Gujarat as well; it is also known as knowledge hub as several educational institutions of national and international repute are situated at Gandhinagar. Gandhinagar city is also known because of popular tourist places like 'Akshardham' and 'Adalaj Stepwell'.

Kadi Sarva Vishvavidyalaya, which was recognized and established in 2009 by Sarva Vidyalaya Kelavani Mandal, a trust with more than 90 years of philanthropic existence. The University is now all set to provide need-based education and develop courses of contemporary relevance. KSV is committed to provide research-based activities which would foster higher economic growth.

IoT4TD 2017 received 212 submissions from around 453 authors of seven different countries such as USA, Afghanistan, Australia, Namibia, Sri Lanka, etc. Each submission has been checked by anti-plagiarism software and was rigorously reviewed by at least two reviewers with an average of 3.07 per reviewer. Even some submissions have more than two reviews. On the basis of these reviews, 35 high-quality papers were selected for publication in this volume, with an acceptance rate of 16.5%.

We are thankful to the speakers, Prof. Rana Singh, University of Jazeera, Dubai, UAE; Dr. T.V. Vijay Kumar, JNU Delhi; delegates and the authors for their participation and their interest in IoT4TD as a platform to share their ideas and innovation. We are also thankful to the Prof. Dr. Janusz Kacprzyk, Series Editor, LNNS, Springer and Mr. Aninda Bose, Senior Editor, Hard Sciences, Springer, India for providing continuous guidance and support. Also, we extend our heartfelt gratitude and thanks to the reviewers and Technical Program Committee Members for showing their concern and efforts in the review process. We are indeed thankful to everyone directly or indirectly associated with the conference organizing team leading it towards the success.

Taichung City, Taiwan   Yu-Chen Hu
Ghaziabad, India   Shailesh Tiwari
Allahabad, India   Krishn K. Mishra
Ghaziabad, India   Munesh C. Trivedi

Chairs

Patrons:
Mr. Vallabh Patel, President, KSV University, Gandhinagar, India
Mr. K.D. Patel, President, Sarva Vidhyalaya Khelvani Mandal, Gandhinagar, India

General Chair:
Prof. D.T. Kapadia, Director, KSV University, Gandhinagar, India

Program Chairs:
Prof. N.N. Jani, KSV-IIIC, KSV University, Gandhinagar, India

Convener:
Dr. Gargi Rajapara (Principal, LDR.P-ITR), KSV University, Gandhinagar, India

Technical Program Committee Chairs:
Prof. Yu-Chen Hu, Professor, Department of Computer Science & Information Management, Providence University, 200, Section 7, Taiwan Boulevard, Shalu District, Taichung City 43301, Taiwan
Dr. Kavi Kumar Khedo, Department of Computer Science and Engineering, Faculty of Engineering, University of Mauritius

Publication Chairs:
Prof. Megha Bhatt, LDR.P-ITR, KSV University, Gandhinagar, India
Prof. J.D. Raol, LDR.P-ITR, KSV University, Gandhinagar, India

Publicity Chairs:
Prof. Maulik Pandya, LDR.P-ITR, KSV University, Gandhinagar, India
Prof. Zalak Shah, LDR.P-ITR, KSV University, Gandhinagar, India

Tutorial Chairs:
Prof. A.R. Patel, LDR.P-ITR, KSV University, Gandhinagar, India
Prof. Nirvesh Mehta, LDR.P-ITR, KSV University, Gandhinagar, India

Organizing Committee:
Dr. H.B. Patel, CE/IT, LDRP-ITR, KSV University, Gandhinagar, India
Dr. S.M. Shah, KSV University, Gandhinagar, India
Dr. Bhadresh Pandya, KSV University, Gandhinagar, India
Prof. Bhavesh Kataria, LDRP-ITR, KSV University, Gandhinagar, India
Prof. J.V. Dave, LDRP-ITR, KSV University, Gandhinagar, India
Prof. Ankita Parikh, LDRP-ITR, KSV University, Gandhinagar, India
Prof. Maulik Modi/Prof Parth Patel, LDRP-ITR, KSV University, Gandhinagar,
India
Dr. Sanjay Vyas, LDRP-ITR, KSV University, Gandhinagar, India
Prof. Sunil Joshi, KSV University, Gandhinagar, India
Prof. Utkarsh Patel, KSV University, Gandhinagar, India

Steering Committee

1. Prof. R. Chandramouli (Mouli), Founder Chair IEEE COMSOC, Stevens
   Institute of Technology, Hoboken, NJ 07030, USA
2. Sanjiv K. Bhatia, Professor, University of Missouri, St. Louis, USA
3. Prof. David Barber, University College London, London
4. Prof. Tariq S. Durrani, University of Strathclyde, Glasgow, Scotland, UK
5. Prof. Khedo Kavi Kumar, University of Mauritius, Mauritius

Technical Program Committee

Prof. Ajay Gupta, Western Michigan University, USA
Prof. Babita Gupta, California State University, USA
Prof. Amit K.R. Chowdhury, University of California, USA
Prof. David M. Harvey, G.E.R.I., UK
Prof. Madjid Merabti, Liverpool John Moores University, UK
Dr. Nesimi Ertugrual, University of Adelaide, Australia
Prof. Ian L. Freeston, University of Sheffield, UK
Prof. Witold Kinsner, University of Manitova, Canada
Prof. Anup Kumar, M.I.N.D.S., University of Louisville, USA
Prof. Prabhat Kumar Mahanti, University of New Brunswick, Canada
Prof. Ashok De, Director, NIT Patna, India
Prof. Kuldip Singh, IIT Roorkee, India
Prof. A.K. Tiwari, IIT, BHU, Varanasi, India
Mr. Suryabhan, ACERC, Ajmer, India
Dr. Vivek Singh, BHU, India
Prof. Abdul Quaiyum Ansari, Jamia Millia Islamia, New Delhi, India
Prof. Aditya Trivedi, ABV-IIITM Gwalior, India
Prof. Ajay Kakkar, Thapar University, Patiala, India

Prof. Bharat Bhaskar, IIM Lucknow, India
Prof. Edward David Moreno, Federal University of Sergipe, Brazil
Prof. Evangelos Kranakis, Carleton University
Prof. Filipe Miguel Lopes Meneses, University of Minho, Portugal
Prof. Giovanni Manassero Junior, Universidade de São Paulo, Brazil
Prof. Gregorio Martinez, University of Murcia, Spain
Prof. Pabitra Mitra, Indian Institute of Technology Kharagpur, India
Prof. Joberto Martins, Salvador University-UNIFACS, Brazil
Prof. K. Mustafa, Jamia Millia Islamia, New Delhi, India
Prof. M.M. Sufyan Beg, Jamia Millia Islamia, New Delhi, India
Prof. Jitendra Agrawal, Rajiv Gandhi Proudyogiki Vishwavidyalaya, Bhopal, MP, India
Prof. Rajesh Baliram Ingle, PICT, University of Pune, India
Prof. Romulo Alexander Ellery de Alencar, University of Fortaliza, Brazil
Prof. Youssef Fakhri, Université Ibn Tofail, Faculté des Sciences, Brazil
Dr. Abanish Singh, Bioinformatics Scientist, USA
Dr. Abbas Cheddad, (UCMM), Umeå Universitet, Umeå, Sweden
Dr. Abraham T. Mathew, NIT, Calicut, Kerala, India
Dr. Adam Scmidit, Poznan University of Technology, Poland
Dr. Agostinho L.S. Castro, Federal University of Para, Brazil,
Prof. Goo-Rak Kwon, Chosun University, Republic of Korea
Dr. Alberto Yúfera, Instituto de Microelectrónica de Sevilla (IMSE), (CNM), Spain
Dr. Adam Scmidit, Poznan University of Technology, Poland
Prof. Nishant Doshi, S.V. National Institute of Technology, Surat, India
Prof. Gautam Sanyal, NIT Durgapur, India
Dr. Alok Chakrabarty, IIIT Bhubaneswar, India
Dr. Anastasios Tefas, Aristotle University of Thessaloniki
Dr. Anirban Sarkar, NIT Durgapur, India
Dr. Anjali Sardana, IIIT Roorkee, Uttarakhand, India
Dr. Ariffin Abdul Mutalib, Universiti Utara Malaysia
Dr. Ashok Kumar Das, IIIT Hyderabad
Dr. Ashutosh Saxena, Infosys Technologies Ltd., India
Dr. Balasubramanian Raman, IIT Roorkee, India
Dr. Benahmed Khelifa, Liverpool John Moores University, UK
Dr. Björn Schuller, Technical University of Munich, Germany
Dr. Carole Bassil, Lebanese University, Lebanon
Dr. Chao MA, Hong Kong Polytechnic University
Dr. Chi-Un Lei, University of Hong Kong
Dr. Ching-Hao Lai, Institute for Information Industry
Dr. Ching-Hao Mao, Institute for Information Industry, Taiwan
Dr. Chung-Hua Chu, National Taichung Institute of Technology, Taiwan
Dr. Chunye Gong, National University of Defense Technology
Dr. Cristina Olaverri Monreal, Instituto de Telecomunicacoes, Portugal

Dr. Chittaranjan Hota, BITS Hyderabad, India
Dr. D. Juan Carlos González Moreno, University of Vigo
Dr. Danda B. Rawat, Old Dominion University
Dr. Davide Ariu, University of Cagliari, Italy
Dr. Dimiter G. Velev, University of National and World Economy, Europe
Dr. D.S. Yadav, South Asian University, New Delhi
Dr. Darius M. Dziuda, Central Connecticut State University
Dr. Dimitrios Koukopoulos, University of Western Greece, Greece
Dr. Durga Prasad Mohapatra, NIT Rourkela, India
Dr. Eric Renault, Institut Telecom, France
Dr. Felipe Rudge Barbosa, University of Campinas, Brasil
Dr. Fermín Galán Márquez, Telefónica I+D, Spain
Dr. Fernando Zacarias Flores, Autonomous University of Puebla
Dr. Fuu-Cheng Jiang, Tunghai University, Taiwan
Prof. Aniello Castiglione, University of Salerno, Italy
Dr. Geng Yang, NUPT, Nanjing, People's Republic of China
Dr. Gadadhar Sahoo, BIT-Mesra, India
Prof. Ashokk Das, International Institute of Information Technology, Hyderabad, India
Dr. Gang Wang, Hefei University of Technology
Dr. Gerard Damm, Alcatel-Lucent
Prof. Liang Gu, Yale University, New Haven, CT, USA
Prof. K.K Pattanaik, ABV-Indian Institute of Information Technology and Management, Gwalior, India
Dr. Germano Lambert-Torres, Itajuba Federal University
Dr. Guang Jin, Intelligent Automation, Inc
Dr. Hardi Hungar, Carl von Ossietzky University Oldenburg, Germany
Dr. Hongbo Zhou, Southern Illinois University Carbondale
Dr. Huei-Ru Tseng, Industrial Technology Research Institute, Taiwan
Dr. Hussein Attia, University of Waterloo, Canada
Prof. Hong-Jie Dai, Taipei Medical University, Taiwan
Prof. Edward David, UFS—Federal University of Sergipe, Brazil
Dr. Ivan Saraiva Silva, Federal University of Piauí, Brazil
Dr. Luigi Cerulo, University of Sannio, Italy
Dr. J. Emerson Raja, Engineering and Technology of Multimedia University, Malaysia
Dr. J. Satheesh Kumar, Bharathiar University, Coimbatore
Dr. Jacobijn Sandberg, University of Amsterdam
Dr. Jagannath V. Aghav, College of Engineering, Pune, India
Dr. JAUME Mathieu, LIP6 UPMC, France
Dr. Jen-Jee Chen, National University of Tainan
Dr. Jitender Kumar Chhabra, NIT Kurukshetra, India
Dr. John Karamitsos, Tokk Communications, Canada
Dr. Jose M. Alcaraz Calero, University of the West of Scotland, UK
Dr. K.K. Shukla, IT-BHU, India

Dr. K.R. Pardusani, Maulana Azad NIT, Bhopal, India

Dr. Kapil Kumar Gupta, Accenture

Dr. Kuan-Wei Lee, I-Shou University, Taiwan

Dr. Lalit Awasthi, NIT Hamirpur, India

Dr. Maninder Singh, Thapar University, Patiala, India

Dr. Mehul S. Raval, DA-IICT, Gujarat, India

Dr. Michael McGuire, University of Victoria, Canada

Dr. Mohamed Naouai, University Tunis El Manar and University of Strasbourg, Tunisia

Dr. Nasimuddin, Institute for Infocomm Research

Dr. Olga C. Santos, aDeNu Research Group, UNED, Spain

Dr. Pramod Kumar Singh, ABV-IIITM, Gwalior, India

Dr. Prasanta K. Jana, IIT, Dhanbad, India

Dr. Preetam Ghosh, Virginia Commonwealth University, USA

Dr. Rabeb Mizouni, (KUSTAR), Abu Dhabi, UAE

Dr. Rahul Khanna, Intel Corporation, USA

Dr. Rajeev Srivastava, CSE, ITBHU, India

Dr. Rajesh Kumar, MNIT, Jaipur, India

Dr. Rajesh Bodade, Military College of Telecommunication, Mhow, India

Dr. Rajesh Kumar, MNIT, Jaipur, India

Dr. Ranjit Roy, SVNIT, Surat, Gujarat, India

Dr. Robert Koch, Bundeswehr University München, Germany

Dr. Ricardo J. Rodriguez, Nova Southeastern University, USA

Dr. Ruggero Donida Labati, Università degli Studi di Milano, Italy

Dr. Rustem Popa, University "Dunarea de Jos" in Galati, Romania

Dr. Shailesh Ramchandra Sathe, VNIT, Nagpur, India

Dr. Sanjiv K. Bhatia, University of Missouri—St. Louis, USA

Dr. Sanjeev Gupta, DA-IICT, Gujarat, India

Dr. S. Selvakumar, National Institute of Technology, Tamil Nadu, India

Dr. Saurabh Chaudhury, NIT, Silchar, Assam, India

Dr. Shijo. M. Joseph, Kannur University, Kerala

Dr. Sim Hiew Moi, University Technology of Malaysia

Dr. Syed Mohammed Shamsul Islam, The University of Western Australia, Australia

Dr. Trapti Jain, IIT Mandi, India

Dr. Tilak Thakur, PED, Chandigarh, India

Dr. Vikram Goyal, IIIT, Delhi, India

Dr. Vinaya Mahesh Sawant, D.J. Sanghvi College of Engineering, India

Dr. Vanitha Rani Rentapalli, VITS, Andhra Pradesh, India

Dr. Victor Govindaswamy, Texas A&M University-Texarkana, USA

Dr. Victor Hinostroza, Universidad Autónoma de Ciudad Juárez

Dr. Vidyasagar Potdar, Curtin University of Technology, Australia

Dr. Vijaykumar Chakka, DAIICT, Gandhinagar, India

Dr. Yong Wang, School of IS & E, Central South University, China

Dr. Yu Yuan, Samsung Information Systems America—San Jose, CA

Eng. Angelos Lazaris, University of Southern California, USA
Mr. Hrvoje Belani, University of Zagreb, Croatia
Mr. Huan Song, SuperMicro Computer, Inc., San Jose, USA
Mr. K.K. Patnaik, IIITM, Gwalior, India
Dr. S.S. Sarangdevot, Vice Chancellor, JRN Rajasthan Vidyapeeth University, Udaipur
Dr. N.N. Jani, KSV University Gandhinagar, India
Dr. Ashok K. Patel, North Gujrat University, Patan, Gujrat, India
Dr. Awadhesh Gupta, IMS, Ghaziabad, India
Dr. Dilip Sharma, GLA University, Mathura, India
Dr. Li Jiyun, Donghua Univesity, Shanghai, China
Dr. Lingfeng Wang, University of Toledo, USA
Dr. Valentina E. Balas, Aurel Vlaicu University of Arad, Romania
Dr. Vinay Rishiwal, MJP Rohilkhand University, Bareilly, India
Dr. Vishal Bhatnagar, Ambedkar Institute of Technology, New Delhi, India
Dr. Tarun Shrimali, Sun rise Group of Institutions, Udaipur, India
Dr. Atul Patel, CU Shah University, Vadhwan, Gujrat, India
Dr. P.V. Virparia, Sardar Patel University, VV Nagar, India
Dr. D.B. Choksi, Sardar Patel University, VV Nagar, India
Dr. Ashish N. Jani, KSV University Gandhinagar, India
Dr. Sanjay M. Shah, KSV University Gandhinagar, India
Dr. Vijay M. Chavda, KSV University Gandhinagar, India
Dr. B.S. Agarwal, KIT, Kalol, India
Dr. Apurv Desai, South Gujrat University, Surat, India
Dr. Chitra Dhawale, Nagpur, India
Dr. Bikas Kumar, Pune, India
Dr. Nidhi Divecha, Gandhinagar, India
Dr. Jay Kumar Patel, Gandhinagar, India
Dr. Jatin Shah, Gandhinagar, India
Dr. Kamaljit I. Lakhtaria, Auro University, Surat, India
Dr. B.S. Deovra, B.N. College, Udaipur, India
Dr. Ashok Jain, Maharaja College of Engineering, Udaipur, India
Dr. Bharat Singh, JRN Rajasthan Vidyapeeth University, Udaipur, India
Dr. S.K. Sharma, Pacific University, Udaipur, India
Dr. Akheela Khanum, Integral University, Lucknow, India
Dr. R.S. Bajpai, Ram Swaroop Memorial University, Lucknow, India
Dr. Manish Shrimali, JRN Rajasthan Vidyapeeth University, Udaipur, India
Dr. Ravi Gulati, South Gujrat University, Surat, India
Dr. Atul Gosai, Saurashtra Univesrity, Rajkot, India
Dr. Digvijai sinh Rathore, BBA Open University, Ahmedabad, India
Dr. Vishal Goar, Government Engineering College, Bikaner, India
Dr. Neeraj Bhargava, MDS University, Ajmer, India
Dr. Ritu Bhargava, Government Womens Engineering College, Ajmer, India
Dr. Rajender Singh Chhillar, MDU, Rohtak, India
Dr. Dhaval R. Kathiriya, Saurashtra Univesrity, Rajkot, India

Dr. Vineet Sharma, KIET, Ghaziabad, India
Dr. A.P. Shukla, KIET, Ghaziabad, India
Dr. R.K. Manocha, Ghaziabad, India
Dr. Nandita Mishra, IMS Ghaziabad, India
Dr. Manisha Agarwal, IMS Ghaziabad
Dr. Deepika Garg, IGNOU, New Delhi, India
Dr. Goutam Chakraborty, Iwate Prefectural University, Iwate Ken, Takizawa, Japan
Dr. Amit Manocha Maharaja Agrasen University, HP, India
Prof. Enrique Chirivella-Perez, University of the West of Scotland, UK
Prof. Pablo Salva Garcia, University of the West of Scotland, UK
Prof. Ricardo Marco Alaez, University of the West of Scotland, UK
Prof. Nitin Rakesh, Amity University, Noida, India
Prof. Mamta Mittal, G.B. Pant Government Engineering College, Delhi, India
Dr. Shashank Srivastava, MNNIT, Allahabad, India
Prof. Lalit Goyal, JMI, Delhi, India
Dr. Sanjay Maurya, GLA University, Mathura, India
Prof. Alexandros Iosifidis, Tampere University of Technology, Finland
Prof. Shanthi Makka, JRE Engineering College, Greater Noida, India
Dr. Deepak Gupta, Amity University, Noida, India
Dr. Manu Vardhan, NIT Raipur, India
Dr. Sarsij Tripathi, NIT Raipur, India
Prof. Wg Edison, HeFei University of Technology, China
Dr. Atul Bansal, GLA University, Mathura, India
Dr. Alimul Haque, V.K.S. University, Bihar, India
Prof. Simhiew Moi, Universiti Teknologi Malaysia
Prof. Vinod Kumar, IIT Roorkee, India
Prof. Christos Bouras, University of Patras and RACTI, Greece
Prof. Devesh Jinwala, SVNIT, Surat, India
Prof. Germano Lambert Torres, PS Solutions, Brazil
Prof. Byoungho Kim, Broadcom Corp., USA
Prof. Aditya Khamparia, LPU, Punjab, India

Contents

Part I Internet of Things

**Privacy Preserving IPv6 Address Auto-Configuration
for Internet of Things** 3
Monali Mavani and Krishna Asawa

**Message Delivery Guarantee and Status Update of Clients
Based on IoT-AMQP** 15
Purvi Bhimani and Gaurang Panchal

**The Role of Internet of Things and Smart Grid for the Development
of a Smart City** 23
Sudeep Tanwar, Sudhanshu Tyagi and Sachin Kumar

**An IoT-Based Portable Smart Meeting Space with Real-Time
Room Occupancy** 35
Jaimin Patel and Gaurang Panchal

IoT Infrastructure: Fog Computing Surpasses Cloud Computing 43
Tasnia H. Ashrafi, Md. Arshad Hossain, Sayed E. Arefin,
Kowshik D.J. Das and Amitabha Chakrabarty

**Sensing–Actuation as a Service in Cloud Centric
Internet of Things** 57
Suchismita Satpathy, Bibhudatta Sahoo, Ashok Kumar Turuk
and Prasenjit Maiti

**Soil Monitoring, Fertigation, and Irrigation System Using
IoT for Agricultural Application** 67
R. Raut, H. Varma, C. Mulla and Vijaya Rahul Pawar

Part II Intelligent Image Processing

Patch-Based Image Denoising Model for Mixed Gaussian Impulse Noise Using L₁ Norm 77
Munesh C. Trivedi, Vikash Kumar Singh, Mohan L. Kolhe,
Puneet Kumar Goyal and Manish Shrimali

Content-Based Image Retrieval Using Multiscale Local Spatial Binary Gaussian Co-occurrence Pattern 85
Prashant Srivastava and Ashish Khare

Importance of Missing Value Estimation in Feature Selection for Crime Analysis 97
Soubhik Rakshit, Priyanka Das and Asit Kumar Das

Image Analysis for the Detection and Diagnosis of Hepatocellular Carcinoma from Abdominal CT Images 107
P. Sreeja and S. Hariharan

A Level-Set-Based Segmentation for the Detection of Megaloblastic Anemia in Red Blood Cells 119
N.S. Aruna and S. Hariharan

Music Playlist Generation Using Facial Expression Analysis and Task Extraction 129
Arnaja Sen, Dhaval Popat, Hardik Shah, Priyanka Kuwor and Era Johri

Digital Forensic Enabled Image Authentication Using Least Significant Bit (LSB) with Tamper Localization Based Hash Function 141
Ujjal Kumar Das, Shefalika Ghosh Samaddar
and Pankaj Kumar Keserwani

Automatic Cotton Leaf Disease Diagnosis and Controlling Using Raspberry Pi and IoT 157
Asmita Sarangdhar Adhao and Vijaya Rahul Pawar

Part III Networks and Mobile Communications

Symmetric Key Based Authentication Mechanism for Secure Communication in MANETs 171
Sachin Malhotra and Munesh C. Trivedi

Performance Comparison of Rectangular and Circular Micro-Strip Antenna at 2.4 GHz for Wireless Applications Using IE3D 181
Mahesh Gadag, Shreedhar Joshi and Nikit Gadag

Optimal Power Flow Problem Solution Using Multi-objective Grey Wolf Optimizer Algorithm 191
Ladumor Dilip, Rajnikant Bhesdadiya, Indrajit Trivedi and Pradeep Jangir

Bisection Logic Based Symmetric Algorithm for the Design of Fault Tolerance Survivable Wireless Communication Networks 203
B. Nethravathi and V.N. Kamalesh

A Bibliometric Analysis of Recent Research on Machine Learning for Cyber Security 213
Pooja R. Makawana and Rutvij H. Jhaveri

A Comparative Study of Various Routing Technique for Wireless Sensor Network with Sink and Node Mobility 227
Tejashri Sawant and Sumedha Sirsikar

Review of Hierarchical Routing Protocols for Wireless Sensor Networks 237
Misbahul Haque, Tauseef Ahmad and Mohd. Imran

A 3.432 GHz Low-Power High-Gain Down-Conversion Gilbert Cell Mixer in 0.18 μm CMOS Technology for UWB Application 247
Gaurav Bansal and Abhay Chaturvedi

Part IV Big Data and Cloud Computing

GCC-Git Change Classifier for Extraction and Classification of Changes in Software Systems 259
Arvinder Kaur and Deepti Chopra

ECDMPSO: A Modified MPSO Technique Using Exponential Cumulative Distribution 269
Narinder Singh and S.B. Singh

Allocation of Resource Using Penny Auction in Cloud Computing 287
Aditya Kumar Naik and Gaurav Baranwal

Optimized Location-Specific Movie-Reviewing System Using Tweets 295
Vijay Singh, Bhasker Pant and Devesh Pratap Singh

Adaptive System for Handling Variety in Big Text 305
Shantanu Pathak and D. Rajeshwar Rao

Storage Size Estimation for Schemaless Big Data Applications: A JSON-Based Overview 315
Devang Swami and Bibhudatta Sahoo

Intelligent Vehicular Monitoring System Integrated with Automated Remote Proctoring 325
Chetan Arora, Nikhil Arora, Aashish Choudhary and Adwitiya Sinha

ObfuCloud: An Enhanced Framework for Securing DaaS Services Using Data Obfuscation Mechanism in Cloud Environment 333
Krunal Suthar and Jayesh Patel

The Interdependent Part of Cloud Computing: Dew Computing 345
Hiral M. Patel, Rupal R. Chaudhari, Kinjal R. Prajapati and Ami A. Patel

**Analysis of Parallel Control Structure for Efficient Servo and
Regulatory Actions** 357
Aarti Varshney, Puneet Mishra and Vishal Goyal

**Fine Grained Privacy Measuring of User's Profile Over Online Social
Network** 371
Shikha Jain and Sandeep K. Raghuwanshi

**Analyzing the Behavior of Symmetric Algorithms Usage in Securing
and Storing Data in Cloud** 381
Ashok Sharma, Ramjeevan Singh Thakur and Shailesh Jaloree

Author Index 391

Editors and Contributors

About the Editors

Dr. Yu-Chen Hu received his Ph.D. degree in Computer Science and Information Engineering from the Department of Computer Science and Information Engineering, National Chung Cheng University, Chiayi, Taiwan in 1999. Currently, Dr. Hu is a Professor in the Department of Computer Science and Information Management, Providence University, Sha-Lu, Taiwan. He is a member of ACM and IEEE. He is also a member of Computer Vision, Graphics, and Image Processing (CVGIP), Chinese Cryptology and Information Security Association, and Phi Tau Phi Society of the Republic of China. He also serves as the Editor-in-Chief of International Journal of Image Processing since 2009. In addition, he is the Managing Editor of Journal of Information Assurance and Security. He is a member of the editorial boards of several other journals. His areas of interest are image and signal processing, data compression, information hiding, and data engineering.

Dr. Shailesh Tiwari is currently working as a Professor in Computer Science and Engineering Department, ABES Engineering College, Ghaziabad, India. He is also administratively heading the department. He is an alumnus of Motilal Nehru National Institute of Technology Allahabad, India. He has more than 16 years of experience in teaching, research, and academic administration. His primary areas of research are Software Testing, Implementation of Optimization Algorithms, and Machine Learning Techniques in Software Engineering. He has also published more than 50 publications in International Journals and in Proceedings of International Conferences of repute. He has served as a program committee member of several conferences and edited Scopus and E-SCI-indexed journals. He has also organized several international conferences under the sponsorship of IEEE and Springer. He is a Senior Member of IEEE, member of IEEE Computer Society and Executive Committee member of IEEE Uttar Pradesh section. He is a member of reviewer and editorial board of several International Journals and Conferences.

Dr. Krishn K. Mishra is currently working as a Visiting Faculty, Department of Mathematics and Computer Science, University of Missouri, St. Louis, USA. He is an alumnus of Motilal Nehru National Institute of Technology Allahabad, India which is also his base working institute. His primary area of research includes Evolutionary Algorithms, Optimization Techniques and Design & Analysis of Algorithms. He has published more than 50 publications in International Journals and in Proceedings of International Conferences of repute. He has served as a program committee member of several conferences and edited Scopus and SCI-indexed journals. He has 15 years of teaching and research experience during which he made all his efforts to bridge the gaps between teaching and research.

Dr. Munesh C. Trivedi is currently working as a Professor in Computer Science and Engineering Department, ABES Engineering College, Ghaziabad, India. He has rich experience in teaching the undergraduate and postgraduate classes. He has published 20 text books and 80 research publications in different International Journals and Proceedings of International Conferences of repute. He has received Young Scientist Visiting Fellowship and numerous awards from different national as well as international forum. He has organized several international conferences technically sponsored by IEEE, ACM, and Springer. He has delivered many invited and plenary conference talks throughout the country. He has also chaired technical sessions in international and national conferences in India. He is on the review panel of IEEE Computer Society, International Journal of Network Security, Pattern Recognition Letter and Computer & Education (Elsevier's Journal). He is an Executive Committee Member of IEEE UP Section, IEEE India Council and also IEEE Asia Pacific Region-10. He is an active member of IEEE Computer Society, International Association of Computer Science and Information Technology, Computer Society of India, International Association of Engineers, and life member of ISTE.

Contributors

Asmita Sarangdhar Adhao Department of Electronics and Tele-Communication, Bharati Vidyapeeth College of Engineering for Women, Savitribai Phule University, Dhankawadi, Pune, India

Tauseef Ahmad Department of Computer Engineering, ZHCET, Aligarh Muslim University, Aligarh, UP, India

Sayed E. Arefin Department of Computer Science and Engineering, BRAC University, Dhaka, Bangladesh

Chetan Arora Department of Electronics, Jaypee Institute of Information Technology, Noida, India

Nikhil Arora Department of Electronics, Jaypee Institute of Information Technology, Noida, India

N.S. Aruna Department of EEE, College of Engineering Trivandrum, Thiruvananthapuram, Kerala, India

Krishna Asawa Jaypee Institute of Information Technology, Noida, UP, India

Tasnia H. Ashrafi Department of Computer Science and Engineering, BRAC University, Dhaka, Bangladesh

Gaurav Bansal Department of Electronics and Communication Engineering, GLA University, Mathura, India

Gaurav Baranwal Department of Computer Science and Engineering, Madan Mohan Malaviya University of Technology, Gorakhpur, Uttar Pradesh, India

Rajnikant Bhesdadiya Lukhdhirji Engineering College, Morbi, Gujarat, India

Purvi Bhimani U & P. U. Patel Department of Computer Engineering, Chandubhai S Patel Institute of Technology, Changa, India

Amitabha Chakrabarty Department of Computer Science and Engineering, BRAC University, Dhaka, Bangladesh

Abhay Chaturvedi Department of Electronics and Communication Engineering, GLA University, Mathura, India

Rupal R. Chaudhari Sankalchand Patel College of Engineering, Visnagar, India

Deepti Chopra University School of Information and Communication Technology, Guru Gobind Singh Indraprastha University, New Delhi, India

Aashish Choudhary Department of Electronics, Jaypee Institute of Information Technology, Noida, India

Asit Kumar Das Department of Computer Science and Technology, Indian Institute of Engineering Science and Technology, Shibpur, Howrah, India

Priyanka Das Department of Computer Science and Technology, Indian Institute of Engineering Science and Technology, Shibpur, Howrah, India

Ujjal Kumar Das Srikrishna College, Bagula, Nadia, West Bengal, India

Kowshik D.J. Das Department of Computer Science and Engineering, BRAC University, Dhaka, Bangladesh

Ladumor Dilip Lukhdhirji Engineering College, Morbi, Gujarat, India

Mahesh Gadag Department of ECE, HIT, Nidasoshi, India

Nikit Gadag Department of Electrical Engineering and Information Technology, Otto Van Gureeke University, Magdeburg, Germany

Puneet Kumar Goyal ABES Engg. College, Ghaziabad, India

Vishal Goyal Department of Electronics and Communication Engineering, GLA University, Mathura, India

Misbahul Haque Department of Computer Engineering, ZHCET, Aligarh Muslim University, Aligarh, UP, India

S. Hariharan Department of EEE, College of Engineering Trivandrum, Thiruvananthapuram, Kerala, India

Md. Arshad Hossain Department of Computer Science and Engineering, BRAC University, Dhaka, Bangladesh

Mohd. Imran Department of Computer Engineering, ZHCET, Aligarh Muslim University, Aligarh, UP, India

Shikha Jain CSE Department, SATI Vidisha, Vidisha, India

Shailesh Jaloree SATI, Vidisha, India

Pradeep Jangir Lukhdhirji Engineering College, Morbi, Gujarat, India

Rutvij H. Jhaveri Department of Information Technology, Shri S'ad Vidya Mandal Institute of Technology, Bharuch, India

Era Johri Department of Information Technology, K. J. Somaiya College of Engineering, Mumbai, India

Shreedhar Joshi Department of ECE, SDMCET, Dharwad, India

V.N. Kamalesh Department of Computer Science and Engineering, Sreenidhi Institute of Science and Technology, Autonomous Under UGC, Hyderabad, Telengana, India

Arvinder Kaur University School of Information and Communication Technology, Guru Gobind Singh Indraprastha University, New Delhi, India

Pankaj Kumar Keserwani Department of Computer Science & Engineering, National Institute of Technology Sikkim, Ravangla, Sikkim, India

Ashish Khare Department of Electronics and Communication, University of Allahabad, Allahabad, India

Mohan L. Kolhe University of Agder, Kristiansand, Norway

Sachin Kumar Department of ECE, Amity University, Lucknow Campus, Noida, UP, India

Priyanka Kuwor Department of Information Technology, K. J. Somaiya College of Engineering, Mumbai, India

Pooja R. Makawana Department of Information Technology, Shri S'ad Vidya Mandal Institute of Technology, Bharuch, India

Prasenjit Maiti National Institute of Technology, Rourkela, India

Sachin Malhotra Department of Information Technology, IMS, Ghaziabad, India

Monali Mavani Jaypee Institute of Information Technology, Noida, UP, India

Puneet Mishra Department of Electronics and Communication Engineering, GLA University, Mathura, India

C. Mulla Department of Electronics and Telecommunication Engineering, Bharati Vidyapeeth's College of Engineering for Women, Pune, India

Aditya Kumar Naik Department of Computer Science and Engineering, Madan Mohan Malaviya University of Technology, Gorakhpur, Uttar Pradesh, India

B. Nethravathi Department of Information Science and Engineering, JSSATE, Affiliated to Visvesvaraya Technological University, Belgaum, Karnataka, India

Gaurang Panchal U & P. U. Patel Department of Computer Engineering, Chandubhai S Patel Institute of Technology, Changa, India

Bhasker Pant Graphic Era University, Dehradun, India

Ami A. Patel Sankalchand Patel College of Engineering, Visnagar, India

Hiral M. Patel Sankalchand Patel College of Engineering, Visnagar, India

Jaimin Patel U & P. U. Patel Department of Computer Engineering, Chandubhai S Patel Institute of Technology, Changa, India

Jayesh Patel Computer Science Department, AMPICS, Kherva, Gujarat, India

Shantanu Pathak CSE Department, K L University (K L Education Foundation), Vijayawada, India

Vijaya Rahul Pawar Department of Electronics and Tele-Communication, Bharati Vidyapeeth College of Engineering for Women, Savitribai Phule University, Dhankawadi, Pune, India

Dhaval Popat Department of Information Technology, K. J. Somaiya College of Engineering, Mumbai, India

Kinjal R. Prajapati Sankalchand Patel College of Engineering, Visnagar, India

Sandeep K. Raghuwanshi CSE Department, MANIT, Bhopal, India

D. Rajeshwar Rao CSE Department, K L University (K L Education Foundation), Vijayawada, India

Soubhik Rakshit Department of Computer Science and Technology, Indian Institute of Engineering Science and Technology, Shibpur, Howrah, India

R. Raut Department of Electronics and Telecommunication Engineering, Bharati Vidyapeeth's College of Engineering for Women, Pune, India

Bibhudatta Sahoo National Institute of Technology Rourkela, Rourkela, Odisha, India

Shefalika Ghosh Samaddar Department of Computer Science & Engineering, National Institute of Technology Sikkim, Ravangla, Sikkim, India

Suchismita Satpathy National Institute of Technology, Rourkela, India

Tejashri Sawant Department of Information Technology, Maharashtra Institute of Technology, Pune, India

Arnaja Sen Department of Information Technology, K. J. Somaiya College of Engineering, Mumbai, India

Hardik Shah Department of Information Technology, K. J. Somaiya College of Engineering, Mumbai, India

Manish Shrimali University of Agder, Kristiansand, Norway; JRN RVD, Udaipur, India

Devesh Pratap Singh Graphic Era University, Dehradun, India

Narinder Singh Department of Mathematics, Punjabi University, Patiala, Punjab, India

S.B. Singh Department of Mathematics, Punjabi University, Patiala, Punjab, India

Vijay Singh Graphic Era University, Dehradun, India

Vikash Kumar Singh IGNTU, Amarkantak, India

Adwitiya Sinha Department of Computer Science, Jaypee Institute of Information Technology, Noida, India

Sumedha Sirsikar Department of Information Technology, Maharashtra Institute of Technology, Pune, India

Ashok Sharma Department of Computer Science, BU, Bhopal, India

Prashant Srivastava Department of Electronics and Communication, University of Allahabad, Allahabad, India

P. Sreeja Department of Electrical Engineering, College of Engineering Trivandrum, Thiruvananthapuram, India

Krunal Suthar Computer Engineering Department, Sankalchand Patel College of Engineering, Visnagar, Gujarat, India

Devang Swami National Institute of Technology Rourkela, Rourkela, Odisha, India

Sudeep Tanwar Department of CE, Institute of Technology, Nirma University, Ahmedabad, India

Ramjeevan Singh Thakur MANIT, Bhopal, India

Indrajit Trivedi GEC Gandhinagar, Gandhinagar, Gujarat, India

Ashok Kumar Turuk National Institute of Technology, Rourkela, India

Sudhanshu Tyagi Department of ECE, College of Technology, GBPUA&T, Pantnagar, India

H. Varma Department of Electronics and Telecommunication Engineering, Bharati Vidyapeeth's College of Engineering for Women, Pune, India

Aarti Varshney Department of Electronics and Communication Engineering, GLA University, Mathura, India

Part I
Internet of Things

Privacy Preserving IPv6 Address Auto-Configuration for Internet of Things

Monali Mavani and Krishna Asawa

Abstract Internet of Things enables every node on a personal network to be managed and monitored remotely over the Internet. Biometric devices, used for access control or as bio-sensors, form a critical part of Internet of Things and are identified using IPv6 address. Malicious users can track activity of these devices by spoofing IPv6 addresses from unsecure wireless communication channels. Tracking device activity and identifying user behavior of the device poses a great threat to device identity and data generated by it. Such a threat can be avoided by keeping the device's IPv6 address hidden from attacker. This study proposes a method to privacy enable IPv6 address configuration for connected devices in general and biometric devices in particular, while connected as a part of Internet of Things. It is proposed that by changing the device's IPv6 address periodically and pseudorandomly, its identity can be kept private to a large extent. These address changes are configured on devices based on congruence classes, which generate non-repeatable integer sequence. It is proposed that the interface identification part of IPv6 address is configured with two-level hierarchy with each level level using a different congruence class. Such configuration generates different identification values to ensure conflict free address configuration. The proposition is analyzed for privacy preserving property and communication cost. The results of performance benchmarking using Cooja simulator show that the method does not impose substantial communication overhead on IPv6 address configuration process.

Keywords IoT · Privacy · IPv6 Address · Congruence relation
Low power devices · 6LoWPAN

M. Mavani (✉) · K. Asawa
Jaypee Institute of Information Technology, Noida, UP, India
e-mail: monamavani@gmail.com

K. Asawa
e-mail: krishna.asawa@jiit.ac.in

© Springer Nature Singapore Pte. Ltd. 2018
Y.-C. Hu et al. (eds.), *Intelligent Communication and Computational
Technologies*, Lecture Notes in Networks and Systems 19,
https://doi.org/10.1007/978-981-10-5523-2_1

1 Introduction

In Internet of Things (IoT), personal devices are connected to each other via Internet. IoT has got numerous applications, e.g., wireless body area networks (WBAN) in medical and healthcare domain, smart homes, smart retail, smart grid, smart cities, etc. In IoT applications, devices need authorized access, therefore, knowledge based access control system or biometric systems are used. Compared to token-based or knowledge-based access control system, biometric systems are more reliable as biometric identification is unique to user. In conventional biometric authentication system, time-invariant, physiological characteristics are first captured and stored. The stored characteristics are referred to enable controlled access. In literature, numerous examples are found which use biometrics for access control, for example, use of biometric footsteps to provide biometric verification system as presented in [21] and [17], voice biometric for entrance door access control [11], personalization of home services [6], etc. There is another use of biometric data which uses biometric sensors to provide physiological characteristics to ensure security as reported in [20] for medical diagnosis, in [19] for health monitoring in WBAN. Authors in [13] have presented the use of bio-sensors in textile networks.

Biometric devices communicating via the Internet require an unique identity like IPv6 network address. But IPv6, built for high speed and reliability of modern communication networks, cannot be used without modification for IoT devices which have constraints like low power and lossy communication channel. As a result, 6LoWPAN has emerged as the most suitable communication protocol for IoT, when sensed data needs to be transferred for remote monitoring and processing via Internet [1, 2, 9, 10]. These low power devices need to configure a unique, global routable IPv6 address in order to communicate via Internet.

This unique, global routable IPv6 address is prone to widely popular and ever-evolving attacks against IPv6 protocol stack. Changing the IPv6 address periodically makes it difficult to associate a device with an address and correlate different activities performed by the said device in the lifetime of a network [15]. This study proposes a method of achieving IPv6 address auto-configuration for maintaining privacy of IoT devices.

Rest of the sections are organized as follows: Sect. 2 describes related work on different addressing schemes. The proposed address auto-configuration method is described in Sect. 1. Section 4 describes analyses, Sect. 5 represents simulation study and finally Sect. 6 concludes the study.

2 Related Work

The biometric devices and authentication systems communicating via Internet present a set of data security and privacy challenges as mentioned in [12] and [24]. By sniffing and spoofing communication messages, the gathered information can be used to track the device's active time. And this activity time can be correlated to deduce time of presence of the user as stated in [27]. If used well, the information can enable more active attacks like modifying biometric data.

Work has been reported in literature to keep IP-based communication secure for different wireless networks. Authors in [27] have proposed cluster-based addressing scheme with privacy support. Privacy is achieved by regularly changing node's address from allocated address set. Size of address set is randomly chosen. Address set is sent by cluster head to every cluster member upon request. In this, IID part is formed using cluster ID and member ID. Member ID changes regularly from the address space, allocated by the cluster head. Cluster ID is assigned by access router. Their scheme eliminates need for DAD as addresses are assigned centrally. However, this scheme causes fragmented address space at cluster head and limits the number of addresses that node can change. It also requires substantial amount of cluster head involvement and increases complexity at cluster head. Authors in [22] have worked on the issue of false claim of ownership of IP address. They have proposed Crypto-ID instead of EUI-64 ID in IID part of address. Crypto-ID is generated using elliptical curve cryptography. Crypto-ID can overcome issue of IP address authentication but does not guarantee IP address privacy. Thus, eavesdropper can still identify node by its IP address. Authors in [26] have modified DAD scheme which protects attacks against DAD. In this, only part of address to be configured is broadcasted so that attacker cannot see the whole address and can not claim it. And address cost is lowered as only part of address to be configured is broadcasted and further DAD is performed at link layer. In this scheme, IID part is formed using DAD ID and Node Id. These are randomly generated. It uses beacon and command frames of IEEE 802.15.4 specification. As these are randomly generated, address conflict probability will be greater than zero. Hence nodes need to perform DAD. In [7], each node can act as authorized proxy and allocates IP address to neighbors securely. Authors have tried to solve the issue of node authorization to obtain IPv4 address in MANET. However, in their scheme, minimum of four messages are exchanged to configure node's IP address which increases communication cost in the network. Moreover once configured, IP address does not change, which makes eavesdropper to identify node from its IP address. Address allocation scheme can be classified as secure vs unsecure. In secure allocation scheme, either confidentiality, integrity, authentication, or privacy is achieved. Authors in [28] have used fundamental property of integers to obtained integer sequence in which repetition of a number comes after long time. Initial random seed is used to generate random number based on a generator function. They have used fundamental theorem of product of prime numbers to generate a sequence. Their scheme defeats the IP spoofing attack, sybil attack and state pollution attack by using two random numbers to generate IP address. However, due to use of random numbers, address duplication can not be avoided. Proxy-based distributed address allocation scheme is presented in [8] known as SD-RAC. Authors have proposed this scheme for MANET. Each node after acquiring IP address can assign IP address to its neighbors who are newly joined. In this scheme, blocks are partitioned among proxy nodes in such a way that it does not overlap, which ensures its uniqueness, and eliminating its need for DAD. Messages exchanged between nodes in address allocation process are authenticated either by digital signature in case of broadcast messages or by MAC in case of unicast messages. This ensures authenticity of messages which lead to correct address assignment. Further nodes are assigned unique IDs before deployment of network and its list is preinstalled in all the nodes. Thus, only authorized nodes can acquire IP address. However, this scheme requires all nodes should store unique Ids of all other nodes in the network due to which storage overhead at all the nodes increases.

It is clear from literature survey that not much work is done in the area of IPv6 address privacy. Though in [27], IPv6 address privacy is achieved but address sets for nodes are not interleaved and are contiguous, so address space fragmentation is observed. The proposed method attempts to achieve device privacy with minimum communication overhead, delay and resource requirements, and yet, without address space fragmentation.

3 IPv6 Address Auto-Configuration

IPv6 address auto-configuration can be stateless or stateful. In stateless mechanism, every device auto-configures address using its Medium Access Control (MAC) address in Interface Identification (IID) part and router advertised global routing prefix. This mechanism requires procedure called Duplicate Address Detection (DAD) [16] to ascertain that the self configured address is unique which incurs heavy communication overhead. Alternatively IPv6 address can be configured using stateful protocol like Dynamic Host Configuration Protocol for IPv6 (DHCPv6) [4]. In this approach, DHCPv6 server uniquely assigns IPv6 address to each device present in the network. Configured either way, the IPv6 address remains unchanged for long period of time.

Keeping device and user security in view, it is required that IPv6 address in 6LoW-PAN based IoT should remain difficult to predict. One way to achieve this is to change device addresses periodically so that it becomes difficult for adversary to track and identify devices. But such address changes come at a cost. If the device address is changed periodically, then DAD process has to run for each address change iteration, causing substantial communication overhead and is resource intensive for address allocator. Failure of address allocator results in loss of state.

128 bits IPv6 address consists of two parts: global routing prefix and Interface Identifier (IID) part [3] as shown in Fig. 1. Global routing prefix is unique and identifies IPv6 subnet globally. IID part is further divided as *Node Id* and *Router Id* part. As shown in Fig. 1, i represents number of bits for *Router Id* part and j represents number of bits for *Node Id* part.

Fig. 1 IPv6 addressing structure

Prefix ID	Router Id (i bits)	Node Id (j bits)
Global routing prefix	Interface ID (IID)	

IID part is generated using two step process.

1. Generation of disjoint congruent integer sequences for *Node Id* and *Router Id* parts.
2. Shuffling the sequences before generating the IPv6 address.

This study considers unique ID of each node *(UID)* and *congruence seed* as address configuration parameters. Every node can be given *an UID* while deployment. 6LoWPAN standard allows use of 16-bit short addresses for intra-PAN communication which should be unique to the PAN [14]. This 16-bit short address can also be taken as *UID* in the proposed algorithm. This study assumes 16 bits for *Node_UID* which is used to calculate 32 bit *Node Id* part in IID. Similarly 16 bit *router_UID* is used to caclulate 32 bit *Router ID* part by routers. In order to achieve privacy, *Node Id* and *Router Id* changes periodically, through non-repeatable congruent integer sequences which are generated automatically. Nodes can pick up any number from generated sequence and forms IID part of IPv6 address. *Node Id* is generated by node itself based on the congruence seed *(CS)* broadcasted by router. Another congruence seed *(cs)* is used to generate *Router Id* which is broadcasted by edge router. This technique is different from all previous stateful approaches found in the literature where IID is allocated by central entity or distributed entity. In those cases, state maintenance becomes more costly when routers or PAN coordinators have to allocate address space periodically to achieve privacy. In proposed method, routers only broadcast congruence seed, instead of address set to the nodes, so communication overhead reduces substantially.

Non-repeatable address sequence for *Node Id* and *Router Id* are created based on congruence relation *node_UID* and *router_UID* respectively.

Let R_m is the congruence relation *modulo m* on the set of all positive integers $b \in \mathbb{Z}$ and $m \in \mathbb{Z}$. Equivalence class of b is denoted by $\bar{b}_m$ or $[b]_m$ and defined as:

$$[b]_m = \{y \ : \ \forall k \in \mathbb{Z}, \ if \ b \equiv y \ mod \ m \ then \ y = b \pm km\}$$

$$[b]_m = \{......b - 2m, b - m, b, b + m, b + 2m, ...\} \tag{1}$$

Equation 1 gives non-repeatable arithmetic sequence. The congruence class $[b]_m$ is also known as residue class of $b \ mod \ m$. The quotient set of *congruence modulo m* denoted as $\mathbb{Z}_m$. Thus, $\mathbb{Z}_m$ is the set of all residue classes *modulo m*.

$$\mathbb{Z}_m = \{[0]_m, [1]_m, [2]_m, ..., [m-1]_m\}$$

The quotient set $\mathbb{Z}_m$ of *congruence modulo m* forms a partition of $\mathbb{Z}$. Here, $\mathbb{Z}$ represents address space for each single PAN and m represents common modulus in single-hop PAN which is congruence seed, generated by router. The fundamental property of equivalence relation that partitions of $\mathbb{Z}$ are disjoint, is used to form disjoint address set for each node in a PAN.

From Eq. 1 *Node Id* takes non zero positive integer values randomly at different time. Address sets for all nodes in single-hop PAN can be generated by considering

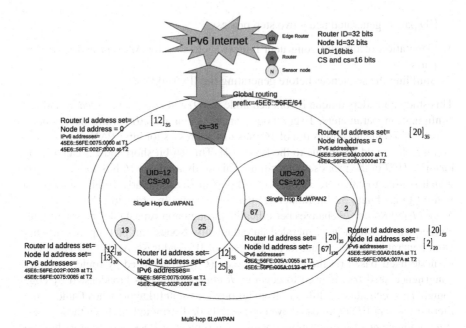

Fig. 2 Address generation process

the sets of all congruence classes of UID *mod CS* where *CS* is congruence seed generated by router and is greater than all *UID* of nodes present in the PAN. With 16-bit *UID* maximum of $2^{16} - 1$ nodes can be in single PAN.

Number of single-hop PAN is characterized by number of routers registered with edge router. After routers register their *UID* with edge router, edge router generates congruence seed greater than the highest *UID* of routers in the subnet. All nodes will choose one of the IDs from sequence generated using congruence seed *(cs)* sent by edge router and *router_UID*. Chosen ID is then combined with global routing prefix to generate 128-bit IPv6 address. This address generation process is explained in Fig. 2.

Figure 2 shows two single hop 6LoWPAN in a IPv6 subnet. Length for global routing prefix is 64 bits and 32 bits each for *Node Id* and *Router Id* part respectively. In single-hop 6LoWPAN1 *router_UID* is 12 and congruence seedgenerated by router is 30 (greater than highest *UID* of nodes in the PAN, i.e., 25). Node with *UID* 13 will generate sequence for *Node Id* part using the members of congruence class 13 mod 30 denoted as $[13]_{30}$. At time *T1* it will generate *Node Id* as 43 i.e. $0 \times 002B$. Router in 6LoWPAN1 is having *UID* as 12 and congruence seed generated by edge router is 35(greater than highest *UID* 20 of routers in the subnet). The *Router Id* part will be generated using the members of congruence class 12 mod 35 denoted as $[12]_{35}$. So at time *T1*it will generate *Router ID* as 47, i.e., $0 \times 002F$. Combining with global routing prefix *45E6::56FE*, which is disseminated by edge router, IPv6 address of node with *UID* 13 will be *45E6::56FE:002F:002B* at time *T1*. When node needs to change its address at time *T2* it will pick up randomly any other member from $[13]_{30}$ for *Node Id* part and any one from $[12]_{35}$ for *Router Id* part and forming IPv6 address as 45E6::56FE:0075:0085. Similar procedure router will follow to generate its address but its *Node Id* part will be zero. Its *Router Id* part will be generated using members of congruence class $[12]_{35}$. At time T1 router in 6LoWPAN1 will have address as *45E6::56FE:0075:0000* and at time T2 it will have as address as *45E6::56FE:002F:0000*. This process picks up the number from congruence classes of same modulus, so numbers generated will not overlap and thus forming conflict free and interleaved addresses at all the times.

However, if nodes change their addresses, it becomes essential to keep track of currently active address of a node. 6LoWPAN-ND [23] mentions registration of IPv6 addresses of a node to the router and IPv6 address of a router to the edge router periodically. This way when node or router changes their IPv6 address, using 6LoWPAN-ND mechanism, they can register their new address to router and edge router, respectively. This facilitates in IP packet forwarding correctly.

4 Analyses

This section presents veracity of privacy preserving property of IPv6 address configuration. Proposed method is analyzed for its privacy-enabled operation and message complexity. Proposed method needs to transmit six types of broadcast messages for address configuration. Those messages are *R_addr_parameter* (address configuration parameters sent by router), *ER_addr_parameter* (address configuration parameters sent by edge router), *node_new* (message sent by new node), *router_new* *(address sent by new router), node_new_ack* (acknowledgement sent in response to *node_new message), router_new_ack* (acknowledgement sent in response to *router_new message)* messages.

Lemma 1 *Node's IP address remains private from eavesdropper.*

Proof Privacy is achieved as nodes change its address periodically from the address set generated locally based on congruence seed received from the router. Address sequence is generated using two step pseudorandomness as explained in Sect. 3. As addresses generated are non-contiguous, i.e., interleaved, eavesdropper cannot determine different addresses belong to different nodes or same nodes. IID part is generated using two level hierarchy in which both the *Router Id* and *Node Id* part will be chosen randomly as explained in Sect. 3. This is superior from similar approach mentioned in [27], where continuous address set for the *Node Id* part is allocated by router which results in fragmentation of address space. Further in their scheme cluster Id remains constant thus from three level hierarchy only last level hierarchy has randomness.

Lemma 2 *Upper bound on number of messages in initialization phase is O(1) when a node configures IP address. Similarly Upper bound on number of messages for* *router addressing is also O(1). Upper bound on total number of messages exchanged in the network when all nodes and routers configures IPv6 addresses are O(N).*

Proof Assume there are no message losses and number of routers *r* are very less than number of nodes *N*. Router transmits *router_new* message and edge router will reply *router_new_ack message*. After *ER_initialization* timer goes off edge router sends *ER_addr_parameter* message. So router configures its address after exchange of three messages. Therefore, growth rate function is a constant *O(1)* for router addressing.

In single-hop PAN node need to transmit *node_new* message and router replies *router_new_ack message*. After *R_initialization* timer goes off, router sends *R_addr_parameter* message. Node configures its address after exchange of *3* messages. Therefore, growth rate function is a constant *O(1)*. With *r router_new* and *r router_new_ack* messages and *1 ER_addr_parameter* message, number of messages required for IP address configuration for all routers are *2r+1*. With *n* number of nodes and 1 router in single PAN, number of messages required for IP address configuration for *n* nodes is *2n+3*. Total number of messages for *r* router and *N* nodes in subnet are *2N+3r+1*. Growth rate function for IP address configuration for whole subnet in initialization phase is *O(N)*.

Lemma 3 *Upper bound on number of messages needed to change address set x times by a single node in operational phase is O(x). Upper bound on number of messages needed to change address set y times by a router is O(y) where x >> y. Upper bound on total number of messages transmitted for N nodes and r routers for changing address sets by is O(x).*

Proof Assuming *y* number of times address set is changed. In each address change, period *y* times *ER_addr_parameter* messages need to be transmitted with different congruence seeds. So number of messages required to generate *y* address sets for a router are *y* messages. Growth rate function for changing number of addresses for router is *O(y)*. Similarly *x* times *R_addr_parameter* messages need to be transmitted with different congruence seeds for *Router Id* part and *Node Id* part respectively. So number of messages required to generate *x* address sets for a node is *x* messages. Growth rate function for changing number of addresses for router is *O(y)*. For *r* routers and *N* nodes, to change address set *x* times, total number of messages transmitted in network is *x+y* where *y* is very small compared to *x*. Therefore neglecting y, growth rate function for IP address configuration during operational phase is *O(x)*.

5 Simulation Experiments

Simulation experiments were carried out in Cooja simulator [18], running Contiki operating system [5] emulating telosb sky motes [25]. Cooja exhibits the same constraints as of real hardware, therefore results produced by cooja are more realistic.

5.1 Simulation Scenarios and Parameters

Cooja simulation parameters are shown in Table 1.

Figure 3 shows address configuration latency for varying node population of 50 to 200 nodes.

It was observed that within 50 s all the 200 nodes were configured with address set. Figure 3 shows that as network population increases, latency to configure address also increases. For 50 node network, delay observed is 8.383 s and for 200 node network, it is 49.03 s. As the number of nodes increase, it results in collisions and subsequently requires retransmissions. This contributes to delayed address configuration for all nodes in initialization phase. Number of messages required for 50 node network is observed to be 115 and for 200 nodes, it is 462 messages (Fig. 4).

Table 1 Simulation parameter

Cooja simulation parameters	Value	Contiki parameters	Value
Radio model	DGRM	NETSTACK_CONF_WITH_IPV6	1
Radio frequency	2.4 Ghz	NETSTACK_CONF_WITH_RIME	1
Transmission range	50 m	NETSTACK_CONF_MAC	csma_driver
Interference range	100 m	NETSTACK_CONF_RDC	nullrdc_driver
Simulation area	100 m x 100 m	NETSTACK_CONF_RADIO	cc2420_driver
Bit rate	250 kbps	NETSTACK_CONF_FRAMER	contikimac_framer
Communication protocol	Rime	NETSTACK_CONF_NETWORK	sicslopan_driver

Fig. 3 Address configuration delay for varying node population (from 50 to 200 nodes)

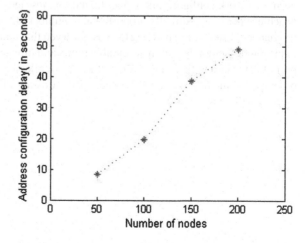

Fig. 4 Address
configuration cost for
varying node population
(from 50 to 200 nodes) and
communication cost

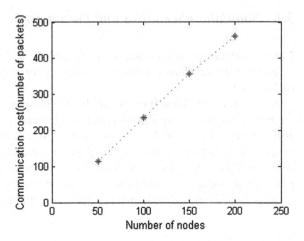

6 Conclusion

This study has presented IPv6 address auto-configuration method for preserving privacy in 6LoWPAN enabled devices forming the IoT. IPv6 addresses give unique identity to IoT devices (nodes) but also introduce threats. By periodically changing IPv6 addresses, these nodes can hide their identity from eavesdropper. The proposed method uses the disjoint property of congruence classes of integers to generate pseudorandom address set for each node. Nodes generate address set by themselves on receiving address configuration parameters from routers. A different address set is generated for *Router Id* and *Node Id part* respectively. Thus, two-level hierarchy ensures uniqueness in generated IPv6 address thereby eliminating the need for DAD. This reduces communication overhead and address configuration latency significantly. Theoretical analyses show that upper bound on address configuration messages is directly proportional to number of nodes, when node configures its address during network configuration. Upper bound on messages for changing address sets, to achieve privacy, is directly proportional to number of times an address set needs to be changed. From simulated results, it is observed that communication overhead due to proposed method is within acceptable limits and offers advantage of privacy. As a part of extension to this study, effect of low power device's sleep/ wake-up schedule on proposed method will be evaluated. Also hardware implementation of proposed method will be evaluated.

References

1. Cao, H., Leung, V., Chow, C., Chan, H.: Enabling technologies for wireless body area networks: A survey and outlook. Comm. Mag. **47**(12), 84–93 (2009). doi:10.1109/MCOM.2009.5350373
2. Cavallari, R., Martelli, F., Rosini, R., Buratti, C., Verdone, R.: A survey on wireless body area networks: Technologies and design challenges. IEEE Communications Surveys Tutorials **16**(3), 1635–1657 (2014). doi:10.1109/SURV.2014.012214.00007
3. Deering, D.S.E., Hinden, R.M.: IP Version 6 Addressing Architecture. RFC 4291 (2015). doi:10.17487/fc4291. https://rfc-editor.org/rfc/rfc4291.txt
4. Droms, R., Bound, J., Volz, B., Lemon, T., Perkins, C., Carney, M.: Dynamic Host Configuration Protocol for IPv6 (DHCPv6). RFC 3315 (Proposed Standard) (2003). http://www.ietf.org/rfc/rfc3315.txt
5. Dunkels, A., Grönvall, B., Voigt, T.: Contiki - A lightweight and flexible operating system for tiny networked sensors. Proc. - Conf. Local Comput. Networks, LCN pp. 455–462 (2004). doi:10.1109/LCN.2004.38
6. Friedewald, M., Da Costa, O., Punie, Y., Alahuhta, P., Heinonen, S.: Perspectives of ambient intelligence in the home environment. Telematics and informatics **22**(3), 221–238 (2005)
7. Ghosh, U., Datta, R.: A secure dynamic ip configuration scheme for mobile ad hoc networks. Ad Hoc Netw. **9**(7), 1327–1342 (2011). doi:10.1016/j.adhoc.2011.02.008.
8. Ghosh, U., Datta, R.: A Secure Addressing Scheme for Large-Scale Managed MANETs. IEEE Trans. Netw. Serv. Manag. **12**(3), 483–495 (2015). doi:10.1109/TNSM.2015.2452292
9. Huq, M.Z., Islam, S.: Home area network technology assessment for demand response in smart grid environment. In: Universities Power Engineering Conference (AUPEC), 2010 20th Australasian, pp. 1–6 (2010)
10. Kamilaris, A., Trifa, V., Pitsillides, A.: Homeweb: An application framework for web-based smart homes. In: Telecommunications (ICT), 2011 18th International Conference on, pp. 134–139 (2011). doi:10.1109/CTS.2011.5898905
11. Lee, K.A., Larcher, A., Thai, H., Ma, B., Li, H.: Joint application of speech and speaker recognition for automation and security in smart home. In: INTERSPEECH, pp. 3317–3318 (2011)
12. Li, T., Ren, J., Tang, X.: Secure wireless monitoring and control systems for smart grid and smart home. IEEE Wireless Communications **19**(3), 66–73 (2012). doi:10.1109/MWC.2012.6231161
13. Lukowicz, P., Kirstein, T., Troster, G.: Wearable systems for health care applications. Methods of Information in Medicine-Methodik der Information in der Medizin **43**(3), 232–238 (2004)
14. Montenegro, G., Hui, J., Culler, D., Kushalnagar, N.: Transmission of IPv6 Packets over IEEE 802.15.4 Networks. RFC 4944 (2015). DOI 10.17487/rfc4944. https://rfc-editor.org/rfc/rfc4944.txt
15. Narten, D.T., Draves, R.P., Krishnan, S.: Privacy Extensions for Stateless Address Autoconfiguration in IPv6. RFC 4941 (2015). DOI 10.17487/rfc4941. https://rfc-editor.org/rfc/rfc4941.txt
16. Narten, T., Nordmark, E., Simpson, W., Soliman, H.: Neighbor Discovery for IP version 6 (IPv6). RFC 4861 (Draft Standard) (2007). http://www.ietf.org/rfc/rfc4861.txt
17. Orr, R.J., Abowd, G.D.: The smart floor: A mechanism for natural user identification and tracking. In: CHI '00 Extended Abstracts on Human Factors in Computing Systems, CHI EA '00, pp. 275–276. ACM, New York, NY, USA (2000). doi:10.1145/633292.633453
18. Osterlind, F., Dunkels, A., Eriksson, J., Finne, N., Voigt, T.: Cross-level sensor network simulation with cooja. In: Local computer networks, proceedings 2006 31st IEEE conference on, pp. 641–648. IEEE (2006)
19. Picard, R.W., Healey, J.: Affective wearables. Personal Technologies **1**(4), 231–240 (1997). doi:10.1007/BF01682026
20. Poon, C.C.Y., Zhang, Y.T., Bao, S.D.: A novel biometrics method to secure wireless body area sensor networks for telemedicine and m-health. IEEE Communications Magazine **44**(4), 73–81 (2006). doi:10.1109/MCOM.2006.1632652

21. Rodríguez, R.V., Lewis, R.P., Mason, J.S., Evans, N.W.: Footstep recognition for a smart home environment. International Journal of Smart Home **2**(2), 95–110 (2008)
22. Sethi, M., Thubert, P., Sarikaya, B.: Address protected neighbor discovery for low-power and lossy networks. Internet-Draft draft-sarikaya-6lo-ap-nd-04, IETF Secretariat (2016). URL http://www.ietf.org/internet-drafts/draft-sarikaya-6lo-ap-nd-04.txt. http://www.ietf.org/internet-drafts/draft-sarikaya-6lo-ap-nd-04.txt
23. Shelby, Chakrabarti, Nordmark, Bormann: Neighbor Discovery Optimization for IPv6 over Low-Power Wireless Personal Area Networks (6LoWPANs). RFC 6775, RFC Editor (2012). http://www.rfc-editor.org/rfc/rfc6775.txt
24. Stankovic, J.A.: Research directions for the internet of things. IEEE Internet of Things Journal **1**(1), 3–9 (2014). doi:10.1109/JIOT.2014.2312291
25. Technologies, W.: Telosb datasheet. Crossbow (2006)
26. Wang, X., Cheng, H., Yao, Y.: Addressing with an improved dad for 6lowpan. IEEE Communications Letters **20**(1), 73–76 (2016). doi:10.1109/LCOMM.2015.2499250
27. Wang, X., Mu, Y., Member, S.: Addressing and Privacy Support for 6LoWPAN. IEEE Sensors **15**(9), 5193–5201 (2015)
28. Zhou, H., Mutka, M.W., Ni, L.M.: Secure prophet address allocation for manets. Security and communication networks **3**(1), 31–43 (2010)

Message Delivery Guarantee and Status Update of Clients Based on IoT-AMQP

Purvi Bhimani and Gaurang Panchal

Abstract Advanced Message Queue Protocol (AMQP) is an open-standard application layer protocol for IoT focusing on message-oriented middleware. It provides asynchronous publish/subscribe communication with messaging. It is store-and-forward feature that ensures reliability even after network disruptions, which is its main advantage. When compared all other IoT protocols with AMQP protocol, it gives better performance. In this paper, we provide features for some cases or situations like when any client disconnected ungracefully or when any client connected and subscribed for a particular topic which it is interested in. This is because these features are used to notify other client(s) about disconnected client and help newly subscribed clients to get a status update immediately after subscribing and do not have to wait until the publishing clients send the new update. So AMQP protocol provides the guarantee of message delivery and provides reliable communication even after a network failure.

Keywords AMQP · Status update · Publish/subscribe
Message delivery guarantee

1 Introduction

A ubiquitous network enables monitoring and control of the physical environment by collecting, processing and analysing of data, which are generated by sensors, actuators or smart objects referred as an Internet of things (IoT) [1–3]. It should be capable of interconnection through the Internet that connects billions and trillions of objects. There are three main parts of IoT: first is a network terminal part that includes sensors, second is the transmission network that refers to information

---

P. Bhimani (✉) · G. Panchal
U & P. U. Patel Department of Computer Engineering, Chandubhai S Patel Institute of
Technology, Changa, India
e-mail: bhimanipurvi56@gmail.com

G. Panchal
e-mail: gaurangpanchal.ce@charusat.ac.in

© Springer Nature Singapore Pte. Ltd. 2018
Y.-C. Hu et al. (eds.), *Intelligent Communication and Computational
Technologies*, Lecture Notes in Networks and Systems 19,
https://doi.org/10.1007/978-981-10-5523-2_2

transmission and third is the data processing centre that controls the processing centre. It grows the world's economy and improves the quality of life [4]. For example, smart home enables their residents to automatically open their garage when reaching home, prepare their coffee or tea, and control climate control systems, TVs and other appliances.

The individual networks are connected together with security, analytics and management that referred as a network of networks for IoT [5, 6]. It is a global concept that provides communication between human-to-human, human-to-machine, machine-to-machine communication [3, 7]. A special set of rules that used by end points when they communicate in telecommunication connection is referred as a Protocol, which specifies the interaction between communication entities. It allows information to be transmitted from one system to another or one device to another device over the Internet [8–16].

2 Brief About AMQP

AMQP [3, 5–7] is an open-standard wire-level application layer protocol for IoT [6]. It is focusing on message-oriented environment. It provides asynchronous publish/subscribe communication with messaging. It is based on reliable transport protocol TCP. It is a store-and-forward feature that ensures reliability even after a network failure. AMQP provides reliable communication via message delivery guarantee that is at least once, at most once, exactly once [17]. As shown in Fig. 1a, at most once guaranteed, delivery referred as the message is sent one time either it is delivered or not. At least once guaranteed delivery referred as the message will be delivered one time and possibly more times (See Fig. 1b). Exactly once guaranteed delivery referred as the message will be delivered only one time (See Fig. 2).

2.1 AMQP Model Structure

To create a desired functionality, there are three main types of components that connected to the processing chain in the server (Fig. 3).

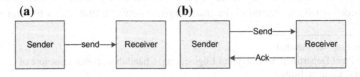

Fig. 1 **a** At most one delivery; **b** At least one delivery

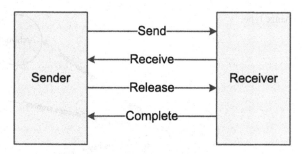

Fig. 2 Exactly once delivery

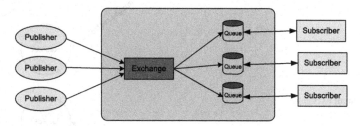

Fig. 3 AMQP model structure

Exchange: Exchange is routing and matching engines. It receives messages from publisher application and forwards it to the message queues according to predefined rules or criteria. It never stores messages. They use routing pattern to route the message to particular queues. There is a number of attributes of exchange namely name and durability. It survives until the broker restarts; auto-delete means it is deleted when all queues have finished to use it, i.e., arguments. There are four types of standard exchange types:

1. Direct exchange: Direct exchange routes message to the queue based on the message routing key. It is suitable for unicast as well as multicast routing. In this type of exchange task, messages are distributed between multiple workers in round robin manner. Direct exchange can be represented graphically as shown in Fig. 4.
2. Fanout exchange: This type of exchange routes message to the entire bounded queue. It is suitable for broadcast routing. If there are N number of queues bounded to fanout exchange, when a new message is published to that exchange by publisher, then a copy of a message delivered to all N queues. It basically delivers a copy of a message to every queue bounded to it, for example, group chat. Fanout exchange can be represented graphically as shown in Fig. 5.
3. Topic exchange: It is suitable for multicast routing of the messages. It matches routing key with routing patterns and based on that it routes the message to the particular queue. The routing pattern follows the same rules as the routing key with addition that single word matches with $*$ and zero or more word matches

Fig. 4 Direct exchange type

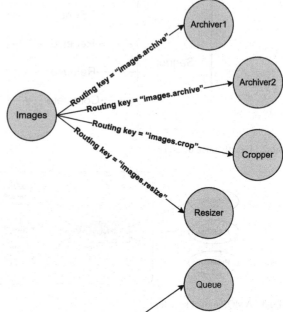

Fig. 5 Fanout exchange type

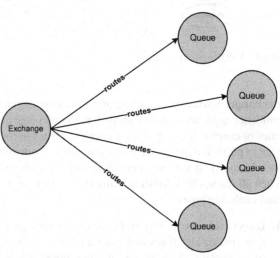

with #. For example, the routing pattern * .*stack*.# matches the routing keys *abc.stack* and *abc.stack.mn* but not *stack.empty* [18, 19].

4. Header exchange: Header exchange is designed for routing on multiple message header attributes than a routing key. It ignores the routing key attribute. If the value of a header is equal to the value of binding, then the message is said to be matching. Header exchange type uses "*x*-match" binding argument. When bind queue to the header exchange using more than one header value, in this case, broker needs one more piece of information that message matches with any of the headers or all of them that referred to the x-match binding argument. If it is set to "any", just one matching header value is sufficient; or if it is set to "all" then all the header values must match.

Message Queue: Message queues are stored and distributed entities. It stores messages on disk or in memory. Message queues are freely created, shared, used and destroyed within the limits to their authority. Messages have some priority level; in the same queue higher priority message is sent first and then the lower priority message. The server will first discard low-priority messages in order to maintain a specific service quality level.

Binding: To route the message from exchange to the message queue based on some specific criteria or rules that referred to as binding, it specifies the relation between message queue and exchange.

Needs of AMQP: AMQP provides full functional communication between client and message middleware server and queueing for later delivery that provides message delivery guarantee.

3   Message Life Cycle

The publisher applications create the message, place the content, set some message properties, label the message with routing information and then send the message to an exchange on the server. The message is received by exchange and routes it to the message queues. If a message is not routable, exchange may be silently dropped it or return the message to the producer.

AMQP provides functionality that a single message can exist on many message queues by doing copies of the message or by reference counting. When a message queue receives the message, it tries to immediately pass it to the subscriber application. If it is not possible, queue stores the message until subscriber is ready. When a message queue delivers the message to subscriber application, it removes the message from its internal buffer immediately or after subscriber's acknowledgement.

4   AMQP Frame Format

In AMQP, information is organised into a frame. As shown in Fig. 6, a frame is divided into three parts: the first is fixed-sized (8 bytes) frame header, the second is variable-sized extended header or payload and the third is a frame end. Frame header includes information necessary to parse the rest of the frames like the type of the frame, channel and the size of the frame. In a frame header, if "type = 1" then it is "Method frame" which carries remote procedural call (RPC) request and response; if "type = 2" then it is "Header frame" that use to establish the connection; if "type = 3" then it is "Body frame" which includes actual content of the message. If a message is too big, then split it into multiple frames and if "type = 4", then it is "Heartbeat frame" that is used to confirm that client is still alive. If frame channel value is zero,

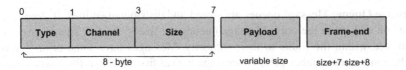

Fig. 6 Frame format

then it is used for global connection; if its value is between 1 and 65535, then it is a frame for specific channels. Payload depends on the type of the frame. Frame end is used to detect the framing error.

5 Proposed Implementation of AMQP

AMQP implements three types of message brokers or servers: RabbitMQ, OpenAMQ and Apache Qpid. Standard AMQP protocol is supported by RabbitMQ. It is an open-source message broker and queueing server. RabbitMQ server is written in Erlang. It is created on the Open Telecom Platform framework for fail over and clustering. When exchanging large amount of data between client and server, the best approach is to use the RabbitMQ server and use a back-end service to consume the messages, process them and send them to the database. In RabbitMQ, data is sent to back-end services and server queue. So, if the back-end service fails, the data is stored in the server. This approach is used to resource saving, prevent data loss and a better organisation of the messages.

OpenAMQ has a faster speed of message routing and high degree of robustness, WireAPI for publish/subscribe application programming. It is in the normal working state and OpenAMQ server opens the direct mode to improve the message forwarding performance. In direct mode, it is five times faster than the normal rate of message forwarding to improve efficiency and reduce delay. Apache Qpid is an open-source messaging system. It provides transaction management, queueing, distribution, security, management, clustering, federation and heterogeneous multi-platform support.

5.1 Status Update of Clients

In future, AMQP provides the feature for some cases or situations like when any client disconnected ungracefully or when any client connects and subscribes for particular topic which it is interested in. These features are used to notify another client about the disconnected client and help newly subscribed clients to get a status update immediately after subscribing and do not have to wait until publishing clients send the new update.

To notify other clients about an ungracefully disconnected client:

Some clients lost the connection; the battery is empty or in any other imaginable case, they will disconnect ungracefully from time to time. In order to take appropriate action, it would be good to know that if a connected client has disconnected gracefully or not. So this feature is used to notify other clients about an ungracefully disconnected client.

When each client is connected to a broker, they specify its message which includes Topic Name, Flag, Quality of Service (QoS) and Payload. Broker detects that the client has disconnected ungracefully till it stores the message. The broker sends the message to all subscribed clients on the topic, when the client has disconnected and stored message will be discarded by sending a DISCONNECT message. It helps to implement strategies when drops the connection of a client or at least informs to other clients about the offline status.

To help newly subscribed clients to get a status update immediately after subscribing to a topic:

A publishing client can only make sure that its message gets delivered safely to the broker. It has no guarantee that a message is actually received by a subscribing client. When a new client is connected and subscribed to the interesting topics, there is no guarantee when the subscriber will get the first message, because it totally depends on a publisher on that topic. It can take a few seconds, minutes or hours until the publisher sends a new message update on that topic. Until then, the new subscribing client is totally in the dark about the current status of that topic.

In the message, if Flag is set to "True", then the broker will store the last message and the corresponding QoS for that topic. So, when a new client is connected and subscribed to interesting topics, it will receive the message immediately after subscribing. For each topic only one message will be stored by the broker. So messages can help newly subscribed clients to get a status update immediately after subscribing to a topic and do not have to wait until a publishing client sends the next update.

6 Conclusion

AMQP provides asynchronous publish/subscribe communication with messaging. It is store-and-forward feature that ensures reliability even after network disruptions, which is its main advantage. It supports reliable communication via message delivery guarantee primitives including at most once, at least once and exactly once delivery. When compared all other IoT protocols with AMQP protocol, it gives better performance compared to others. Therefore, AMQP protocol provides the guarantee of message delivery and it is a reliable communication even after a network failure.

References

1. Al-Fuqaha, Ala, et al. *Internet of things: A survey on enabling technologies, protocols, and applications*, IEEE Communications Surveys & Tutorials 17.4 (2015): 2347–2376.
2. Karagiannis, Vasileios, et al., *A survey on application layer protocols for the internet of things*, Transaction on IoT and Cloud Computing 3.1 (2015): 11–17.
3. Luzuriaga, Jorge E., et al., *A comparative evaluation of AMQP and MQTT protocols over unstable and mobile networks*, 2015 12th Annual IEEE Consumer Communications and Networking Conference (CCNC). IEEE, 2015.
4. Maciej, Krzysztof Grochla, and Aleksander Seman, *Evaluation of highly available and fault-tolerant middleware clustered architectures using RabbitMQ*, Computer Science and Information Systems (FedCSIS), 2014 Federated Conference on. IEEE, 2014.
5. Xiong, Xuandong, and Jiandan Fu, *Active Status Certificate Publish and Subscribe Based on AMQP*, Computational and Information Sciences (ICCIS), 2011 International Conference on. IEEE, 2011.
6. Subramoni, Hari, et al., *Design and evaluation of benchmarks for financial applications using Advanced Message Queuing Protocol (AMQP) over InfiniBand*, High Performance Computational Finance, 2008. WHPCF 2008. Workshop on. IEEE, 2008.
7. Fernandes, Joel L., et al., *Performance evaluation of RESTful web services and AMQP protocol*, 2013 Fifth International Conference on Ubiquitous and Future Networks (ICUFN). IEEE, 2013.
8. Vinoski, Steve, *Advanced message queuing protocol*, IEEE Internet Computing 10.6 (2006): 87.
9. G. Panchal , A. Ganatra, Y. Kosta, D. Panchal, "Forecasting Employee Retention Probability using Back Propagation Neural Network Algorithm", *IEEE 2010 Second International Conference on Machine Learning and Computing (ICMLC)*, Bangalore, India, pp. 248–251, 2010.
10. G. Panchal, A. Ganatra, P. Shah, D. Panchal, "Determination of over-learning and over-fitting problem in back propagation neural network", *International Journal on Soft Computing*, vol. 2, no. 2, pp. 40–51, 2011.
11. G. Panchal, A. Ganatra, Y. Kosta, D. Panchal, "Behaviour analysis of multilayer perceptrons with multiple hidden neurons and hidden layers," *International Journal of Computer Theory and Engineering*, vol. 3, no. 2, pp. 332–337, 2011.
12. G. Panchal and D. Panchal, "Solving np hard problems using genetic algorithm," *International Journal of Computer Science and Information Technologies*, vol. 6, no. 2, pp. 1824–1827, 2015.
13. G. Panchal, D. Panchal, "Efficient attribute evaluation, extraction and selection techniques for data classification," *International Journal of Computer Science and Information Technologies*, vol. 6, no. 2, pp. 1828–1831, 2015.
14. G. Panchal, D. Panchal, "Forecasting electrical load for home appliances using genetic algorithm based back propagation neural network," *International Journal of Advanced Research in Computer Engineering & Technology (IJARCET)*, vol. 4, no. 4, pp. 1503–1506, 2015.
15. G. Panchal, D. Panchal, "Hybridization of Genetic Algorithm and Neural Network for Optimization Problem," *International Journal of Advanced Research in Computer Engineering & Technology (IJARCET)*, vol. 4, no. 4, pp. 1507–1511, 2015.
16. Y. Kosta, D. Panchal, G. Panchal, A. Ganatra, "Searching most efficient neural network architecture using Akaikes information criterion (AIC)," *International Journal of Computer Applications*, vol. 1, no. 5, pp. 41–44, 2010.
17. *AMQP: Advanced Message Queuing, version 0.8, AMQP working group protocol specification*, June 2006 [Online] Available. https://www.iona.com/opensource.
18. Programming WireAPI, http://www.openamq.org/
19. Pivotal Software, Inc., Messaging that just works, [Online] Available: https://www.rabbitmq.com, 2014

The Role of Internet of Things and Smart Grid for the Development of a Smart City

Sudeep Tanwar, Sudhanshu Tyagi and Sachin Kumar

Abstract The smart grid (SG) and Internet of things (IoT) are arguably most important feature in any smart city (SC). If energy unavailable for a significant period of time, all other associated functions will be eventually affected. On the other side, IoT ensures integration of diversified networks, wireless technologies, and appliances. This enables smart communication between various devices, so that they can smartly communicate with each other. Smart city is one of the applications of IoT. Smart city encompasses improved infrastructure for better economic growth and quality life style. Recent technological trends have led to the rise in demand for efficient communication between devices without any human intervention, thus creating a huge market for IoT. With the abundance of opportunities for growth, every developed country, even some of the developing countries, are now come up with their own fully funded IoT projects. This paper summarized the role of IoT for the development of smart city and also presents architecture for the smart city for efficient utilization of infrastructure and challenges of IoT infrastructure development with their solution; further it enlists applications of IoT based on their domain.

Keywords Smart city · Internet of things · Smart grid · Future internet

1 Introduction

Over the last couple of years, a rapid technological development and research is going on in the field of IoT, especially targeting smart city-based applications. Two main reasons for these developments are as follows: first, exponential growth in population in urban area, and second global increment in a number of smart

S. Tanwar (✉)
Department of CE, Institute of Technology, Nirma University, Ahmedabad, India
e-mail: sudeep.tanwar@nirmauni.ac.in

S. Tyagi
Department of ECE, College of Technology, GBPUA&T, Pantnagar, India
e-mail: sudhanshutyagi123@gmail.com

S. Kumar
Department of ECE, Amity University, Lucknow Campus, Noida, UP, India
e-mail: skumar3@lko.amity.edu

© Springer Nature Singapore Pte. Ltd. 2018   23
Y.-C. Hu et al. (eds.), *Intelligent Communication and Computational
Technologies*, Lecture Notes in Networks and Systems 19,
https://doi.org/10.1007/978-981-10-5523-2_3

devices/objects/things connected to the internet, which can participate in any IoT-based application. According to the United Nations, currently 50% of world population lives in cities and by year 2050, this percentage is expected to increase and become 70% of world population, i.e., 2.3 billion [1]. In addition to this, as per the report given in [2], the number of interconnected devices is expected to increase by 50 billion in the next 10 years. Nowadays, cities consume 60% of all water demand and are forecasted to increase by 6 times in the next 50 years, even though in present scenario some municipalities lose up to 50% of their water resource because of leaky infrastructures [3] and similarly some portion of electricity gets wasted due to line losses. Increasing city population has brought the focus toward the development of cities in a sustainable manner for improving the quality of lives in cities. Thus, due to these facts, major range of problems can be solved by the use of IoT and smart grid (SG) for the development of a smart city. In September 2009, the European Union passed a strategic research road map of IoT that was proposed by CERP (Cluster of European Research Project). In January 2009, [4] CEO of IBM proposed the idea of "Smart Earth", in the meeting with the U.S. President along with several business leaders. A proposal was presented there recommending installation of sensors in rivers, railways, power grid, forest, roads, medical bottles, trucks, and other objects.

The energy management is undoubtedly a crucial factor for any city. Power failure cannot be tolerable at any cost while planning to build the SC. Therefore, SC needs to perform the following tasks: First, it supports existing power systems through latest designs, automation of services and supply, monitoring and controlling of remote services, and must include setup for charging and discharging process. Second, it gives awareness among users about how to use energy in a cost-effective manner and suggest alternative options, to enable them to make decisions. Third, it integrates all energy resources at a single platform. Thus, SG forms an integral part of the smart city, without which full optimization of the resources is not possible. The concept of smart city is similar to human organ system that works coherently even when put under extreme conditions. So in smart city diverse applications like energy, water, transportation, public health, and safety work in tandem resulting in better services for the inhabitants.

A city usually consists of a very complex system of different types of infrastructure like ICT infrastructure, civil engineering infrastructure, social networks, financial network, etc. All of these systems require management effort like monitoring, reporting, and interaction to ensure efficient performance of all activities. The example of such system is water distribution system, gas distribution system, electricity supply system, public transport system, maintenance of city infrastructure (parks and roads), and waste management. The aim of an SC is to provide long-term solution to the city transportation system, public administration, health, and public utility.

The paper is organized as follows: Sect. 2 highlights the IoT platform classification at application level. The challenges in smart cities are described in Sect. 3. The generic architecture of IoT for smart city is given in Sect. 4. Section 5 highlights the IoT services to be used in SC. Section 6 gives open issues which need to be taken care while thinking for smart city. Finally, we conclude the article in Sect. 7 with future directions.

2 IoT Platform Classification at Application Level

This section provides a brief classification of IoT at application level. IoT brings communication paradigm to the next level of 'anytime-anywhere-anyhow' concept, required mainly for the development [5–7] of an SC, as shown in Fig. 1. At application layer IoT is divided into various platforms like e-Governance platform, manufacturing industry platform, service industry platform, and business-oriented IoT services platform.

2.1 e-Governance IoT Platforms

In order to improve government information system and management, the government (local, state, national level) should use the IoT services in all areas of important concern. For example, smart city developments include traffic control, environment protection, citizen security, education, health, water conservation, etc. The e-Governance through IoT will support the foundation of city-, state-, and national-level economic development and management.

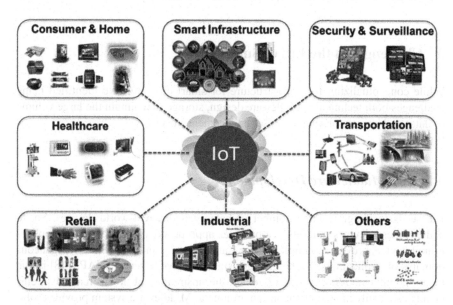

Fig. 1 High-level application view of IoT

2.2 Manufacturing-Based and Service-Based Company IoT Platform

In order to increase market share and improve service assurance, the manufacturing industry requires an independently funded and developed IoT project for distribution, transport, and supply chain management systems. The service-based companies may also use IoT for internal management in company, which should be developed independently.

2.3 Business-Oriented IoT Services

The business-oriented IoT structure plays an important role in making of growing economy because it attracts investment from development industries. It boosts economy by optimization and integration of all end users and equipments in modern industry. For example, RFID (Radio Frequency Identification) in product logistics is useful for customized secure delivery; further tracking of product is possible. The service providers for electricity, gas, water, and health can employ services of IoT, which can convert them into major players of their industry.

3 Challenges in the Era of Smart City

While conceptualizing IoT infrastructure for smart city, we have identified many challenges both technical (Middleware design, storage to maintain the large volume of data) and citizen (privacy and user engagement) level.

3.1 IoT Middleware Development

IoT needs to integrate large number of heterogeneous real-world devices. So, to address these devices, middleware provides appropriate abstraction to the application of device, provides multiple services, and connects and joins the heterogeneous domains of applications. Naming, addressing, and profile server (NAPS) [8] is one of the popular middlewares that handled the upstream and downstream data efficiently especially in heterogeneous environment. Middleware system provides solution of device heterogeneity, service interoperability, device discovery, and device self-management. In the layered architecture of IoT middleware layer works in the bidirectional mode. Middleware works like an intermediate between the application and physical layer [9]; one can implement this by any of these: SensorGird4Env approach [10] or out smart approach [11]. SensorGird4Env has a core collection of

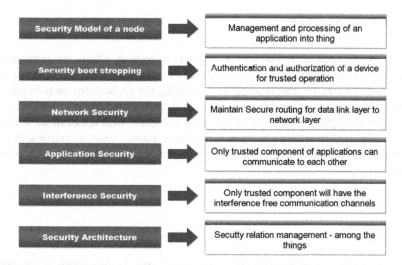

Fig. 2 The class of security

services with semantic sensor web architecture from where discovery, integration, and publication of data have been done. Out smart addresses four necessary services of smart city, i.e., waste management, water and sewage, street lighting, smart metering, and environment monitoring.

3.2 Privacy and Anonymity

Security is one of the challenging issues for IoT; in [12], several conventional security requirements have been discussed. The class of security for IoT [9] is shown in Fig. 2. Encryption was used to maintain the confidentiality; message authentic codes (MAC) were used to maintain the integrity. When an individual interacts on IoT platform, then large amounts of data are being recorded, stored, processed, and shared. So the anonymity and privacy of user should be ensured. To solve this problem there are two approaches: privacy trust mechanism and security mechanism. Privacy and trust-based mechanisms were proposed for WSN (wireless sensor network) like SAS (Simple Anonymity Scheme) and CAS (Cryptographic Anonymity Scheme). Security mechanisms use cryptography and key management like ECC (Elliptic curve Cryptography), and PBC (Pairing-Based Cryptography) is feasible with constrained IoT devices because they need less computation as compared to other [13–15] cryptography technologies.

3.3   Crowd Sourcing

Social and technological innovations need collaborative efforts by companies, researcher, municipalities, and users. User needs to participate in smart city experiments and services, by downloading and installing the applications on their smartphones. Through them developers can access data collected by integrated sensors present in smartphones. Most of the users do not install such applications without any benefits, and in order to increase user's participation, some incentive schemes are required. Those incentive schemes have to match with the interest and needs of users.

3.4   Storage

Cloud storage provides the flexibility and the security in the system at low cost. Advanced generations of the wireless communication system are also supporting the cloud-based storage. Mobile cloud computing (MCC) [16] is one of the attractive issues of cloud computing and very popular to solve the dynamic storage problem and computing competences. In order to use the same under the applications of SC, security issues like authentication and authorization related to MCC must be taken care. Why are we integrating the security issues to the MCC? Several users want to use remote storage on MCC, which creates the potential data storage on cloud and will be targeted from other adversaries. MCC will also support to reduce the delay and jitter to perform the task if proper connectivity is maintained in the application areas.

3.5   Demand-Based Service Delivery

Due to potential use of IoT devices in the SC scalable, suitable, reliable, and on-time services will be required from either main IoT equipment supplier or third-party servicing center. This is one of the challenging issues to maintain the SC so that resident will have QoS for their IoT-based equipments. Domain-oriented services [17] are very popular and handled by the third-party service provider; work will be done under the application development environment (ADE). Predefined components of the domain-specific application may use the drag and drop facility of GUI-based ADE. With the inclusion of service-oriented readers, RFID technology can also be used to provide the better service. In [18] detailed discussion was given on resource-oriented architecture (ROA) to provide the service for IoT equipments under business perspective.

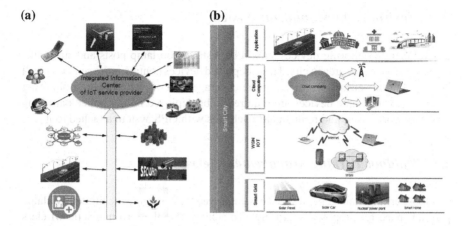

Fig. 3 Proposed architectures of IoT for smart city **a** generic architecture, **b** layered architecture

4 Proposed Generic Architecture of IoT for Smart City

The core element of the proposed generic architecture of IoT is the integrated information center run by IoT service provider shown in Fig. 3a and the layered architecture of SC is shown in Fig. 3b. The need of this integrated information center is to collect, process, distribute, and publish the data of every sensor node in the city. User that subscribes the IoT services can get information from information center through mobile service provider and also from Internet service provider. The city administration connects integrated information center via emergency command. Every government office is attached to integrated information center for e-Government platform. The other facilities like electricity services, traffic monitoring and control, smart parking, medical services, environment monitoring, security services (fire and citizen security), and water and gas supply are also directly attached to the city integrated information center for their data storing and processing. This data may be collected and stored by using cloud computing and data centers.

5 IoT Services for Smart Cities

There are diversified applications of IoT that can be employed for smart city projects, and these can be summarized as follows.

5.1 Environment Monitoring System

Environment monitoring devices should be installed on lamp posts and street like poles of across every area in city. These deployed devices contain noise sensor as well as air pollution sensors. To increase the coverage area of environment monitoring system network, some devices are required to be deployed on the police patrolling cars, transport buses, and municipality vehicles with GPS system attached to it.

5.2 Outdoor Parking Management Services

Outdoor parking management service is employed for managing the limited available parking space for proposed smart city. To achieve this the ferromagnetic wireless sensors are installed to each parking space of the places that have high foot count like Airport, Railway Stations, Court House, etc. Collected data of these sensors are stored in integrated information center of city and then end user can use this data through application of smartphone or through public displays installed at entry gate of each parking that shows a number of free parking places in different areas. Automated multilevel underground and on ground parking systems are already in existence where this technology can be further incorporated.

5.3 Participatory Sensing System

This service requires user's mobile phone to sense environment data such as noise, temperature, GPS coordinates, etc. [19, 20]. The information is collected and stored in integrated information center (IIC). The information provides important data regarding area which needs attention and is underdeveloped. IP multimedia sub-system (IMS), an IoT-based architecture [20], was used in the applications those were focused on ecosystem. Short-range communications of smartphones like blue-tooth, RFID, and near-field communication (NFC) were the basic interest of IMS architecture. IoT was included in two layers, application layer and perception layer. Validation of the IoT-based IMS architecture was performed on the net field programmable gate array (Net FPGA) which is a low-cost open-source platform. By the use of same application of smartphone, user can subscribe special services from which he can get notification of specific kind of events like traffic jam, protest, car accident, etc., currently happening in city.

5.4 Augmented Reality Service

This service can be utilized to tag a place of city as a point of interest (POI) like parks beaches, touristic points, public services, and shops. User can get additional information about tagged places and also leave feedback for city administration with

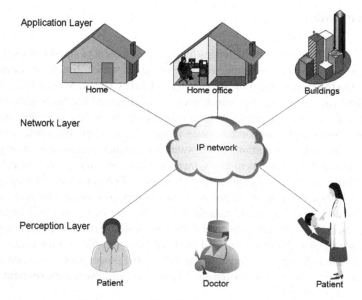

Fig. 4 IoT architecture for the persons with disabilities

the use of application of smartphone. This knowledge will help authorities to adjust their resource like public transport, garbage collection, etc.

5.5 Garden Irrigation Monitoring System

The garden irrigation monitoring system [21] provides real-time data for parks and garden authorities regarding the current status of parks [22]. IoT device should deploy underground sensors to measure the moisture content in soil and other relevant parameters. Apart from providing data to concerned authority, this service may alert other users when irregular condition gets detected. This service will reduce water wastage, energy, and labor cost for park maintenance.

5.6 Supportive Devices for the Persons with Disabilities

Special attention and specific care are required to the persons with physical disabilities. Hence, appropriate supportive IoT devices are required to those persons. Nature of IoT device should be very specific as per the requirement. In most of the situations physically disable persons are dependent on others. Therefore, SC should provide the special look after to those persons, and a possible IoT-based architecture is shown in Fig. 4.

6 Open Issues

As per the classifications and challenges for the development of smart city explained in multiple sections, we have selected different areas which directly or indirectly contribute to the development of SC. Services of IoT, CC, and SG can be used but one can find various open issues which need to be addressed and investigated properly. There are number of proposals exist addressing about the role of IoT and SG in SC [13]. Some of the important issues those cannot be ignored while developing the SC are traffic congestion, waste water management, awareness among citizens, unplanned developments in cities, increased level of heterogeneity, scalability, interoperability, mobility, efficient energy utilization of sensor nodes, data analytics, security, selection of appropriate protocol, and quality of service. Due to potential data production by IoT devices, data fusion is also one of the key challenges for the researchers. Data fusion supports the improvement in processing efficiency and updated intelligence. Management and manipulations are the two important parameters for fusion. Smartly, energy utilization from the power grid is also one of the open issues; a combination of SG and electric vehicle can solve the above-mentioned problem to some extent.

7 Conclusions

In this paper, the role of IoT and SG for the development of smart city has been presented. Different areas have been identified which will relate directly or indirectly to the development of smart city; however, there are various issues which need to be addressed and investigated properly. Critical issues while planning for smart city are traffic congestion, waste water management, awareness among citizens, unplanned developments in cities, increased level of heterogeneity, scalability and Interoperability, mobility, need to support any device with at least basic communication and computing capabilities, wide levels of data producers and consumers, disturbance occurred by haphazard development and finally environmental degradation. Due to limited resources and lack of awareness among citizens about positive impacts of smart city, the development of smart city has not been fully explored.

References

1. United Nations, "World Urbanization Prospects", 2009.
2. Ericsson, "More than 50 billion Connected Devices", Report, Feb. 2011.
3. D. Toppeta, "The Smart City Vision: How Innovation and ICT can build Smart, Livable, Sustainable Cities", *iThink, Report* 005, 2010.
4. P. Guillemin, and P. Friess. "Internet of things strategic research roadmap", *CERPIoT Project*, 2009. [Online]. Available - http://www.internet-of-things-research.eu/.

5. M. O'Droma, and I. Ganchev, "The creation of a ubiquitous consumer wireless world through strategic ITU-T standardization", *Communications Magazine, IEEE*, 48(10), pp. 158–165, 2010.
6. F. Xia, L. T. Yang, L. Wang, and A. Vinel, "Internet of Things", *International Journal of Communication Systems*, 25(9), pp. 1101–1102, 2012.
7. J. Gubbi, R. Buyya, S. Marusic and M. Palaniswami, "Internet of Things (IoT): A vision, architectural elements, and future directions", *Future Generation of Computer Systems*, 29, pp 1645–1660, 2013.
8. C. H. Liu, B. Yang and T. Liu, "Efficient naming, addressing and profile services in Internet-of-Things sensory environment", *Ad Hoc Networks*, 18, pp 85–101, 2014.
9. T. Heer, O. G. Morchon, R. Hummen, S. L. Keoh, S. S. Kumar and K. Wehrle, "Security challenges in the IP-based Internet of Things", *Wireless Pers Commun*, 61, pp 527–542, 2011.
10. Andrea Zanella, Nicola Bui, Angelo Castellani, Lorenzo Vangelista & Michele Zorzi, Internet of Things for Smart Cities, *IEEE Internet of Things Journal*, 1(1), 2014, pp 22–32.
11. Qiang Tang, Kun Yang, Dongdai Zhou, Yuansheng Luo, & Fei Yu, A Real-Time Dynamic Pricing Algorithm for Smart Grid With Unstable Energy Providers and Malicious Users, *IEEE Internet of Things Journal*, 3(4), 2016, pp 554–562.
12. S. Alam, Moh. M. R. Chowdhury, and J. Noll, "Interoperability of Security Enabled Internet of Things", *Wireless Pers Commun*, 61, pp 567–586, 2011.
13. Maninder Jeet Kaur, Piyush Maheshwari, "Building smart cities applications using IoT and cloud-based architectures", *International Conference on Industrial Informatics and Computer Systems (CIICS)*, 2016, pp 1–5.
14. L. B. Oliveira, R. Dahab, J. Lopez, F. Daguano, and A. A. F. Loureiro, "Identity-based encryption for sensor networks", in Proceedings of the 5th Annual *IEEE International Conference on Pervasive Computing and Comm., PERCOM* '07, pp. 223–229.
15. Charith Perera, Canberra, Chi Harold Liu, Srimal Jayawardena, The Emerging Internet of Things Marketplace From an Industrial Perspective: A Survey, *IEEE Transactions on Emerging Topics in Computing*, 3(4), 2015, pp 585–598.
16. T. Shon, J. Sho, K. Han and H. Choi, "Towards Advanced Mobile Cloud Computing for the Internet of Things: Current Issues and Future Direction", *Mobile Netw Appl*, 19, pp 404–413, 2014.
17. J. P. Wang, Q. Zhu, and Y. Ma, "An agent-based hybrid service delivery for coordinating internet of things for 3rd party service providers", *Journal of Networks and Computer Applications*, 36, pp 1684–1695, 2013.
18. K. Dar, A. Taherkordi, H. HanBaraki, F. Eliassen and K. Geihs, "A resource oriented integration architecture for internet of things: A business processes prospective ", *Pervasive and Mobile Computing*, doi:10.1016/j.pmcj.2014.11.005.
19. N. Maisonneuve, M. Stevens, and B. Ochab, "Participatory noise pollution monitoring using mobile phones," *Info. Pol.*, 15, pp. 51–71, Apr. 2010.
20. E. Kanjo, J. Bacon, D. Roberts, and P. Landshoff, "Mobsens: Making smart phones smarter," *IEEE Pervasive Computing*, vol. 8, pp. 50–57, Oct. 2009.
21. V. Ramya, B. Palaniappan and B. George, "Embedded System for Automatic Irrigation of Cardamon Field using XBee-PRO Techonology" *International Journal of Computer Applications*, 53(14), pp. 36–43, September 2012.
22. Y. Kim, R. Evans and W. Iversen, "Remote Sensing and Controlo of na Irrigation System Using a Distributed Wireless Sensor Network", *IEEE Transactions on instrumentation and measurement*, 57(7), pp. 1379–1387, July 2008.

An IoT-Based Portable Smart Meeting Space with Real-Time Room Occupancy

Jaimin Patel and Gaurang Panchal

Abstract The collaboration is the main aspect of very large organizations. Meeting rooms are the main enablers of collaboration. However, most organizations face challenges in using their meeting rooms effectively due to lack of efficient and affordable systems. All the traditional purely software-based solutions are not enough to meet the demands of today's organization. So our aim is to incorporate the latest Internet of Things (IoT) technologies into the meeting rooms based on many different types of research. Our proposed system will help every organization those who want to convert their traditional meeting spaces into a smart meeting space without spending too much money and space. Our proposed system is based on sensors, microcontrollers and minicomputers. We have selected best of available devices keeping the price and portability in mind. This proposed system will be capable enough to display real-time room occupancy status, room booking and management, generate notifications and the control of room environment from a remote location.

Keywords IoT · Smart meeting space · Room occupancy · PIR sensor

1 Introduction

If any organization want to hold more effective meetings, they have to invest in them [1–3]. Over the last few decades, corporations have invested heavily in technology designed to increase individual employee efficiency. But, fewer companies have extended this investment into the meeting rooms/spaces. Plush leather chairs and polished boardroom tables are no longer enough to furnish a meeting room, tools are essential to creating a fully functional workspace [3–6]. A smart meeting room uses various sensors to detect and sense activity and room environment of meeting

J. Patel (✉) · G. Panchal
U & P. U. Patel Department of Computer Engineering, Chandubhai S Patel
Institute of Technology, Changa, India
e-mail: jaiminpatel6890@gmail.com

G. Panchal
e-mail: gaurangpanchal.ce@charusat.ac.in

© Springer Nature Singapore Pte. Ltd. 2018
Y.-C. Hu et al. (eds.), *Intelligent Communication and Computational
Technologies*, Lecture Notes in Networks and Systems 19,
https://doi.org/10.1007/978-981-10-5523-2_4

rooms [6, 7]. This is done in order to provide real-time support to participants and to provide real-time occupancy status and conveniences to operate meeting room appliances using any web-enabled devices [8–29].

2 Related Work

2.1 Smart Meeting Rooms: A Survey and Open Issues [1]

Freitas et al. have proposed a mechanism for smart meeting rooms. This paper gives an overview of some of the latest technologies and different approaches used for the meeting rooms and labs and also states some of the key challenges in constructing smart spaces. This paper suggests the use of ambient intelligence. According to this paper, Smart Meeting Rooms (SMR) is a sub-discipline of Ambient Intelligent Environments.

2.2 SencePresence: Infrastructure-Less Occupancy Detection for Opportunistic Sensing Applications [2]

This paper represents the mobile technology to detect and count the human presence at a particular place. Sense Presence is a technology that combines smartphone-based acoustic and motion sensing to determine the number of people in a partially conversational and non-conversational environment. This system works well when all the participants bring their cell phone and keep it turned on.

2.3 IoT-Based Occupancy Monitoring Techniques For Energy-Efficient Smart Buildings [3]

This paper states that with the proliferation of Internet of Things (IoT) devices such as smart phones, sensors, cameras and RFIDs, it is possible to collect a massive amount of data for localization and tracking of people within commercial buildings. Enabled by such occupancy monitoring capabilities, there are extensive opportunities for improving the energy consumption of buildings. The paper concludes the use of various future technologies in this emerging era to save energy in buildings. The paper shows us to various technologies that are very useful in detecting human presence and are listed below:

1. Wi-Fi/Bluetooth-based occupancy detection
2. Camera- and OpenCV-based presence detection

3. RFID Tag-based presence detection
4. Sensor-based presence detection.

2.4   The Internet of Things: A Survey [4]

This paper gives a brief idea of Internet of Things, its applications, limitations, architecture, IoT protocol stack and different light weight IoT protocols.

2.5   A Smart Meeting Room Scheduling and Management System with Utilization Control and Ad-Hoc Support Based on Real-Time Occupancy Detection [5]

This paper gives a brief introduction to smart meeting room system and its implementation. The proposed system consists of good presence detection sensor with higher detection accuracy and microcontrollers and minicomputers. But they use typical caballed/wired connection in entire set-up. This could be a major limitation of the project. This shows the complexity of implementation. It is quite complex and costly process to set up wired connection in geographically distributed organization.

3   Limitations of the Existing Work

In this section, we describe the limitations of the existing work.

3.1   Most of the Corporate Do Not Have Any Systems that Show Real-Time Room Occupancy and Booking System

Many corporate offices do not have modern meeting spaces and they require a human to manage meetings for them or they buy third-party services to manage their meetings/gathering. So they have to rely on other for such services. In the new era of technology, they are facing many problems to manage meetings and utilization of free spaces as the frequency of meetings have been increased due to the expansion of business.

3.2 Many Corporate Has Obsolete and Costly Systems

Many corporate offices have a biometric or ID card based system. Such systems are quite costly and large in volume. Systems like Biometric machines are not portable and quite annoying. The installation of such systems is quite complex and handling such systems requires knowledge in particular field.

3.3 Many Room Booking Systems Are Purely Based on Calendar System

It is a quite outdated approach to using calendar-based meeting room booking system. It does not show the real-time occupancy of a room and it is a major disadvantage of this system. One more disadvantage of this system is that all the participants should be notified through an email or any other means of communication. If the meeting gets cancelled, again they have to notify all the participants regarding meeting status.

3.4 Many Systems Are Based on Video Camera and OpenCV Software

This is the modern approach to detect room occupancy and yet better utilization of spaces. Even meetings can be recorded for future use.

But no technology is 100% secure. The system with a camera can be hacked and the privacy of an organization or participants can be compromised. The attacker can take advantage of it and might use it for own good or against the organization.

Since it uses OpenCV, the system requires higher GPU (Graphical Processing Unit) and computational power and it is quite costly to develop a software based on openCV. So the implementation of this system is quite complex and costly. More processing means more power consumption. The disadvantages over advantage are very high so such systems should not be implemented in offices where privacy is the main concern.

4 Proposed Approach

In this section, we have mentioned the details architecture of the proposed approach and its aim.

4.1 Architecture of Proposed System

The architecture of Smart Meeting Space consists of following components:

1. Sensor Node (PIR Sensor and Arduino UNO Microcontroller)
2. Gateway (Raspberry Pi 3 Model B)

Figure 1 shows that each room is equipped with a sensor node. Sensor node is a combination of a PIR Sensor, Arduino UNO Microcontroller and a zigbee (XBee Series 2). These sensor nodes are capable of detecting the presence in a room and capable controlling the room environment. These sensor nodes are connected to a common gateway-cum-local server via zigbee. The gateway collects the data from all these sensor nodes. Same time gateway acts as web server where a database and web service reside so a user can access these services remotely using any web-enabled devices. The user will see a good GUI on user end and can easily see room occupancy status in real time. And based on the room availability one can even book the meeting room remotely.

4.2 Aim of the Project

The aim of this project is to build a low-cost, low-power and portable Smart Meeting Space with real-time room occupancy status and a booking facility.

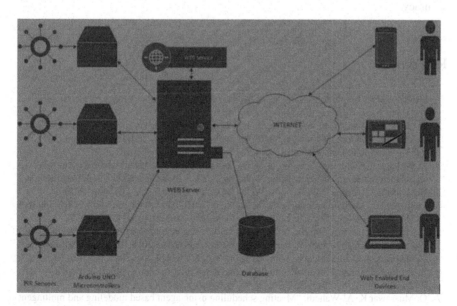

Fig. 1 Architecture of proposed system

1. Converting any traditional meeting room will become easier and affordable using this proposed system.
2. This proposed system is very small in size yet easy to hide in a room.
3. The proposed system is very easy to setup and manage.

5 Conclusion and Future Work

We have analysed key protocols, technologies and services that can be incorporated into a smart meeting space.

The proposed system will be helpful for the companies with traditional meeting spaces who seek a productive and affordable meeting room solution. This system would be helpful for conducting meetings in office, industry using any web-enabled devices such as mobile or laptops from anywhere and anytime that will help a user to save his/her time and money. This solution for meeting room can be useful for schools and colleges too.

Adding IoT capabilities into meeting rooms can help converting old boring meeting rooms into affordable and manageable smart meeting rooms. And the installation is quite simple and management of these devices is quite simple. It also consumes very less amount of energy and occupies a very less amount of space in the room.

The future task of the proposed system will be to integrate more features like AI and machine learning to make this system real smart. And for accurate presence detection in a room multiple/same sensors can be fused together to achieve great accuracy.

References

1. Freitas, Carlos Filipe, Joao Barroso, and Carlos Ramos, "A Survey on Smart Meeting Rooms and Open Issues". International Journal of Smart Home 9.9 (2015): 13–20.
2. M. A. A. H. Khan, H. M. S. Hossain and N. Roy, "SensePresence: Infrastructure-Less Occupancy Detection for Opportunistic Sensing Applications," 2015 16th IEEE International Conference on Mobile Data Management, Pittsburgh, PA, 2015, pp. 56–61.
3. K. Akkaya, I. Guvenc, R. Aygun, N. Pala and A. Kadri, "IoT-based occupancy monitoring techniques for energy-efficient smart buildings," 2015 IEEE Wireless Communications and Networking Conference Workshops (WCNCW), New Orleans, LA, 2015, pp. 58–63.
4. Luigi Atzori, Antonio Iera, Giacomo Morabito c (2010) 'Computer Networks', "The Internet of Things: A Survey." SpringerLink. N.p., n.d. Web. 3 Nov. 2016, 54(15), pp. 2787–2805.
5. L. Tran, "A smart meeting room scheduling and management system with utilization control and ad-hoc support based on real-time occupancy detection," 2016 IEEE Sixth International Conference on Communications and Electronics (ICCE), Ha Long, 2016, pp. 186–191.
6. Z. Yu Y. Nakamura, "Smart meeting systems: A survey of state-of-the-art and open issues", ACM Computing Surveys (CSUR), vol. 42, p. 8, 2010.
7. O. Mussawar K. Al-Wahedi, "Meeting scheduling using agent based modeling and multiagent decision making", 2013 Third International Conference on Innovative Computing Technology (INTECH), pp. 252–257, 2013.

8. A. Radloff, A. Lehmann, O. Staadt and H. Schumann, "Smart Interaction Management: An Interaction Approach for Smart Meeting Rooms," 2012 Eighth International Conference on Intelligent Environments, Guanajuato, 2012, pp. 228–235.
9. The ZigBee Alliance | Control your World. (n.d.). Retrieved October 25, 2016, from http://www.zigbee.org/.
10. Smart Connected Communities. (n.d.). Retrieved October 10, 2016, from http://www.cisco.com.
11. A. Ganatra, G. Panchal,Y. Kosta, C. Gajjar, "Initial classification through back propagation in a neural network following optimization through GA to evaluate the fitness of an algorithm", *International Journal of Computer Science and Information Technology*, vol. 3, no. 1, pp. 98–116, 2011.
12. G. Panchal, A. Ganatra, Y. Kosta, D. Panchal, "Forecasting Employee Retention Probability using Back Propagation Neural Network Algorithm", *IEEE 2010 Second International Conference on Machine Learning and Computing (ICMLC)*, Bangalore, India, pp. 248–251, 2010.
13. G. Panchal, A. Ganatra, P. Shah, D. Panchal, "Determination of over-learning and over-fitting problem in back propagation neural network", *International Journal on Soft Computing*, vol. 2, no. 2, pp. 40–51, 2011.
14. G. Panchal, A. Ganatra, Y. Kosta, D. Panchal, "Behaviour analysis of multilayer perceptrons with multiple hidden neurons and hidden layers," *International Journal of Computer Theory and Engineering*, vol. 3, no. 2, pp. 332–337, 2011.
15. G. Panchal and D. Panchal, "Solving np hard problems using genetic algorithm," *International Journal of Computer Science and Information Technologies*, vol. 6, no. 2, pp. 1824–1827, 2015.
16. G. Panchal, D. Panchal, 11Efficient attribute evaluation, extraction and selection techniques for data classification," *International Journal of Computer Science and Information Technologies*, vol. 6, no. 2, pp. 1828–1831, 2015.
17. G. Panchal, D. Panchal, "Forecasting electrical load for home appliances using genetic algorithm based back propagation neural network," *International Journal of Advanced Research in Computer Engineering & Technology (IJARCET)*, vol. 4, no. 4, pp. 1503–1506, 2015.
18. G. Panchal, D. Panchal, "Hybridization of Genetic Algorithm and Neural Network for Optimization Problem," *International Journal of Advanced Research in Computer Engineering & Technology (IJARCET)*, vol. 4, no. 4, pp. 1507–1511, 2015.
19. G. Panchal, D. Samanta, "Comparable features and same cryptography key generation using biometric fingerprint image," *2nd IEEE International Conference on Advances in Electrical, Electronics, Information, Communication and Bio-Informatics (AEEICB)*, pp. 1–6, 2016.
20. G. Panchal, D. Samanta, "Directional Area Based Minutiae Selection and Cryptographic Key Generation Using Biometric Fingerprint," *1st International Conference on Computational Intelligence and Informatics (Springer)*, pp. 1–8, 2016.
21. G. Panchal, D. Samanta, S. Barman, "Biometric-based cryptography for digital content protection without any key storage," *Springer (Multimedia Tools and Application)*, pp. 1–18, 2017.
22. G. Panchal, Y. Kosta, A. Ganatra,D. Panchal, "Electrical Load Forecasting Using Genetic Algorithm Based Back Propagation Network," *in 1st International Conference on Data Management*, IMT Ghaziabad. MacMillan Publication, 2009.
23. G. Patel, G. Panchal,"A chaff-point based approach for cancelable template generation of fingerprint data," *International Conference on ICT for Intelligent Systems (ICTIS 2017)*, p. 6, 2017.
24. J. Patel, G. Panchal, "An iot based portable smart meeting space with real-time room occupancy," *International Conference on Internet of Things for Technological Development (IoT4TD-2017)*, pp. 1–6, 2017.
25. K. Soni, G. Panchal, "Data security in recommendation system using homomorphic encryption," *International Conference on ICT for Intelligent Systems (ICTIS 2017)*, pp. 1–6, 2017.
26. N. Patel, G. Panchal, "An approach to analyze data corruption and identify misbehaving server," *International Conference on ICT for Intelligent Systems (ICTIS 2017)*, pp. 1–6, 2017.

27. P. Bhimani, G. Panchal, "Message Delivery Guarantee and Status Up- date of Clients based on IOT-AMQP," *International Conference on Internet of Things for Technological Development (IoT4TD-2017)*, pp. 1–6, 2017.
28. S. Mehta, G. Panchal, "File Distribution Preparation with File Retrieval and Error Recovery in Cloud Environment," *International Conference on ICT for Intelligent Systems (ICTIS 2017)*, p. 6, 2017.
29. Y. Kosta, D. Panchal, G. Panchal, A. Ganatra, "Searching most efficient neural network architecture using Akaikes information criterion (AIC)," *International Journal of Computer Applications*, vol. 1, no. 5, pp. 41–44, 2010.

IoT Infrastructure: Fog Computing Surpasses Cloud Computing

Tasnia H. Ashrafi, Md. Arshad Hossain, Sayed E. Arefin,
Kowshik D.J. Das and Amitabha Chakrabarty

Abstract The Internet of Things can be defined as an expansion of Internet network further off conventional gadgets like computers, smart phones, and tablets to an assorted scope of gadgets and ordinary things that use embedded technology to communicate with the outer environment, all by means of the Internet. With the increasing time, data access and computing are proceeding towards more complications and hurdles requiring more efficient and logical data computation infrastructures but the vision of future IoT seems foggy as it is a matter of concern that our current infrastructure may not be able to handle large amount of data efficiently involving growing number of IoT devices. To get diminished from this issue, Fog computing has been introduced which is an expansion of cloud computing working as a decentralized infrastructure in which reserving or computing data, service, and applications are distributed in the most efficient and logical position between source and the cloud. In this paper, the current infrastructure has been depicted and proposed another model of IoT infrastructure to surpass the difficulties of the existing infrastructure, which will be a coordinated effort of Fog computing amalgamation with Machine-to-Machine (M2M) intelligent communication protocol followed by incorporation of Service-Oriented Architecture (SOA) and finally integration of agent-based SOA. This model will have the capacity to exchange data by breaking down dependably and methodically with low latency, less bandwidth, heterogeneity in less measure of time maintaining the Quality of Service (QoS) precisely.

T.H. Ashrafi (✉) · Md.A. Hossain · S.E. Arefin · K.D.J. Das · A. Chakrabarty
Department of Computer Science and Engineering, BRAC University,
66, Mohakhali C/A, Dhaka 1212, Bangladesh
e-mail: tasnia.heya@gmail.com

Md.A. Hossain
e-mail: arshad.antu@gmail.com

S.E. Arefin
e-mail: erfanjordison@gmail.com

K.D.J. Das
e-mail: koushikjay66@gmail.com

A. Chakrabarty
e-mail: amitabha@bracu.ac.bd

© Springer Nature Singapore Pte. Ltd. 2018   43
Y.-C. Hu et al. (eds.), *Intelligent Communication and Computational
Technologies*, Lecture Notes in Networks and Systems 19,
https://doi.org/10.1007/978-981-10-5523-2_5

Keywords Internet of things (IoT) · Cloud computing
Fog computing · M2M communication · Service-oriented architecture (SOA)
Agent-based SOA · Quality of service (QoS) · Heterogeneous devices

1   Introduction

The more we are heading towards the future with ever-growing number of IoT
devices which are expected to take place in a giant number almost doubling or
tripling by reaching near trillions within a couple of years. Handling such amount of
data will be a matter of challenge for present infrastructure as it is developing in a
discrete manner and is not well structured. IoT does not follow any specific
infrastructure yet as Internet of Things is a growing field and many architectural
models have been proposed by researchers which are on the verge of getting
implemented with successful results though a complete functional model is still
needed which can be effectuated in real world as each of this architecture individually lacks behind on some prospects on which other architectures can perform
better. Therefore, those are unable to be considered as a complete and perfectly
workable infrastructure which IoT demands for such a massive number of data.

Having a significant consequence, cloud computing is an anarchic technology as
despite everything, it has a few issues in regards to service-level agreements
(SLA) with security, privacy, and energy efficiency. Cloud has three conveyance
models which are Software as a Service (SaaS), Platform as a Service (PaaS), and
Infrastructure as a Service (IaaS) with various level of security issues [1]. These
security issues of service models of cloud computing can be decreased surprisingly
through applying trust management in the agent-based SOA level (third level) of
the proposed infrastructure [2].

Moreover, transferring data from IoT to the cloud for analyzing would require
inconceivable measures of bandwidth. Present cloud models are not intended for
the volume, variety, and velocity of data that the IoT produces [3]. Edge computing
is a model that empowers extensive variety of applications and services to the end
users by amplifying distributed cloud computing model towards the edge of network. Data transactions over the network by means of Internet without
human-to-human connections or human-to-machine associations are the strong
points of IoT which incorporates components, for example, mobility support,
extensive variety of geo-distribution, wireless accessibility, and a superior stage for
a specific number of IoT services [4].

Here, new and efficient infrastructure for IoT using distributed Fog computing
model has been proposed which will be able to transfer data by reliable, systematic,
and efficient analysis with low latency, less bandwidth, heterogeneity in less
amount of time maintaining the Quality of Service (QoS) befittingly.

2 Literature Review

From the standpoint of IoT, devices have the capacity to communicate within themselves without any interventions. A wireless sensor network contains expansive number of mobile devices, RFID and sensors which are considered as the end point of the system. For every device to be recognized, it stores their necessary information in a local database with a name, model number, hardware type, unit, version, and timestamp of the sensor values which makes metadata for every gadget [5]. An initial configuration of the device and its endpoints is done making an API by JSON containing the static description [6]. This API uses HTTP GET and PUSH request to read and write the configuration of the device. The attributes of this API include Location, Id, Name, Value, URI, and Protocol [7].

A middleware is required for the nodes in the end points to accomplish the distributed architecture where the primary concern is the communication process of the devices.

Services of IoT are represented by five-tuple differing IoT from other services. Where Ini = Identification, Pui = Purpose, PIi = Provided Interface, RIi = Required Interface, SAi = Set of Attributes[8] and defined as,

$$\text{IoT } service_i = \langle In_i, Pu_i, PI_i, RI_i, SA_i \rangle \tag{1}$$

A kind of URI (Uniform Resource Identifier) named URN (Uniform Resource Name) is used for distinguishing each IoT service nodes individually with name or id [8].

Service-oriented architecture is one of the most used architectures for heterogeneous devices. A light-weighted distributed service composition model can be used for data procurement which will convert existing basic devices into better software units along with complex functionality added with corresponding QoS features following soft-real-time restrictions with most appropriate time of specific services [9]. Being a lightweight model, it can be used in the lower levels of the fog computing nodes with lower resources.

In the upper middleware, an agent-based composition model has been used for the nodes to make complex compositions. These nodes have high resources, thus, it does not need to be lightweighted. An agent-based composition has three actors: service provider, business process manager, and users for making autonomous agents to collaborate to attain its goals by being able to do activities which are, building workflows, compose the external web services, and monitor execution. This architecture is based on Society of Agents and mostly made up of two components, component manager and workflow manager. Here, each component manager is in charge of interacting with one or more web services and can communicate with web services by converting into JSON formatted messages within the agent society and workflow manager supports users to build the workflows, compose external web services and monitor their executions.

Local names and the tokens are the main factors for the Trust Management Principles. These work as a distributed system building peer-peer networks with each node responsible for security as it provides proper credentials to access resources of other nodes. The authorization is very critical and important as it helps building up the trusted network without a central control following "least privilege" [10].

IoT applications run real-time control and analytics by data transmission in real time ensuring fast response time. These nodes have temporary storage for saving data locally. Receiving data from Fog nodes, data are collected and analyzed to generate business insights for new application rules to be carried out to the other fog nodes [11].

3 Proposed Infrastructure

The Fog model is separated in several layers of nodes. Here, as shown in the figure, the proposed infrastructure consists of four layers. In the first layer, there are end devices and they are communicating within themselves via M2M protocols and directly to their upper layer. The next two layers, second and the third, are considered as the core of the fog model defined as Middleware. Now, in the second layer, which is the lower fog layer, service-oriented architecture is used by utilizing service composition model which is like a local repository for services and a very lightweight and low resource model. After that, in the third layer or upper layer of fog, the agent-based SOA (a modified version of SOA) is implemented which is capable of autonomous service functionalities and computations by learning which service to request in order to get the local data which is possible by using the proper agent from the society of agents. Finally, at the end, the main cloud server for higher computations has been used. This gives this model a novel solution leading to the support for a wide range of devices (Fig. 1).

In brief, at all part of the Fog model, the time scales of these interactions range from seconds to minutes (real-time analytics), and even days (transactional analytics) which results in, the Fog supporting several types of storage, from short-lived at the lowest layer to semipermanent at the highest layer. Wider geographical coverage, and longer time scale can be obtained in higher layers. The ultimate global coverage is provided by the cloud which is used as repository for data that has a permanence of months and years, and which is the basis for business intelligence analytics [4].

IoT Infrastructure

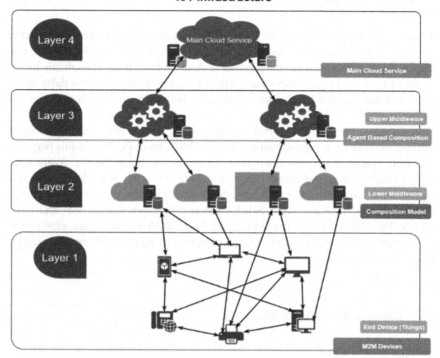

Fig. 1 Proposed infrastructure

4 Advantages of Proposed Infrastructure

1. Geo-Redundant Replicable Local Backup: In this proposed model, there will be geo-redundant replicable local backups that means the data will be stored and requested to local storages which will be geographically very close to end devices so that even if the main server gets down or busy, end devices can have backups to get response.

2. Less latency, less request redundancy, less bandwidth and traffic: By the backups mentioned above, it focuses on less data and network latency where it will take less time for sending data or packets from one place to another. Another concern was less bandwidth and traffic by reducing the pressure from main server by establishing local storages which also reduce the request redundancy where same requests should not hit the server again and again.

5   Result and Implementation

Datadog, a monitoring service for tracking the virtual machine's network has been used for monitoring purpose. Datadog provides monitoring as a service and to use that, Datadog agents need to be integrated in Azure VMs which sends metric of the Azure VMs to the Datadog dashboard. Here, Datadog agents can have delay up to 2 min to send the data to Datadog dashboard. This may cause a bit delay in the generated graphs.

In the experimental setup, end devices, SOA architecture, and agent-based architecture have been represented using Virtual Machines (VM). For this purpose, Microsoft Azure as an implementation structure has been chosen. Azure data centers were situated in different geographical positions, this is really efficient and convenient to perform some test runs. From the start, the plan was to use two different geographical positions: North Central US, South Central US, and Central US. The VMs represent the SOA, agent-based SOA and machines which were in the same geographically available data centers. The main cloud service could be deployed in any region.

5.1   Flowchart of Algorithms

See Figs. 2, 3, 4 and 5.

5.2   Result Graph in Individual VMs

A small-scale experiment was performed to monitor the network usage of each of the VMs for the proposed infrastructure by requesting and responding with JSON upto few hundred kilobytes. It is to be considered that, for a large-scale deployment, the requests and responses will exceed by millions and network will be adjusted to cope up with delivering terabytes of data.

5.3   Result Graph of South Central US

The graphs below show the network usage of the VMs of South Central US, which was involved with the test environment while the experiment was conducted.

For a trial within the first layer, request for the same service was sent from every device, "SCUSL1M1" at 2:31:40am (Fig. 6), "SCUSL1M2" at 2:36:00am and "SCUSL1M3" at 2:36:00am. Initially, VM of the second layer "SCUSL2M1" received the request but it did not have the service. So, request of the service was

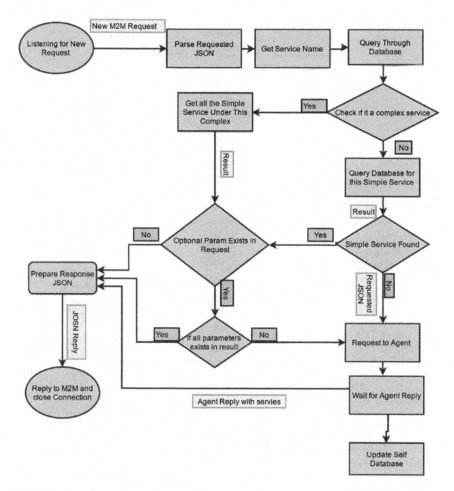

Fig. 2 SOA request and response

sent to "SCUSL3M1" at 2:31:00am (Fig. 7) in the third layer. When the service was not even found in the third layer it was sent to the main server "CUSMAIN" at 2:30:40am (Fig. 8). Later, from the main server, the result was saved and sent back to the third layer and after that in the second layer.

Observing "SCUSL2M1" (Fig. 7) and "SCUSL3M1" (Fig. 8) where only request from "SCUSL1M1" was sent but not from the other two devices as the result was already saved in the second layer while processing for "SCUSL2M1". In the main server, request received and requested at 2:36:00am was not sent from "SCUSL1M2" and "SCUSL1M3" but from North Central.

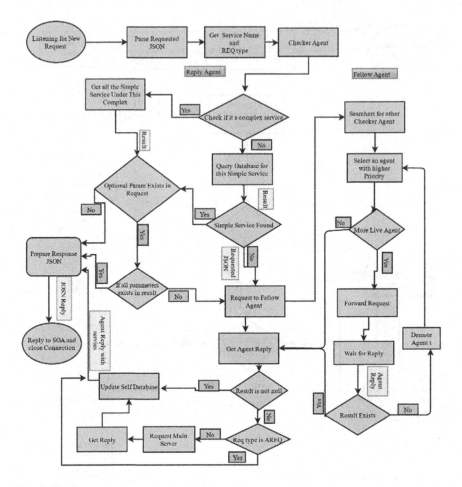

Fig. 3 Checker agent

5.4 Result Comparison

As mentioned before, two experiments were conducted in two different test environments. Among them, one represented the proposed infrastructure and the other one represented the conventional infrastructure. For the sake of computing the data transactions between the VMs and main cloud server and comparing the proposed and present infrastructure, same service was requested from four end devices of different regions as shown in Table 1.

After finishing the data processing, gathered result was projected through different graphs. Figure 9 shows the results of the conventional infrastructure, where four devices have requested for the same service in between 3:33:00 am and

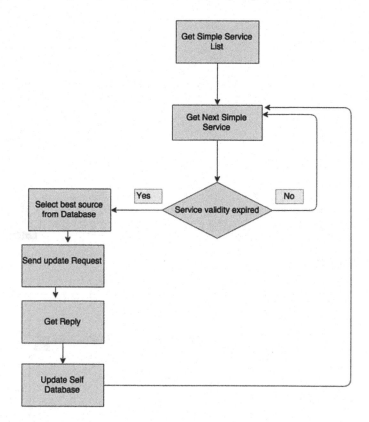

Fig. 4 Update agent

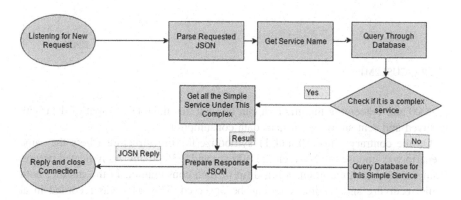

Fig. 5 Main server request and response

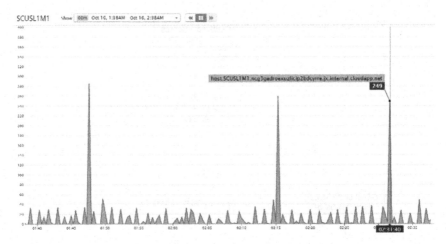

Fig. 6 SCUSL1M1

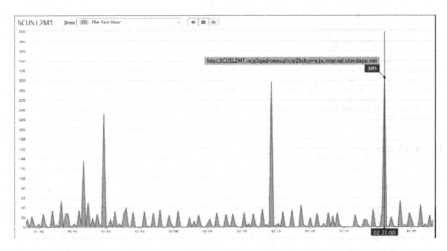

Fig. 7 SCUSL2M1

3:38:00 am. Perceiving the graph, it can be seen that the total amount of data both received and sent shows a constant data consumption.

On the contrary, Figs. 10 and 11 symbolize the infrastructure where the same scenario was imposed in between 2:31:00 am and 2:36:00 am. As indicated before, from the kilobyte/time graph a big drop of data consumption in the middle, both while receiving and sending data can be observed. This data was recorded in at most 15 s interval, which gives this inconsistent growth of the graph. Now, comparing the graphs it can be distinctly comprehended that the proposed infrastructure has a very low amount of network usage as it has a highest usage of 14–11 KB

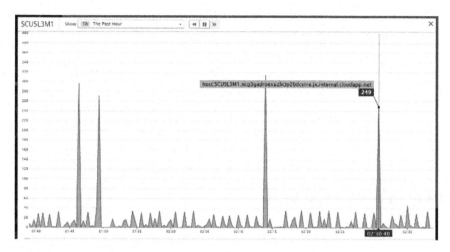

Fig. 8 SCUSL3M1

Table 1 VM information

VM name	Layer	Location
SCUSL1M1	Layer 1	South Central US
SCUSL1M2		
SCUSL1M3		
SCUSL2M1	Layer 2	
SCUSL2M2		
SCUSL3M1	Layer 3	
NCUSL1M1	Layer 1	North Central US
NCUSL2M1	Layer 2	
NCUSL3M1	Layer 3	

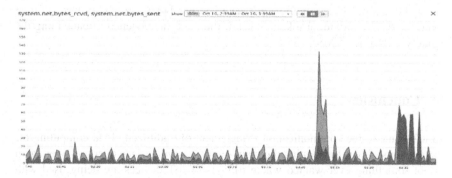

Fig. 9 Conventional infrastructure result

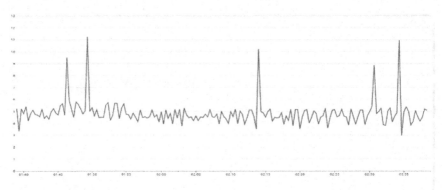

Fig. 10 Fog model's total received data

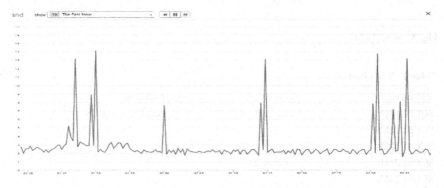

Fig. 11 Fog model's total sent data

whereas in the conventional infrastructure it reaches 70–60 KB within that time limit.

From these results of the described algorithms along with the comparison with present infrastructure, it can be ensured that the proposed infrastructure surpasses the traditional one in less traffic along with less bandwidth, reliability through trust management by providing token authentication and heterogeneity maintaining the Quality of Service (QoS).

6   Conclusion

In IoT context, the communication is inefficient of traditional cloud computing for having some flaws. Requesting the main server, a same service over and over again is not something ever wanted where trillions of IoT devices are being used with multiple requests from each of them. Therefore, this infrastructure is relied upon based on Fog computing and using edge networking, data is provided to the local nodes to run further computation. With running lightweight SOA in the fog nodes, very low resources are used and agent-based SOA provides autonomous service providence which is basically an artificial intelligence becoming mature over time. These make the infrastructure more efficient and scalable for future researches and open for additional improvements. Finally, to finish up, detailed descriptions, proper definitions, proposed model's implementation, and its pseudocodes and the final results to verify the infrastructure's advantages over the currently used infrastructure have been described in this paper.

References

1. B. Azimdoost, H. R. Sadjadpour and J. J. Garcia-Luna-Aceves. 2013. *Capacity of Wireless Networks with Social Behavior*. In IEEE Transactions on Wireless Communications, vol. 12, no. 1, pp. 60–69. doi:10.1109/TWC.2012.121112.111581.
2. K. P. Birman, R. van Renesse and W. Vogels. 2001. Spinglass: secure and scalable communication tools for mission-critical computing. In DARPA Information Survivability Conference & Exposition II, 2001. DISCEX '01. Proceedings, Anaheim, CA, vol. 2, pp. 85–99. doi:10.1109/DISCEX.2001.932161.
3. F. Boccardi, R. W. Heath, A. Lozano, T. L. Marzetta and P. Popovski. 2014. *Five disruptive technology directions for 5G*. In IEEE Communications Magazine, vol. 52, no. 2, pp. 74–80. doi:10.1109/MCOM.2014.6736746.
4. F. Bonomi, R. Milito, J. Zhu, and S. Addepalli. 2012. *Fog computing and its role in the internet of things*. In Proceedings of the first edition of the MCC workshop on Mobile cloud computing (MCC '12). ACM, New York, NY, USA, 13–16. doi:10.1145/2342509.2342513.
5. S. Subashini, V. Kavitha. 2011. A survey on security issues in service delivery models of cloud computing. In Journal of Network and Computer Applications, vol. 34, no. 1, pp. 1–11. doi:10.1016/j.jnca.2010.07.006.
6. S. K. Datta, C. Bonnet and N. Nikaein. 2014. An IoT gateway centric architecture to provide novel M2M services. In Internet of Things (WF-IoT), 2014 IEEE World Forum on, Seoul, 2014, pp. 514–519. doi:10.1109/WF-IoT.2014.6803221.
7. F. H. Bijarbooneh, W. Du, E. C. H. Ngai, X. Fu and J. Liu. 2016. Cloud-Assisted Data Fusion and Sensor Selection for Internet of Things. In IEEE Internet of Things Journal, vol. 3, no. 3, pp. 257–268, June 2016. doi:10.1109/JIOT.2015.2502182.
8. R. Garcia-Castro, C. Hill, and O. Corcho. 2011. SemserGrid4Env Deliverable D4.3 v2 Sensor network ontology suite.
9. Sandra Rodríguez-Valenzuela, Juan A. Holgado-Terriza*, José M. Gutiérrez-Guerrero and Jesús L. Muros-Cobos, "Distributed Service-Based Approach for Sensor Data Fusion in IoT Environments".
10. Giancarlo Fortino and Wilma Russo, "Towards a Cloud-assisted and Agent-oriented Architecture for the Internet of Things", 2013.
11. Cisco. 2015. Fog Computing and the Internet of Things: Extend the Cloud to Where the Things Are.

Sensing–Actuation as a Service in Cloud Centric Internet of Things

Suchismita Satpathy, Bibhudatta Sahoo, Ashok Kumar Turuk and Prasenjit Maiti

Abstract Trillions of connections and interconnection without cloud are just impossible. Therefore, Sensing–Actuation as a Service is a cloud centric service delivery model which authorizes access to the IoT Architecture (IoT-A). Sensed, actuated, and computed data from various existing mobile devices can be accessed by end user or consumer through IoT Platforms on a pay-as-you-go fashion of cloud which may enhance the usage of existing sensors deployed in the IoT network. Proper sensor management, virtual sensor management web servers, and IoT platforms are the emerging components of this architecture. Opportunistic sensors, participatory sensors along with sensor owners claim for various challenges like cost, reliability, trustworthiness, etc. Similarly, the expectations of the end users also appear as a big challenge. In this paper, we present a Sensing–Actuation as a Service Delivery Model (SAaaSDM) framework consisting of sensing, actuating, and computing with open issues and future directions for the researchers in this field. This paper has used the standard UML notations to represent our proposed model for Cloud centric Internet of Things.

Keywords Internet of Things · Cloud centric IoT · S^2aaS · Sensors SAaaSDM

1 Introduction

All possible kinds of data, documents, images, recordings, and games everything were created for people and it is all about people. The Internet of Things consists

S. Satpathy (✉) · B. Sahoo · A.K. Turuk · P. Maiti
National Institute of Technology, Rourkela, India
e-mail: satpathy.suchi@gmail.com

B. Sahoo
e-mail: bibhudatta.sahoo@gmail.com

A.K. Turuk
e-mail: akturuk@nitrkl.ac.in

P. Maiti
e-mail: pmaiti1287@gmail.com

© Springer Nature Singapore Pte. Ltd. 2018
Y.-C. Hu et al. (eds.), *Intelligent Communication and Computational Technologies*, Lecture Notes in Networks and Systems 19,
https://doi.org/10.1007/978-981-10-5523-2_6

of sensors, software, hardware, and connectivity to accredit it to produce enormous amount of service by exchanging information with the producer, manufacturer, and all other connected devices. An enactment of utilizing a network of faraway servers provided on Internet to stock, control, organize, and supply sensory data, rather than a bounded server or any personal computer is called cloud computing which enables companies to absorb a computer resource as a utility rather than creating an infrastructure in home [1]. Cloud computing delivers an ample amount of storage capacity and process capability using which huge amount of sensory data could be collected by linking the cloud through two types of gateways from both the sides those are physical sensor gateway from sensor side and cloud gateway from cloud side. The physical sensor gateway gathers all the information from the physical sensor nodes. Then after abbreviating and compressing the information, it transmits them to the cloud gateway that decomposes and stocks them in sufficiently large storage of cloud which is basically the concept of service delivery model. Recently, the IoT and cloud integration seems to be one of the most demanding and winning solutions in order to opportunely manage the proliferation of both data and devices. The principal idea is to employ the cloud computing platform to stock and exercise the sensed information and then using edge computing, accelerate, and improve the performance of cloud computing which remits to sensory data processing at the edge of a wireless network as the substitute to cloud network or any centralized data warehouse and finally provide the sensor data and information to the end users or consumers after the processing is being completed. Sensing and actuation accessories accept to be allotment of the cloud infrastructure and accept to be managed by afterward the circumscribed cloud approach, i.e., through a set of Appliance Programming Interfaces fortifying limited ascendancy of hardware and software resources admitting their bounded position. The main actors in the scenario are the mobile device owners and the end users. Mobile device owners accommodate sensing analysis and actuation assets architecture up the Sensing–Actuation as a Service Delivery Model (SAaaSDM). The sensor networks owners, accessory owners, device owners, general peoples alms their Personal Digital Assistants as an antecedent of data in a crowd-sourcing model could be the examples of mobile device owner [2–6]. We presume that the resources of sensing and actuation are allotted to the basement via an amount of hardware-constrained units, which we accredit as the physical sensor nodes. The physical sensor nodes provide resources of sensing and actuation and act as intermediary in affiliation to the infrastructure of cloud which charge to accept connectivity to the Internet for the applicability of our approach.

1.1 Motivations

At this moment, the smart phones even consist of various sensors like accelerometer, gyroscope, digital compass, barometer, ambient light sensor, heart, humidity, temperature, and proximity sensors that one can use as participatory sensors. The population density in London exceeds 4000 inhabitants per square kilometer and the UK smart phone penetration reaches 55% [7]. The density of smart phones' sensors in London today exceeds 14,000 sensor per square kilometer. By 2020, the global number of sensor-equipped and location-aware devices (e.g., wearable, smart home, and fleet management devices) will reach tens of Billions, potentially creating dense, dynamic, location-aware to manage networks of devices that can realize the vision of providing a versatile remote sensing services, known as "Sensing as a Service" (S^2aaS) [7–10]. We conjecture that employing IoT devices sensing resources in a cloud computing like platform to support remote sensing applications may be an effective approach to realize the Sensing–Actuation as a Service vision.

The further part of this paper is arranged as follows. Section 2 introduces the system architecture of SAaaS Delivery Model with the objectives and the assumptions. A use case scenario is discussed in Sect. 3 which highlights the various aspects of the SAaaSDM in cloud-centric IoT. Finally, Sect. 4 ends the paper with some closing consideration.

2 Sensing–Actuation as a Service Delivery Model (SAaaSDM) in Cloud centric IoT

2.1 Objectives and Assumptions

We are trying to propose a novel model to organize and manage the existing sensing data in our SAaaSD model which is in cloud-centric IoT system. We are trying to utilize the non-cloud gateway that is the Physical Sensor Management (PSM) to perform various data processing tasks. For-bye, the gateway of the cloud side supervises data recommendation. Then the physical sensor gateway and the cloud sensor gateway authenticate the functionalities of sensory data security. This section of SAaaSDM is focused to mediate the appropriate sensed data from the physical sensors to the cloud through the cloud gateway in a swift, quick, authenticate, and assured way [11]. Sensory data processing may elongate the lifespan and existence of the network, decrease the requirement of the storage, and enhance the service delivery performance of the model. This model may help to decrease the traffic and bandwidth needed for the data transmission of the sensory data and storage in the cloud [12, 13]. Also, it is competent of latency reduction, security enhancement, and speedy data processing by giving the appropriate and desired sensory data and information to the mobile users [14]. We are assuming heterogeneous sensors in our service delivery model which may be opportunistic or participatory in nature.

2.2 SAaaSDM System Architecture

Mobile Device Owner A Mobile Device Owner is an actor who owns the physical sensors which are participatory in nature and the owner of the smart phone as per the entity name. A Mobile Device Owner whether allows or denies others to use those personal sensors for specific purposes through Sensing–Actuation as a Service Delivery Model (SAaaSDM). The all possible advantages for the Mobile Device Owner would be the chartering fees for sharing the Participatory Sensor's information or any reward or discount in some specific convenient services. First of all the Mobile Device Owner has to register the participatory sensors with theirs specific properties to SAaaSDM infrastructure.

Participatory Sensors Participatory sensing is the abstraction of contributing sensory information to anatomy a physique of knowledge. Participatory sensors depend on the owner's approval whether to share the information or not. That is why such sensors are more secure than the opportunistic sensors in this dynamic cyber world.

Physical Sensor Management The Physical Sensor Management (PSM) acts as a sensor gateway. Each participatory sensor gathers the sensory data and the information and sends to the PSM. After the PSM receives the sensory data, it processes the data to check whether it is applicable for the cloud demand services or not. If applicable then it sends the processed data to the cloud gateway else it discards the faulty data and sends a warning message to the Mobile Device Owner to increase the feasibility or to disconnect the sharing.

Cloud Sensor Management This is the cloud gateway which accepts the compressed encrypted information of sensed data from the PSM. It decrypts and decompresses the received data by decryption and decompression units respectively. Now the ready to serve sensory data are deposited, kept, organized, and managed by robust servers in Cloud Sensor Management (CSM). After receiving the requested service details from the Virtual Sensor Management, the vital culpability of CSM is to spot and notice obtainable sensors. When CSM receives the service request, it tries to collect all possible and feasible sensory information fulfilling the demand of VSM. Then it sends a request to the Sensor Quality Review Management (SQRM) to get all the information, properties, and functionalities details of the selected sensors. Once the CSM gathers all according to the request of VSM, it sends the selected sensory data with proper details to the Virtual Sensor Management.

Sensor Quality Review Management This is another cloud gateway which receives the properties details of the sensors, functionality details of the selected processed sensory data rental fees information given by the Mobile Device Owner from the PSM. All the details are deposited, kept, organized, and managed by robust servers in SQRM. Once the CSM sends a request to get all the information, properties, and functionalities details of the selected sensors, it provides as per the requirements. After the delivery of the service, the end user sends the quality improvement request if he/she is not satisfied with the quality of the service and sensor details. Once the SQRM is informed from ESTP, it checks the sensor quality whether the quality review is valid or not. If the sensor details deemed faulty SQRM sends the warning message to the Mobile Device Owner through PSM.

Virtual Sensor Management The encrypted service forwarded from the ESTP with some service level agreements are being decrypted, processed, and executed in this section of SAaaSDM. The VSM converts the decided SLAs to certain parameters. The service details with the decided parameters are sent to the Cloud Sensor Management for the discovery and selection of the IoT Device sensors. Once the CSM delivers the appropriate device sensor information and sensor details using the above parameters, Virtual Sensor Management creates a virtual sensor network and virtualizes the device sensors efficiently and tries to create optimum solutions for the service request sent from the ESTP. VSN is a concept which provides protocol solace for the origination, configuration, utilization, adaptation, maintenance, and preservation of subsets of sensors collaborating on specific tasks [15]. The arrangement of the applications, intensification of the performances and scalability, smoothing the resource sharing are the real job of Virtual Sensor Management.

Extended Sensing Task Provider Extended Sensing Task Provider (ESTP) acts as a gateway between the request and the service as it handles the Sensing Task Request (STR) initiated by the End User (EU). The objectives of the STR would be to retrieve the information about the existing sensor's sensory data and information and to execute distributed algorithms on Virtual Sensor Networks deployed on multiple interconnected devices. After receiving the STR, the ESTP analyzes it and translates it to a particular corresponding service. Now, the service consists of several number of virtual sensors which will create a Virtual Sensor Set (VS). The VS needs to be deployed on virtual sensors connected devices which all are located with in a certain geographic radius from the center. After finalizing the service from the STR, the ESTP sends the encrypted service to the VSM. ESTP creates a SLA according to the service created from the STR before it is handing over the service to the VSM. Once the VSM virtualizes the device sensors efficiently according to the parameters and creates an optimum solution for the service request, it delivers the STR service to the ESTP within decided time.

End User An End User may be an actor, any business organization, any institution, any scientific research communities, or the government who does not have the Physical Sensors of its own but need the sensory data for various purposes and applications. All the End Users require to enroll themselves and procure a well founded viable bona fide certificate from the authorization so that they will be able to utilize sensory data and information. End Users having minimum technical capabilities may also obtain needed sensory data because all the onerous tasks like integrating sensory data from number of CSMs, choosing suitable sensors as per the End User's need and demand are managed by various managements who finally deliver the optimum service back to the end user. After the successful delivery of the service, the end user pays the rent to the ESTP as decided earlier which will be forwarded to the Mobile Device Owner later. Also the end user sends the quality review to the ESTP in case of any violation.

Figure 1 shows the block diagram of the complete system model. Figure 2 shows the flow of the process of sensory data before submitting to the cloud. Figure 3 describes the sequence diagram of service request and delivery in the cloud. Finally

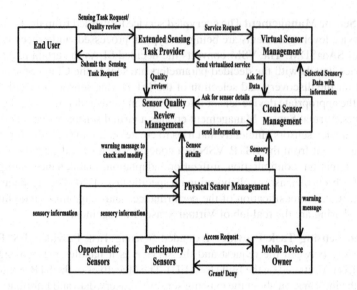

Fig. 1 SAaaSDM system architecture

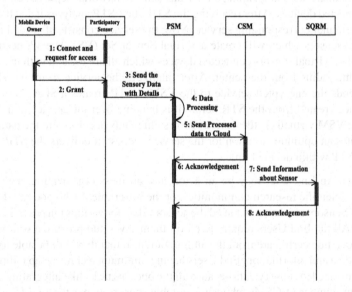

Fig. 2 Sequence diagram for sending sensory information to the Service Provider

the Fig. 4 depicts the quality measurement and improvement in the SAaaS Delivery Model.

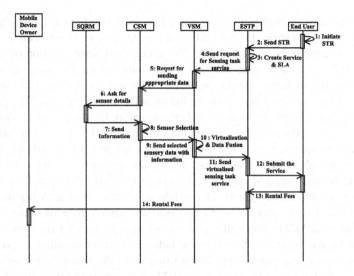

Fig. 3 Sequence diagram of service request and delivery

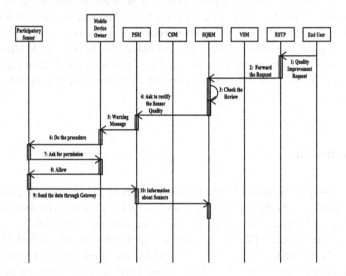

Fig. 4 Sequence diagram for quality improvement

3 SAaaSDM in Action: The Hereafter

3.1 Smart Institute

The researchers are passionately implicated in planning, creating, evolving accessible open platforms for sensory data collection, distribution, administration with assorted domains through abounding altered projects like OpenIoT and phenonet [11, 16]. In this scenario, the SAaaSDM allows to conduct so abounding altered things more effectively and efficiently. The educational administration of a country may analyze the intelligent adequate sensor nodes having the power to observe and monitor the students' attendance, performance information and institute environment, classroom decorum, employee information, and many more. The primary objective of positioning the sensors and accumulating the sensory data and information is to maintain everything within the institute, but the aforementioned set of sensors can be activated and utilized to accomplish an array of altered analysis and research proposals in various arena and disciplines. The sensory data and information could be distributed among altered corporations, federations, and institutions amid about various altered places. Due to bound funding, short-term requirements, a lot of the institutions may not be able to maintain ample calibration sensor deployments. The SAaaSDM allows all such kinds of institutions who are clumsy intellectual and not adequate to structure their individual sensor installation and placements, to accomplish various application using real information and sensory data with decidedly beneath cost. Again SAaaSDM creates opportunities beyond altered domains. The aloft mentioned sensory data can be used to understand the education standard and related issues of a country. Also, they can be used for food supply in the institution for canteen, mesh purposes. Employment issues can be discussed using the above sensor information. Added chiefly the SAaaSDM allows advisory observers to share information all over the world and comprehend the circumstances and experiences which are in accessible and inconvenient in their respective territories.

4 Conclusion

We have shown the potential Cloud centric Internet of Things to scale cloud computing vertically by exploiting sensing resources of IoT devices to provide Sensing–Actuation as a Service. Our paper describes the Sensing–Actuation as a Service Delivery Model (SAaaSDM) using some standard notations of Unified Modeling Language. As the procedure of UML, it clarifies and unravels the design of complicated procedure of the service delivery model and delivers enormous support at each and every stage of the required service delivery. In this paper, several sequence diagrams decorate the effective processes and flows of the complete model. The described delivery model may discover and virtualize the sensing resources of physical IoT devices to increase the usability of the existing sensors which will ultimately decrease the complexity of IoT world.

References

1. Pradhan, S., Satpathy, S., Tripathy, A.K., Mantri, J.K.: Research issues on WSBS performance evaluation. In: 2nd International Conference on Computational Intelligence and Networks (CINE) IEEE, doi:10.1109/CINE.2016.29, ISSN. 2375-5822, pp. 123–129, January (2016).
2. Hu, X., Li, X., Ngai, E.C.H., Leung, V.C., Kruchten, P.: Multidimensional context-aware social network architecture for mobile crowdsensing. In: IEEE Communications Magazine. doi:10.1109/MCOM.2014.6829948, ISSN. 0163-6804, vol. 52, no. 6, pp. 78–87, June (2014).
3. Kazemi, L., Shahabi, C., Chen, L.: GeoTruCrowd: Trustworthy query answering with spatial crowdsourcing. In: Proceedings of the 21st acm sigspatial international conference on advances in geographic information systems, SIGSPATIAL'13, ACM. doi:10.1145/2525314.2525346, ISBN. 978-1-4503-2521-9, acmid. 2525346, pp. 314–323 (2013).
4. Longo, F., Bruneo, D., Distefano, S., Merlino, G., Pulia to, A.: Stack4things: An openstack-based framework for IoT. In: 3rd International Conference on Future Internet of Things and Cloud (Fi-Cloud) IEEE. doi:10.1109/FiCloud.2015.97, pp. 204–211, August (2015).
5. Sheng, X., Tang, J., Xiao, X., Xue, G.: Leveraging GPS-less sensing scheduling for green mobile crowd sensing. In: IEEE Internet of Things Journal. doi:10.1109/JIOT.2014.2334271, ISSN. 2327-4662, vol. 1, no. 4, pp. 328–336, August (2014).
6. Yang, D., Xue, G., Fang, X., Tang, J.: Crowdsourcing to Smartphones: Incentive mechanism design for mobile phone sensing, Mobicom '12. In: Proceedings of the 18th annual international conference on Mobile computing and networking, ACM. doi:10.1145/2348543.2348567, ISBN. 978-1-4503-1159-5, acmid. 2348567, no. 12, pp. 173–184 (2012).
7. Abdelwahab, S., Hamdaoui, B., Guizani, M., Znati, T.: Cloud of Things for Sensing as a Service: Sensing resource discovery and virtualization. In: IEEE Global Communications Conference (GLOBECOM). doi:10.1109/GLOCOM.2015.7417252, pp. 1–7, December (2015).
8. Distefano, S., Merlino, G., Puliato, A.: A utility paradigm for IoT: The sensing cloud. In: Pervasive and mobile computing. doi:10.1016/j.pmcj.2014.09.006, ISSN. 1574-1192, vol. 20, pp. 127–144, July (2015).
9. Sheng, X., Tang, J., Xiao, X., Xue, G.: Sensing as a Service: Challenges, solutions and future directions. In: IEEE Sensors journal. doi:10.1109/JSEN.2013.2262677, ISSN. 1530-437X, vol. 13, no. 10, pp. 3733–3741, October (2013).
10. Sheng, X., Xiao, X., Tang, J., Xue, G.: Sensing as a Service: A cloud computing system for mobile phone sensing. In: IEEE Sensors, doi:10.1109/ICSENS.2012.6411516, ISSN. 1930-0395, pp. 1–4, October (2012).
11. Perera, C., Zaslavsky, A., Liu, C.H., Compton, M., Christen, P., Georgakopoulos, D.: Sensor search techniques for Sensing as a Service architecture for the Internet of Things. In: IEEE Sensors Journal. doi:10.1109/JSEN.2013.2282292, ISSN. 1530-437X, vol. 14, no. 2, pp. 406–420, February (2014).
12. Zhu, C., Leung, V.C., Yang, L.T., Shu, L.: Collaborative location-based sleep scheduling for Wireless Sensor Networks integrated with Mobile Cloud Computing. In: IEEE Transactions on Computers. doi:10.1109/TC.2014.2349524, ISSN. 0018-9340, vol. 64, no. 7, pp. 1844–1856, July (2015).
13. Zhu, C., Wang, H., Liu, X., Shu, L., Yang, L.T., Leung, V.C.M.: A novel sensory data processing framework to integrate sensor networks with mobile cloud. In: IEEE Systems Journal. doi:10.1109/JSYST.2014.2300535, ISSN. 1932-8184, vol. 10, no. 3, pp. 1125–1136, September (2016).
14. Chang, W., Wu, J.: Progressive or conservative: Rationally allocate cooperative work in mobile social networks. In: IEEE Transactions on Parallel and Distributed Systems, doi:10.1109/TPDS.2014.2330298, ISSN. 1045-9219, vol. 26, no. 7, pp. 2020–2035, July (2015).
15. Jayasumana, A.P., Han, Q., Illangasekare, T.H.: Virtual Sensor Networks-A resource efficient approach for concurrent applications. In: Fourth International Conference on Information Technology, ITNG'07. doi:10.1109/ITNG.2007.206, pp. 111–115, April (2007).

16. Perera, C., Zaslavsky, A., Christen, P., Georgakopoulos, D.: Sensing as a Service model for smart cities supported by Internet of Things. In: Transactions on Emerging Telecommunications Technologies. doi:10.1002/ett.2704, ISSN. 2161-3915, vol. 25, no. 1, pp. 81–93 (2014).

Soil Monitoring, Fertigation, and Irrigation System Using IoT for Agricultural Application

R. Raut, H. Varma, C. Mulla and Vijaya Rahul Pawar

Abstract India's population reached beyond 1.2 billion and the population rate is increasing day-by-day; then after 25–30 years there will be a serious problem for food, so the development of agriculture is necessary. Today, the farmers are suffering from the lack of rains and scarcity of water. The main objective of the project is to provide an automatic irrigation system and also to check the amount of the three major macronutrients, nitrogen (N), phosphorus (P), and potassium (K), in the soil thereby saving time, money, and power of the farmer. The N, P, and K amounts in the soil sample are determined by comparing the solution with color chart. This will describe the amount of N, P, and K as high, medium, and low. The traditional farm-land techniques require manual intervention. With the automated technology of irrigation the human intervention can be minimized. Whenever there is a change in temperature and humidity of the surroundings, these sensors sense the change in temperature and humidity and give an interrupt signal to ARM 7 Processor, thereby initiating the irrigation. All this functioning will be updated to the user by e-mail sent by the system PC through IoT.

Keywords NPK · Temperature and humidity check · Fertilizer dispensing
Irrigation · E-mailing through IoT

R. Raut (✉) · H. Varma · C. Mulla · V.R. Pawar
Department of Electronics and Telecommunication Engineering, Bharati Vidyapeeth's
College of Engineering for Women, Pune 411043, India
e-mail: ranirautb@gmail.com

H. Varma
e-mail: heenavarma99@gmail.com

C. Mulla
e-mail: chandbibimulla@gmail.com

V.R. Pawar
e-mail: pawarvr74@gmail.com

© Springer Nature Singapore Pte. Ltd. 2018
Y.-C. Hu et al. (eds.), *Intelligent Communication and Computational Technologies*, Lecture Notes in Networks and Systems 19,
https://doi.org/10.1007/978-981-10-5523-2_7

1 Introduction

India's major source of income is from agriculture sector and 70% of farmers and general people depend on the agriculture. In India most of the farming systems are operated manually. The available traditional techniques are like sowing, digging, and irrigation system.

Due to long-established methods of agricultural process, the Indian farmer faces many problems about productivity of agricultural product than others. It is due to unbalance feeding of fertilizer without knowing the actual requirement of nutrient to a particular crop.

Nowadays, soil is checked in the laboratory and proper examine of soil is done; their proportion in soil is checked.

1.1 Need of Automatic Irrigation

- Farmers would be able to smear the right quantity of water at the right time of day by automating farm or nursery irrigation.
- Avoiding irrigation at the inexact time of day, reduce runoff from overwatering saturated soils which will improve crop performance.
- Automated irrigation system uses valves to turn motor ON and OFF. Motors can be automated easily by using controllers and no need of labor to turn motor ON and OFF. It is a precise method for irrigation.

The system will check the soil nutrient level by the built-in standard color chart, and accordingly dispense fertilizers in the water reservoir. It will also check the water level in the reservoir simultaneously, and if not full, will get automatically filled. Later the system will check the soil moisture level as well as the temperature; accordingly, it will initiate irrigation in the fields

All this functioning will be updated to the user through Internet of Things (IoT) by sending an e-mail from the system PC.

2   Background Details and Related Work

Many research papers from reputed national and international journals are surveyed and few are presented here:

Suresh et al. [1] proposed a system in which N, P, K, and PH values of soil sample are measured in real time and compared with the pre-stored values received from agricultural department. The system also provides the information about the crops that can be grown in respective soils.

Shaligram and Singh [2] proposed a system that has an NPK kit to test the soil and also dispensing fertilizers required.

Parameswaran and Sivaprasath [3] proposed an automated irrigation system based on soil humidity. The irrigation status is updated to the server or local host using PC. This connectivity is carried by IoT.

Kim et al. [4] proposed remote sensing and control of an irrigation system using a distributed wireless sensor network.

Londhe and Galande [5] proposed an automated irrigation system using ARM processor. The system monitors and controls all the activities of irrigation system.

The valves are turned ON or OFF and automatically provide the systematic data regarding the soil pH and nutrients like nitrogen along with the proper recommendations and provide the communication interface.

Various limitations like only the irrigation and not the NPK values are updated to the user through IoT [3] in the work reported at national and international level; that is why the present system is designed for soil parameter monitoring and fertilizer dispensing along with irrigation using iot for agricultural application.

3 Proposed Approach

Present system consists of

A. *Hardware*:

(1) *Processor:*

The processor used in the system is ARM7 LPC2138. It has an 8-channel 10-bit ADCs that provide a total of up to 16 analog inputs, with conversion times as low as 2.44 ms per channel.

Multiple serial interfaces are used including two UARTs (16C550), two Fast I2C-bus (400 Kbit/s), SPI and SSP with buffering, and variable data length capabilities.

It also has 60 MHz max. CPU clock available from programmable on-chip PLL with settling time of 100 ms; also an on-chip integrated oscillator operates with external crystal in the range of 1–30 MHz and with external oscillator up to 50 MHz.

It plays a key role; it processes the data received by color sensor and compares data with the data which is already stored in controller according to the sensor output signal.

(2) *Sensors*:

- Color sensor: The color sensor used is 1185 SunRom color sensor which will sense the color of the chemicals in which the soil sample is mixed to check the soil nutrient level. The resultant color sensed by the sensor is matched with the values in the color chart programmed in the processor.

Table 1 Specifications of color sensor

Parameter	Valve	Unit	Notes
Operating voltage	5	VDC	Provide regulated 5 V supply
Current	20	mA	
Color detecting capacity	16.7 millions	RGB	R = 8 bit(2^8 = 256 levels) G = 8 bit(2^8 = 256 levels) B = 8 bit(2^8 = 256 levels) 256 * 256 * 256 = 16.7 million shades detection
Color measuring range	350–750	Nm	
Luminance range	100	Lux	
Response time	500	ms	
Output data baud	9600	Bps	5 V level output UART properties (8-N-1) Start bit: 1 bit Data bit: 8 bits Parity: none Stop bit: 1 bit

Table 1 shows the technical specifications of color sensor:

- Moisture sensor: HSY 220 senses the humidity in the surrounding environment and accordingly commands the processor.
- Temperature sensor: LM 35 series with operating temperature range from −55 °C to 150 °C. Its output voltage is proportional to the temperature. The output voltage is converted to digital form by the built-in DAC in the ARM processor.

(3) *Solenoid Valves*:

The system uses four solenoid valves, three for dispensing the N, P, and K fertilizers each, and one for irrigation through the reservoir.

B. *Software*:

- OrCAD used for designing PCB layout.
- MikroC PRO for AVR used for coding the microcontroller in embedded C.
- Flash Magic used to burn the program onto the µC.
- Visual Basics 6.0 used to create GUI, front-end and back-end acquisition, and sending e-mails.

C. *Working*:

The logic of the entire system is described here: The soil sample is first to be mixed with the chemicals; this is carried out manually. The resultant color of the chemical is sensed by the color sensor and fed to the processor through the serial port [6] (Fig. 1).

The processor compares the sensed color with the color chart which is programmed; and with reference to the table the processor decides whether the soil nutrient (N, P, K) level is appropriate or not; and if not appropriate, the processor

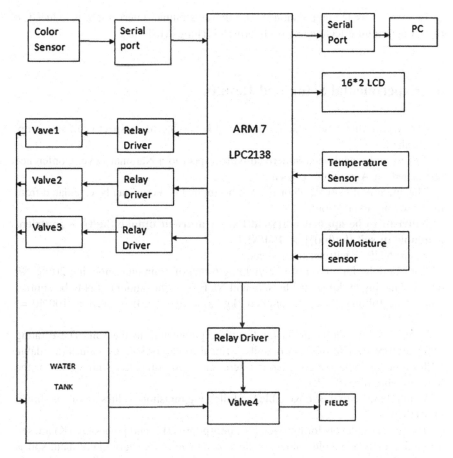

Fig. 1 System block diagram

commands the respective N, P, and K valve to open and dispense the required amount of fertilizer in the water reservoir.

Meanwhile, the system checks the water level in the reservoir; if not full, the reservoir will be filled.

The system also checks the soil moisture level as well as the surrounding temperature [7] by the moisture sensor (HSY 220) and temperature sensor (LM35), respectively. These analog signals are fed to the built-in ADC of the processor and then to the processor.

The processor then decides whether the humidity level and the temperature level are appropriate or not by the program fed; if the humidity and temperature are inappropriate, then it commands the valve of the reservoir to be open and irrigate the fields.

All this functioning will be updated to the user by an **e-mail** sent by the system through IoT [8].

In such a way the automation works and thus human intervention is reduced in digital agricultural techniques by Internet of Things (IoT) [9].

4 Experimental Setup and Results

In this system, threshold value for N, P, and K is selected irrespective of soil, crop, or weather.

The formula mentioned is derived with respect to a National Level Conference conducted on Agriculture at Delhi.

The amount of N or P2O5 or K2O to be applied to a crop can be calculated from the following expression:

Nutrient to be applied = [{yield(t/ha) x nutrient uptake (kg/t)} – {nutrient available in soil (kg/ha)}] × 100/NUE,

where NUE = nutrient use efficiency

For example, for a crop of rice yielding 6 t/ha of grain and removing 20 kg N/t on a soil having 60 kg available N/ha and NUE 40%, the amount of N to be applied will be as follows: N to be applied (kg/ha) = [(6 × 20) – (60)] × 100/40 = 150 kg N/ha

Table 2 shows rating limits for the three parameters in the soil. These rating limits are irrespective of crops or soils. Critical limits for soil test values (available NPK) used in India are summarized here and then compared with values in the look-up table in controller.

With reference to the above table the following threshold values are set as shown in Table 3.

The threshold levels for nitrogen (N), phosphorus (P), and potassium (K) are set. Table 3 depicts threshold values for three parameters. The agricultural land soil is tested and above results are obtained

The user present on-field or off-field gets the system update by an e-mail generated by the on-field PC and transmitted over IoT (Internet of Things).

Table 2 Rating limits for soil test values used in India

Nutrients	Low	Medium	High
Nitrogen (kg/ha)	<280	280–560	>560
Phosphorus (kg/ha)	<10	10–24.6	>24.6
Potassium (kg/ha)	<108	108–280	>280

Table 3 Threshold values for soil NPK level

Nutrients	Low	Medium	High
Nitrogen	<15	15–25	20–25
Phosphorus	16–20	20–35	35–50
Potassium	20–25	25–40	50–60

5 Future Scope

The system has a vast scope for future expansion including the addition of more parameters to be monitored and controlled. The system can also be implemented for sowing the seeds and also reaping the crops. We can also update the user about the system functioning through a mobile application.

In this paper, IoT platform is a viable solution in improving the agricultural techniques in the run of making a *Digital India*.

6 Conclusions

The system will measure the values of N, P, and K from the soil and also monitor the level of *soil nutrients* contents and accordingly dispense the required quantity of the *fertilizers* through *irrigation system*. All the data will be updated to the user through e-mail.

References

1. D. S. Suresh, Jyothi Prakash K. V., Rajendra C. J. "Automated Soil Testing Device", ITSI Transactions on Electrical & Electronics Engineering (ITSI-TEEE) ISSN (PRINT): 2320–8945, Vol. 1, Issue 5, 2013.
2. Dr. A. D. Shaligram, Nishant Singh, "NPK Measurement in Soil & Automatic Soil Fertilizer Dispensing Robot", International Journal of Engineering Research & Technology (IJERT) Vol. 3, Issue 7, July. 2014.
3. G. Parameswaran, K. Sivaprasath, "Arduino Based Smart Drip Irrigation System using Internet of Things" IJESC, Vol. 6, Issue 10, April. 2016.
4. Yunseop (James) Kim, Robert G. Evans, and William M. Iversen, "Remote Sensing and Control of an Irrigation System Using a Distributed Wireless Sensor Network", IEEE transactions on instrumentation and measurement, VOL. 57, NO. 7, PP 1379–1387, JULY 2008.
5. Gayatri Londhe, Prof. S.G. Galande, "Automated Irrigation System By Using ARM Processor", IJSRET, ISSN 2278–0882, Vol. 3 Issue 2, May 2014.
6. Leenata Vedpathak, Pooja Salape, Snehal Naik, "An Automated Agricultural Robot", IJARCCE, Vol. 4, Issue 3, March 2015.
7. Shweta S. Patil, Ashwini V. Malviya, "Agricultural Field Monitoring System Using ARM", IJAREEIE, Vol. 3, Issue 4, April 2014.
8. Nisha Mary Lemos, Shruti Narayan Nair, Sonali Sanjay Yadav, Prof. Dr. Vijaya Rahul Pawar, "Building a Smart City through an of Internet of Things (IoT)', IJSRD, Vol. 4, Issue 02, 2016| ISSN (online): 2321–0613.
9. Tanmay Baranwal, Nitika, Pushpendra Kumar Pateriya, "Development of IoT based Smart Security and Monitoring Devices for Agricultural" IEEE, Vol., pp 592–602, Issue 2016.

Part II
Intelligent Image Processing

Patch-Based Image Denoising Model for Mixed Gaussian Impulse Noise Using L_1 Norm

Munesh C. Trivedi, Vikash Kumar Singh, Mohan L. Kolhe, Puneet Kumar Goyal and Manish Shrimali

Abstract Image denoising is the classes of technique used to free the image form the noise. The noise in the image may be added during the observation process due to the improper setting of the camera lance, low-resolution camera, cheap, and low-quality sensors, etc. Noise in the image may also be added during the image restoration, image transmission through the transmission media. To obtain required information from image, image must be noise free, i.e., high-frequency details must be present in the image. There are number of applications where image denoising is needed such as remote location detection, computer vision, computer graphics, video surveillance, etc. In last two decades, numbers of models have been published for image denoising. In this research work, we proposed patch-based image denoising model for mixed impulse, Gaussian noise using L_1 norm. Mat lab 2014a on the Intel i5 with 4 GB RAM platform is used to simulate the proposed model. Simulation results show the effectiveness of our proposed model for image denoising as compared to state-of-the-art methods.

Keywords Image denoising · Patch-based · L1 norm · Mixed gaussian impulse noise

M.C. Trivedi (✉) · P.K. Goyal
ABES Engg. College, Ghaziabad, India
e-mail: munesh.trivedi@gmail.com

P.K. Goyal
e-mail: puneet.goyal@abes.ac.in

V.K. Singh
IGNTU, Amarkantak, India
e-mail: drvksingh76@gmail.com

M.L. Kolhe · M. Shrimali
University of Agder, Kristiansand, Norway
e-mail: mohan.l.kolhe@uio.no

M. Shrimali
e-mail: manishshrimali2009@gmail.com

M. Shrimali
JRN RVD, Udaipur, India

© Springer Nature Singapore Pte. Ltd. 2018   77
Y.-C. Hu et al. (eds.), *Intelligent Communication and Computational Technologies*, Lecture Notes in Networks and Systems 19,
https://doi.org/10.1007/978-981-10-5523-2_8

1 Introduction

Nowadays, videos and pictures play a major role in the entertainment. With the development of new technologies in the field of telecommunication, computing, and processing, the applications of image and videos are increasing rapidly. Now the world is moving from voice to video calling, and hence this needs high quality of videos, pictures, etc. With increase in the number of applications, the demand of image quality improvement techniques like image denoising, and image super resolution is also increasing. In recent years, many methods have published to fullfil these requirements [1–6].

During the image processing, steps like image observation, image transmission, image storing process the quality of image may be degraded. During the observation process, the quality of image is degraded due to the cheap and low-quality observing devices like camera, sensors, etc. In the observation process, image quality also degrades due to the factors like: if the lance of the camera/sensor not set properly, low-resolution camera/sensors, object may be far away from the observing device, object in motion, etc. In the image, video transmission process image, video quality is also degrading due to the noise (unwanted information added during the message passing through the various Chanel). The process of image restoration and fetching may also degrade the quality (adding noise) of the images.

In practical, the noisy image is the outcome of the imaging system uses to observe the image. But in simulation, we have to use degradation model to generate the degraded noisy image from original high-resolution and noise-free image. Let I_h is the original high-resolution noise-free image. I_{nl} image generating by image degradation model as follows:

$$I_{nl} = I_h + MGIN \tag{1}$$

where I_{nl} is the noisy low-resolution image, $I_h$is the original noise-free high-resolution image, and *MGIN* is the mixed Gaussian impulse noise.

Denoising of the image is ill-posed inverse problem. The objective of image denoising is to obtain I_h from the I_{nl}. To obtain I_h, numbers of research have been published that have used the some regularization technique to estimate the I_h form I_{nl} and prior are also used to select the best estimated value from the all estimated values. Some of the known priors are total variation (TV) regularization [7], edge prior [8], gradient profile prior [9], and Bayesian prior model [10, 11]. Few regularization techniques use the fact that we can sparsely represent most image signals using de-correlation transforms, e.g., discrete wavelet transform (DWT) or discrete cosine transform (DCT), so that the signal can be well separated from the noise [12, 13].

The remaining part of this paper is organized as follows: Sect. 2 gives the brief overview about the works which have done in this field. Section 3 describes the proposed denoising formulation and objective function of model. In this section, algorithm pseudocode is given. The mathematical model formulation and the objective function of proposed model is also given in this section. Section 4 explains simulation results (in the form of table and visual image) and discussion, and finally the summary of proposed model given in Sect. 5.

2 Related Work

G. Yu and G. Sapiro [12] used the discrete cosine transformation for image denoising. In this work, authors divided the impute images into the block and applied proposed method onto each block separately. This is the block wise processing. Soft thresholding based image denoising model published by D.L. Donoho [13]. In this image is partitioned into some subband and then the noise is reduced by applying soft thresholding to the coefficient of its. C. Zuo [14] proposed the image denoising model based on the quadtree-based nonlocal means (NLM) and locally adaptive principal component analysis. This model successfully removed the high-level noise from the image and gives better performance as compared to state-of-the-art methods in term of numeric result as well as visual quality.

3 Proposed Denoising Model: Algorithm and Mathematical Formulation

To formulate our model, we use following abbreviations

1. I_h: Original high-resolution noise-free image
2. I_{nl}: Noisy low-resolution image
3. UI_{nl}: Updated noisy low-resolution image
4. I_{dn}: Do-noised Image
5. $I_{h\ i}$: ith patch of original high-resolution noise-free image
6. $I_{nl\ i}$: ith patch of noisy low-resolution image
7. $UI_{nl\ i}$: ith patch of updated noisy low-resolution image
8. $I_{dn\ i}$: ith patch of do-noised Image

Let's start with the background of image denoising, the optimal solution to the problem defined in Eq. (1) is the MAP (maximum a posterior probability estimator) given in Eq. (2)

$$I_{dn} = arg\ max_{I_h}\ Pr(I_h|I_{nl}) \tag{2}$$

In the proposed model, we are targeting mixed Gaussian impulse noise (*MGIN*). For removing the *MGIN*, the objective function of the proposed model is defined as follows in Eq. (3).

$$I_{dn} = arg\ min_{I_h}\ I_{nl} - I_{h1}^1 + \lambda . \Psi(I_h) \tag{3}$$

Here we are taking the absolute distance between the noisy low-resolution image and original noise-free image. This process is done patch wise but for reading convenient we representing as image instead of patch. L_1 is taken to balancing the reconstruction error. $\Psi(I_h)$ Is the regularization term and λ is the regularization parameter. The value of λ is select according to the noise level in the input image. Regularization term is used to find out high-quality estimating value of the given input. Here $I_{nl} - I_{h1}^1$ is the fidelity term used to measure the goodness of the estimation. As we are using L_1 norm for the regularization purpose, the regularization function $\Psi(I_h)$ in our proposed model is defined as follows (given in Eq. 4)

$$\Psi(I_h) = \Phi I_{h1}^1 \tag{4}$$

Algorithm pseudocode for our proposed model given as following (Algorithm 1):

ALGORITHM 1: Denoising Algorithm using L_1 norm

Step1: Given original high resolution image I_h.
Step2: Obtain corresponding noisy low resolution imagefor testing I_{nl} Using model specified in **Eq. (1)**
Step 3: Let I_{dn}= De-noised image and UI_{nl}= updated noise image is
Set $I_{dn} = I_{nl}$ and $UI_{nl} = I_{nl}$;
Step4: do {
        Update UI_{nl} according to **Eq. (5)**
        Update the noise variance σ_n^2
For i = 1, 2, -----, N
        {
        Measure the distance of patch by **Eq. (6);**
        Make a group of similar patches of UI_{nli} into Y_i
        Find covariance and PCA matrix for $I_{h\,i}$
        Patches in Y_i analyses using PCA
        Calculate mean and standard deviation for
        Apply soft-thresholding operation given in **Eq. (7)**
        Calculate $I_{dn\,i}$ by using **Eq. (8)**
        }
    Combined all the patches by **Eq. (9)**
}while (condition full fill)
Step5: Result: obtain **Denoised** image I_{dn} from I_{nl} which is almost equivalent to I_h

---

Rest of the mathematical equations needed to compute the proposed algorithm given in equation from Eq. (5) to Eq. (9)

After every iteration, updated noisy image is calculated by Eq. (5)

$$UI_{nl} = I_{dn} + \rho(I_{nl} - I_{dn}) \tag{5}$$

Here updated noisy image is the sum of the denoised image obtained after the iteration and diffidence of noisy and denoisy image with some multiplying factor before the iteration.

Distance between two patches of noise low-resolution image is calculated using the Eq. (6)

$$d(i, N) = \frac{\|I_{nli} - I_{nlN}\| \cdot_1^1}{S^1} \tag{6}$$

Here $d(i, N)$ is the absolute distance between $(I_{nli} - I_{nlN})$ ith and Nth patch of noisy low-resolution image. Hare absolute distance is calculated instate of Euclidean distance because in the model impulse and Gaussian noisy image is considering.

Rest of the work done by the Eqs. (7), (8), and (9)

$$\tilde{\alpha}._{i,j} = \mu_{i,j} + soft\left(\beta_{i,j} - \mu_{i,j}, \frac{\lambda \omega_{i,j}}{2}\right) \tag{7}$$

$$I_{dn} = \Phi_i^T . \tilde{\alpha}_i \tag{8}$$

$$I_{dn} = \left(\sum_i R_i^T R_i\right)^{-1} . \sum_i \left(R_i^T I_{dn\,i}\right) \tag{9}$$

$I_{h\,i}$ is fetch from I_h using matrix R_i. The transform domain representation of is I_h is fetch from I_{nl} is α and β.

4 Simulation and Result Analysis

We have implemented proposed model using Matlab 2014a on the Intel i5 with 4 GB RAM platform. Parameter setting used in this simulation is given in Table 1.

Numeric results of our proposed model with comparison to state-of-the-art methods in terms of PSNR and SSIM given in the Table 2.

Figure 1 shows the six given input images. Figure 2 shows the noisy image with mixed impulse Gaussian noise and Fig. 3 shows six images as results of our proposed model.

Table 1 Simulation parameters

Simulator	Matlab 2014a
Noise	Impulse (paper and salt), gaussian
Gaussian noise	Standard deviation = 10
Impulse noise	Random value with 5% and the salt-and-pepper impulse noise with 10%)
Patch size	8 × 8
Performance majoring factors	PSNR, SSIM
Test images	Barbara, monarch, flowers, cameraman, coins, pout

Table 2 Simulation results

Methods parameters	Median filter		DWT		Proposed	
	PSNR	SSIM	PSNR	SSIM	PSNR	SSIM
Barbara	24.4	0.905	26.03	0.91	27.07	0.92
Flowers	24.9	0.91	26.03	0.901	27.12	0.925
Cameraman	24.3	0.902	26.03	0.901	27.03	0.901
Monarch	24.6	0.905	26.03	0.89	27.4	0.92
Pout	24.8	0.91	26.1	0.904	27.21	0.91
Coins	24.9	0.915	26.1	0.904	27.01	0.90

Barbara (256 x 256)   Flowers (256 x 256)   Pout (256 x 256)

Cameraman (256 x 256)   Monarch (256 x 256)   Coins (256 x 256)

Fig. 1 Input test images of size (256 × 256)

Barbara (256 x 256)   Flowers (256 x 256)   Pout (256 x 256)

Cameraman (256 x 256)   Monarch (256 x 256)   Coins (256 x 256)

Fig. 2 Images with mixed impulse gaussian noise of size (256 × 256)

Barbara (256 x 256)   Flowers (256 x 256)   Pout (256 x 256)

Cameraman (256 x 256)   Monarch (256 x 256)   Coins (256 x 256)

Fig. 3 De-noised images (Results of our proposed model)

5 Conclusion

In this paper, we have proposed image denoising model for *MIGN* noise using L_1 norm based regularization technique. A simulation result shows that L_1 efficiently deal with the impulse noise. Two parameters, PSNR and SSIM, are used to test the effectiveness of our proposed model to remove the noise from the image. Results show that our proposed model effectively removes the noise in terms of numeric value (PSNR, SSIM) as well as visual quality.

Acknowledgements The Indira Gandhi National Tribal University, Amarkantak (India) and the University of Agder (Norway) gratefully acknowledge partial support of the India—Norway Cooperation Program (INCP), Project No. INCP-2014/10110, for this collaborative work with the Indian partners.

References

1. C. Liu, R. Szeliski, S. B. Kang, C. L. Zitnick, and W. T. Freeman, "Automatic estimation and removal of noise from a single image," *IEEE Trans. Pattern Anal. Mach. Intell.*, vol. 30, no. 2, pp. 299–314, Feb. 2008.
2. H. Takeda, S. Farsiu, and P. Milanfar, "Kernel regression for image processing and reconstruction," *IEEE Trans. Image Process.*, vol. 16, no. 2, pp. 349–366, Feb. 2007.
3. D. Zoran and Y. Weiss, "From learning models of natural image patches to whole image restoration," in *Proc. IEEE Int. Conf. Comput. Vis.*, Nov. 2011, pp. 479–486.
4. P. Chatterjee and P. Milanfar, "Patch-based near-optimal image denoising," *IEEE Trans. Image Process.*, vol. 21, no. 4, pp. 1635–1649, Apr. 2012.
5. M. Zhang and B. K. Gunturk, "Multiresolution bilateral filtering for image denoising," *IEEE Trans. Image Process.*, vol. 17, no. 12, pp. 2324–2333, Dec. 2008.
6. F. Luisier, T. Blu, and M. Unser, "Image denoising in mixed Poisson–Gaussian noise," *IEEE Trans. Image Process.*, vol. 20, no. 3, pp. 696–708, Mar. 2011.
7. L. I. Rudin, S. Osher, and E. Fatemi, "Nonlinear total variation based noise removal algorithms," *Phys. D, Nonlinear Phenomena*, vol. 60, nos. 1–4, pp. 259–268, 1992.
8. Y. -W. Tai, S. Liu, M. S. Brown, and S. Lin, "Super resolution using edge prior and single image detail synthesis," in *Proc. IEEE Conf. Comput. Vis. Pattern Recognit., Jun. 2010, pp. 2400–2407.*
9. Beck and M. Teboulle, "Fast gradient-based algorithms for constrained total variation image denoising and deblurring problems," *IEEE Trans. Image Process.*, vol. 18, no. 11, pp. 2419–2434, Nov. 2009.
10. R. O. Lane, "Non-parametric Bayesian super-resolution," IET Radar, Sonar, Navigat., vol. 4, no. 4, pp. 639–648, Aug. 2010.
11. L. C. Pickup, D. P. Capel, S. J. Roberts, and A. Zisserman, "Bayesian image super-resolution, continued," in *Proc. Adv. Neural Inf. Process. Syst., 2006, pp. 1089–1096.*
12. G. Yu and G. Sapiro, "DCT image denoising: A simple and effective image denoising algorithm," *Image Process. On Line*, vol. 1, Oct. 2011. [Online]. Available: http://dx.doi.org/10.5201/ipol.2011.ys-dct.
13. D. L. Donoho, "De-noising by soft-thresholding," *IEEE Trans. Inf. Theory*, vol. 41, no. 3, pp. 613–627, May 1995.
14. Chenglin Zuo, Ljubomir Jovanov, Bart Goossens, Hiep Quang Luong, Wilfried Philips, Yu Liu, and Maojun Zhang, "Image Denoising Using Quadtree-Based Nonlocal Means With Locally Adaptive Principal Component Analysis" IEEE SIGNAL PROCESSING LETTERS, VOL. 23, NO. 4, PP: 434–438, APRIL 2016

Content-Based Image Retrieval Using Multiscale Local Spatial Binary Gaussian Co-occurrence Pattern

Prashant Srivastava and Ashish Khare

Abstract With the invention of low-cost smartphones and other image capturing devices, image acquisition is no longer a difficult task. This has created a huge amount of unorganized images which require proper indexing for easy access. The field of Content-Based Image Retrieval (CBIR) proposes algorithms to solve this problem. This paper proposes a new multiresolution descriptor, named Multiscale Local Spatial Binary Gaussian Co-occurrence Pattern (MLSBGCP) for CBIR. Grayscale image is subjected to three-level Gaussian filtering process to perform multiresolution processing of the image. Local Spatial Binary Pattern (LSBP) of resulting filtered image is computed to gather local features. Finally, a feature vector is constructed using Gray-Level Co-occurrence Matrix (GLCM) which is then utilized as the feature vector to retrieve visually similar images. Performance of the proposed method is tested on Corel-1 K dataset and measured in terms of precision and recall. The experimental results demonstrate that the proposed method achieves better retrieval accuracy than some of the other state-of-the-art CBIR methods.

Keywords Content-based image retrieval · Gaussian filter
Gray-level co-occurrence matrix · Local spatial binary pattern

1 Introduction

The invention of low-cost image capturing devices such as digital camera and smartphones has made the task of image capturing quite easy. Nowadays, getting an image is quite easy than it was earlier. This has created huge repository of different

P. Srivastava (✉) · A. Khare
Department of Electronics and Communication, University of Allahabad,
Allahabad, India
e-mail: prashant.jk087@gmail.com

A. Khare
e-mail: ashishkhare@hotmail.com

© Springer Nature Singapore Pte. Ltd. 2018
Y.-C. Hu et al. (eds.), *Intelligent Communication and Computational Technologies*, Lecture Notes in Networks and Systems 19,
https://doi.org/10.1007/978-981-10-5523-2_9

types of images. In order to access images from such huge repository easily, it is essential to arrange and organize the images. This problem is solved by image retrieval systems. Image retrieval systems can be broadly classified into two types—text-based image retrieval and content-based image retrieval systems. Text-based image retrieval system retrieves images based on keywords and text. Such systems require manual tagging of a large number of images and fail to retrieve visually similar images. The other type of image retrieval system is Content-Based Image Retrieval system which retrieves images based on features present in the image. No manual annotation is required in such systems unlike text-based retrieval systems.

Content-Based Image Retrieval (CBIR) refers to indexing and retrieval of images based on the features that are inbuilt in the image. Typical CBIR systems accept query in the form of image and extract features present in the image to construct feature vector. The feature vectors of query image and database images are matched to retrieve images that are visually similar to query image [1].

A number of algorithms have been proposed for CBIR since the term came into existence. Early CBIR systems were mostly based on primary features such as colour, texture and shape. Colour feature has been mostly used in the form of colour histogram [2]. The use of texture feature for image retrieval has been done through various techniques such as Gabor transform [3] and local pattern [4]. Shape feature has been exploited through polygonal shape [5] and moments [6].

As the complexity of images increased, single feature started proving to be insufficient. The combination of features overcame this drawback. The combination of features extracts more details from an image than single feature. The combination of colour, texture and shape in [7] proves this fact. Most of the above techniques have been exploited on single resolution of image. An image is a complex structure and single resolution of image fails to gather varying level of details. Multiresolution analysis tends to overcome this drawback. The main objective of multiresolution analysis is to analyze images at more than one resolution so that features left undetected at one scale get detected at another scale. A number of techniques based on multiresolution processing of image have been proposed for image retrieval [8]. This paper proposes a new descriptor, Multiscale Local Spatial Binary Gaussian Co-occurrence Pattern (MLSBGCP) for CBIR. The proposed descriptor combines multiresolution technique and local pattern for image retrieval. Multiresolution processing has been performed using three-level Gaussian filtering of image. Local information has been extracted through Local Spatial Binary Pattern proposed by Xia et al. [9] which computes local patterns in four directions. The proposed descriptor extracts directional local information at multiple scales of image by computing LSBP of three- level filtered image. The feature vector has been constructed through GLCM which analyzes how frequently adjacent pixel pairs occur in an image.

The remaining part of the paper is organized as follows. Section 2 discusses some of the Related Work in the field of image retrieval. Section 3 discusses Gaussian filtering and LSBP. The proposed method has been discussed in Sect. 4. Section 5 discusses Experiment and Results. The concluding part of the paper is discussed in Sect. 6.

2 Related Work

The field of image retrieval has caught a lot of attention of scientists across the world in the past decade. A number of feature descriptors such as Local Binary Pattern (LBP) [4] and Local Ternary Pattern (LTP) [10] have been proposed in the past. Recently, some of the new feature descriptors such as Microstructure Descriptor (MSD) [11] which combines colour and texture feature, Colour Difference Histogram (CDH) [12] which combines colour space and co-occurrence matrix, Hybrid Information Descriptor (HID) [13] which combines low level features and high level features, Correlation Descriptor [14] which combines colour and texture features, have been proposed for image retrieval. These feature descriptors combine multiple features for constructing an efficient feature vector for CBIR. These descriptors have been used either as a single feature or combined with other image features [15] for image retrieval. Most of these descriptors have been exploited on single resolution of image. Use of these features on single resolution of image fails to exploit complex level of details. Multiresolution analysis overcomes this drawback. Xia et al. [9] proposed multiscale Local Spatial Binary Pattern which exploited multiple scales of image for image retrieval. Multiresolution processing of image attempts to gather details at more than one resolution of image which is advantageous as features left undetected at one scale of image get detected at another scale. This paper proposes a multiresolution descriptor, Multiscale Local Spatial Binary Gaussian Co-occurrence Pattern (MLSBGCP) for image retrieval.

3 Gaussian Filtering and Local Spatial Binary Pattern

3.1 Gaussian Filter

The Gaussian filter is used for performing smoothing or blurring of images. It helps in removing noise from the image. Gaussian function for two dimensional signals is defined as

$$G(x, y) = \frac{1}{2\pi\sigma^2} \exp\left(- \frac{(x^2 + y^2)}{2\sigma^2} \right) \tag{1}$$

where σ denotes standard deviation of the Gaussian function. Standard deviation plays an important role in noise removal. If the value of σ is very small, smoothing effect will be very little. In case of very large standard deviation, loss of important details of image such as edges will occur. Hence the value of σ should be chosen such that it should remove maximum noise without losing much details of an image [16].

Apart from noise removal from an image, Gaussian filter has numerous applications in image processing. An important application of Gaussian filter is multiresolution processing. In order to perform multiresolution processing of image, convolution operation between different kernel size and image is performed. This produces a series of images at different levels of scale. Image features can be extracted from these images for constructing the feature vector for image retrieval.

3.2 Local Spatial Binary Pattern

The concept of Local Spatial Binary Pattern (LSBP) was proposed by Xia et al. [9]. LSBP combines LBP with the spatial distribution information of gray-level variation between centre pixel and neighbourhood pixels. LSBP between two gray values P and Q is defined as

$$LSBP(P, Q) = \begin{cases} 00, if (P - Q) \geq T \& P \geq Q \\ 01, if (P - Q) < T \& P \geq Q \\ 10, if (Q - P) < T \& P < Q \\ 11, if (Q - P) \geq T \& P < Q \end{cases} \tag{2}$$

where T denotes threshold which differentiates between two pixels.

LSBP computes binary codes at four different angles, $0°, 45°, 90°, 135°$ in a 3×3 window. LSBP value is obtained by combining binary codes of two angles. For example, binary codes of $0°$ and $90°$ can be combined to form eight bit binary code varying 00000000 to 11111111. In the proposed method, LSBP codes have been computed by combining codes obtained at $0°$ and $90°$ angles and $45°$ and $135°$ angles.

3.3 Gray-Level Co-occurrence Matrix

The concept of Gray-Level Co-occurrence Matrix (GLCM) was originally proposed by Haralick et al. [17]. GLCM determines how frequently adjacent pixel pairs of specified values and at specified directions occur in an image. This helps in determining spatial distribution of pixel values which other features such as histogram fail to provide.

3.4 Advantages of the Proposed Multiscale Local Spatial Binary Gaussian Co-occurrence Pattern (MLSBGCP) Descriptor

Following are the advantages of the proposed MLSBGCP descriptor-

1. It extracts local directional information at multiple scales of image so that features left undetected at one scale get detected at another.

2. LSBP extract local information in spatial directions which other features such as LBP fail to gather.
3. Presence of noise in an image may bring variations in local pattern descriptors. Three-level Gaussian filtering helps in removal of existing noise and hence improves computation of local pattern.
4. Gaussian filtering of image performs multiresolution processing of image. However, this fails to extract directional information. Combination of LSBP, which generates directional local information, with three-level Gaussian filtering, helps in extracting directional information.
5. Use of GLCM for feature vector construction helps in getting information about spatial distribution of pixels which other features such as histogram fail to provide.

4   The Proposed Method

The proposed method consists of the following steps:

1. Smoothing of grayscale image through Gaussian filtering.
2. Computation of Local Spatial Binary Pattern descriptors of resulting filtered image.
3. Construction of Gray-Level Co-occurrence Matrix (GLCM) of LSBP descriptors.
4. Similarity measurement.

Figure 1 shows the schematic diagram of the proposed method.

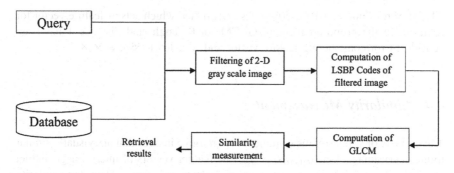

Fig. 1 Schematic diagram of the proposed method

4.1   Gaussian Filtering

Multiresolution processing of grayscale image is done through Gaussian filtering. Multiresolution processing of image decomposes image into multiple resolutions to gather the varying level of details which single resolution processing fails to do. Grayscale image is subjected to Gaussian filtering by performing convolution operation between Gaussian filter of different size and grayscale image. In the proposed method, three-level Gaussian filtering is performed using three different filter size and standard deviation values. For level 1, size of kernel is taken as 3×3, and standard deviation value as 1, for level two, size of kernel is 5×5 and standard deviation as 1.25 and for level 3, size of kernel is 7×7, and value of standard deviation as 1.5. The three-level filtering operation not only performs multiresolution processing of image but also helps in noise removal from grascale image.

4.2   Computation of LSBP

The next step of the proposed method is computation of LSBP descriptors of resulting three-level filtered images. The resulting descriptors are then stored in three separate matrices. The advantage of using LSBP is that it extracts local information in different directions.

4.3   Construction of Gray-Level Co-occurrence Matrix (GLCM)

GLCM of resulting LSBP descriptors is constructed which acts as feature vector for retrieval. In the proposed method, GLCM for 0° angle and distance 1 has been considered for constructing feature vector and rescaled to size 8×8.

4.4   Similarity Measurement

The purpose of similarity measurement is to retrieve images that are visually similar to the query image. Let $(f_{Q1}, f_{Q2}, \ldots f_{Qn})$ be feature vectors of query image and let $(f_{DB1}, f_{DB2}, \ldots f_{DBn})$ be the feature vectors of database images. Then, the similarity between the query image and database image is computed using the following formula:

$$\sum_{i=1}^{n} \left| \frac{f_{DBi} - f_{Qi}}{1 + f_{DBi} + f_{Qi}} \right| \quad i = 1, 2, \ldots n \tag{3}$$

5 Experiment and Results

In order to carry out the experiment using the proposed technique, images from Corel-1 K dataset [18] has been used. This dataset consists of 1000 images divided into ten categories and each category consists of 100 images. The size of each image is either 256 × 384 or 384 × 256.

Each image of Corel-1 K dataset has been rescaled to size 256 × 256 to ease the computation. Each image of dataset is taken as query image. If the retrieved images belong to the same category as that of the query image, the retrieval is considered to be successful. Otherwise, the retrieval fails.

5.1 Performance Evaluation

The performance of the proposed method is measured using precision and recall metrics. Precision refers to the ratio of the total number of relevant images retrieved to the total number of images retrieved. Mathematically, precision can be expressed in the following way

$$P = \frac{I_R}{T_R} \tag{4}$$

where I_R denotes total number of relevant images retrieved and T_R denotes total number of images retrieved. Recall refers to the ratio of total number of relevant images retrieved to the total number of relevant images in the database. Mathematically, recall can be expressed in the following way

$$R = \frac{I_R}{C_R} \tag{5}$$

where I_R denotes total number of relevant images retrieved and C_R denotes total number of relevant images in the database. In this experiment, $T_R = 10$ and $C_R = 100$.

5.2   Retrieval Results

For conducting experiment, each image of Corel-1 K dataset has been rescaled to size 256 × 256 to ease the computation. Grayscale image is filtered using Gaussian filter up to three levels and Local Spatial Binary Pattern of resulting filtered image is computed. Finally, feature vector is constructed using GLCM.

Three-level Gaussian filtering using different filter size and standard deviation values results in three smoothed images. LSBP, which produces directional local patterns, of resulting filtered images is computed. This produces three LSBP matrices. GLCM of these matrices is constructed for computing feature vectors. Each of these feature vectors is used separately to perform similarity measurement. This results in three sets of similar images. Final set of similar images is then produced by taking union of these sets. Computation of recall is done by counting total number of relevant images in the final set. For computing precision values, the top n matches for each set is counted followed by performing union of these sets. This results in final image set. The top n matches in the final set are considered for evaluating precision. The relevant image set of a level is obtained by considering relevant image set of previous level along with the relevant image set of that level. If the values of precision and recall are high, the retrieval is considered to be good.

The average values of precision and recall obtained for three levels of resolution are shown in Table 1. The plot between average values of precision and recall for three levels of resolution is shown in Fig. 2. From Table 1 and Fig. 2, it can be clearly observed that there is an increase in the values of precision and recall with the increase in the level of resolution. It is due to the multiresolution analysis that the features left undetected at one level get detected at another. The increase in the values of precision and recall of the proposed method with the level of resolution is due to this phenomenon.

5.3   Performance Comparison

The performance of the proposed descriptor MLBGCP has been compared with other state-of-the-art methods, Srivastava et al. [15], CDH [12, 14], Xia et al. [9] and MSD [11, 13] in terms of precision. These descriptors exploit single resolution

Table 1 Average recall and precision values of the proposed method for three levels of resolution

Level of resolution	Recall (%)	Precision (%)
Level 1	42.65	68.93
Level 2	45.21	73.84
Level 3	46.98	76.46

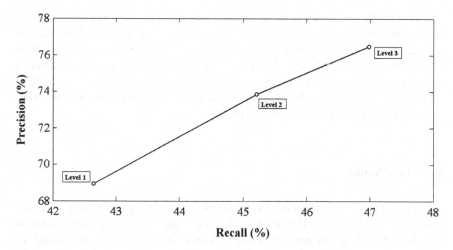

Fig. 2 Average recall versus precision plot of the proposed method for three level of resolutions

Table 2 Performance comparison of the proposed method with other state-of-the-art methods

Methods	Precision (%)
Srivastava et al. [15]	53.70
CDH [12, 14]	65.75
Xia et al. [9]	75.39
MSD [11, 13]	75.67
Proposed method	**76.46**

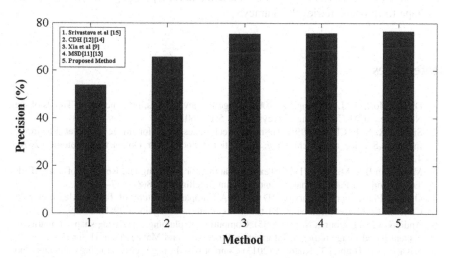

Fig. 3 Performance comparison of the proposed method with other state-of-the-art methods

of image for feature vector construction and hence produce low retrieval accuracy as compared to the proposed descriptor. The proposed descriptor exploits multiple resolutions of image for constructing feature vector and hence outperforms other state-of-the-art methods. Table 2 and Fig. 3 show the performance comparison of the proposed descriptor with other state-of-the-art methods. Table 2 and Fig. 3 clearly demonstrate that the proposed descriptor MLBGCP outperforms other state-of-the-art techniques in terms of precision.

6   Conclusion

This paper proposed a novel descriptor Multiscale Local Spatial Binary Gaussian Co-occurrence Pattern (MLSBGCP). This descriptor combines multiresolution technique and local pattern for image retrieval. Multiresolution processing of grayscale image was done with the help of a three-level Gaussian filter. The local information from filtered image was gathered through Local Spatial Binary Pattern (LSBP). Finally, feature vector was constructed through GLCM of resulting LSBP descriptors which were used to retrieve visually similar images. The advantage of this technique is that it helps in extracting directional local information at multiple resolutions of image so that features left undetected at one scale get detected at another. The performance of the proposed descriptor was measured using precision and recall metrics. As demonstrated by experimental results, the proposed method outperformed some of the other state-of-the-art methods in terms of precision. The proposed method can be further improved by using different filters such as Laplacian filter for multiresolution processing of image along with other directional local features such as Local Derivative Pattern and incorporating more features such as shape to improve retrieval accuracy.

References

1. Dutta R, Joshi D, Li J, Wang J Z (2008) Image Retrieval: Ideas, Influences, and Trends of the New Age. ACM Computing Surveys 40(2): 5:1–5:60.
2. Smith J R, S. F. Chang (1996) Tools and Techniques for Color Image Retrieval. Electronic Imaging, Science and Technology, International Society for Optics and Photonics 2670: 426–437.
3. Manjunath B S, Ma W Y (1996) Texture Features for Browsing and Retrieval of Data. IEEE Transactions on Pattern Analysis and Machine Intelligence 18(8): 837–842.
4. Ojala T, Pietikainen M, Harwood D (1996) A Comparative Study of Texture Measures with Classification Based on Feature Distributions. Pattern Recognition 29(1): 51–59.
5. Andreou I., and Sgouros N. M. (2005). Computing, explaining visualizing shape similarity in content-based image retrieval. Information Processing and Management 41: 1121–1139.
6. Srivastava P, Binh N T, Khare A (2013) Content-based image retrieval using moments. 2nd International Conference on Context-Aware Systems and Applications, Phu Quoc, Vietnam pp. 228–237.

7. Wang X., Yu Y., and Yang H. (2011). An Effective Image Retrieval Scheme Using Color, Texture And Shape Features, Computer Standards & Interfaces 33(1): 59–68.

8. Srivastava P, Prakash O, Khare A (2014) Content-Based Image Retrieval using Moments of Wavelet Transform. International Conference on Control Automation and Information Sciences, Gwangju, South Korea pp. 159–164.

9. Xia Y, Wan S, Jin P, Yue L (2013) Multi-scale Local Spatial Binary Patterns for Content-Based Image Retrieval. Active Media Technology, Springer International Publishing 423–432.

10. Tan X, Triggs B (2010) Enhanced Local Texture Feature Sets for Face Recognition Under Difficult Lighting Conditions. IEEE Transactions on Image Processing 19(6): 1635–1650.

11. Liu G, Li Z, Zhang L, Xu Y (2011) Image retrieval based on microstructure descriptor. Pattern Recognition 44(9) 2123–2133.

12. Liu G H and Yang J Y (2013) Content-based image retrieval using color difference histogram. Pattern Recognition, 46(1): 188–198.

13. Zhang M, Zhang K, Feng Q, Wang J, Jun K, Lu Y (2014) A novel image retrieval method based on hybrid information descriptors. Journal of Visual Communication and Image Representation 25(7): 1574–1587.

14. Feng L, Wu J, Liu S, and Zhang H (2015) Global correlation descriptor: a novel image representation for image retrieval. Journal of Visual Communication and Image Representation 33: 104–114.

15. Srivastava P, Binh N T, Khare A (2014) Content-Based Image Retrieval using Moments of Local Ternary Pattern. Mobile Networks and Applications 19: 618–625.

16. Forsyth D A, Ponce J Computer Vision- A Modern Approach, Prentice Hall of India.

17. Haralick R M and Shanmugam K (1973). Textural features for image classification. IEEE Transactions on systems, man, and cybernetics 6: 610–621.

18. http://wang.ist.psu.edu/docs/related/ Accessed April 2014.

Importance of Missing Value Estimation in Feature Selection for Crime Analysis

Soubhik Rakshit, Priyanka Das and Asit Kumar Das

Abstract Missing values are most likely to be present in voluminous datasets that often lead to poor performance of the decision-making system. The present work carries out an experiment with a crime dataset that deals with the existence of missing values in it. The proposed methodology depicts a graph-based approach for selecting important features relevant to crime after estimating the missing values with the help of a multiple regression model. The method selects some features with missing values as important features. The selected features subsequently undergo some classification techniques that help in determining the importance of missing value estimation without discarding the feature for crime analysis. The proposed method is compared with existing feature selection algorithms and it promises a better classification accuracy, which shows the importance of the method.

Keywords Crime records · Missing value estimation · Correlation coefficient
Feature selection · Multiple regression · Rough set theory

1 Introduction

The existence of missing values is a common phenomenon in the field of data mining and information retrieval. Occurrence of the missing values can take place due to different reasons. It may occur as a result of some flawed system or often the collected data from surveys containing opinions of a number of people result in incomplete datasets. For example, clinical datasets often endure the presence of missing data occurring due to human error and noise, while sometimes the police and witness

S. Rakshit (✉) · P. Das · A.K. Das
Department of Computer Science and Technology, Indian Institute of Engineering
Science and Technology, Shibpur, Howrah, India
e-mail: soubhik.dd2015@cs.iiests.ac.in

P. Das
e-mail: priyankadas700@gmail.com

A.K. Das
e-mail: akdas@cs.iiests.ac.in

© Springer Nature Singapore Pte. Ltd. 2018   97
Y.-C. Hu et al. (eds.), *Intelligent Communication and Computational
Technologies*, Lecture Notes in Networks and Systems 19,
https://doi.org/10.1007/978-981-10-5523-2_10

narrative reports for crime may contain missing values due to lack of information collected. Predicting the missing values is an important task for further crime analysis. Few missing data are trivial to analyse and some of them can prove to be manageable, but the presence of some of the missing values in relevant features can have an unsparing effect on the system accuracy. So, proper imputation of the missing values definitely fulfil the prerequisites for generating a modified dataset leading to pertinent feature selection.

A handful of schemes suggesting the prediction of missing data exists in the literature. An experiment with 12 different datasets and an evaluation of the impact on miss-classification error rate of some prior work handling missing values were performed in [1]. Another scheme suggesting online approach of feature extraction along-with classification is suggested in [2]. A habitual course at first estimates the missing values followed by selection of features. But a lot of techniques have been proposed that directly extracts the features from incomplete datasets. Such a work is represented in [3]. According to [4], high-dimensional microarray data are often incomplete and error-prone. It involves K-nearest neighbour (KNN) feature selection technique comprising the notion of feature selection and estimation of new values. Sun et al. [5] described an efficient model-based study accentuating on the impact of missing value prediction on classification for DNA microarray gene data showing better results for missing value estimation provided a high noise level. The presence of missing values of influential attributes can give underperforming results in feature selection methods. An adaptation of nearest neighbours based mutual information estimator is suggested in [6] to get rid of the missing data. This method shows results for both artificial and real-world datasets and selects crucial features without any imputation algorithm. An approach to the feature selection problem with incomplete datasets is suggested in [7]. Here, apart from the concept of mutual information estimator, a partial distance strategy and a greedy forward search technique have been adopted to fill the missing data. A comparative study in [8] analyses the treatments for few common cases of missing values at prediction time. This analysis deals with some of the problematic issues faced during missing value estimation. The method described in the above-mentioned work of overcoming the issues also suggests better accuracy compared to other techniques. Eugene et al. suggested an algorithm using tree-based ensembles to generate a compact subset of important features in [9]. This method concentrates on the elimination of some inconsequential attributes using statistical comparisons with artificial contrasts by fixing an edge for importance estimated from the parallel ensemble of trees capable of scoring a very large number of features.

Predicting the crime patterns with the presence of missing values in criminal datasets often becomes a challenging task. An improving classification accuracy with missing data imputation algorithm has been considered in [10] for crime datasets. The method combines KNN imputation with Grey Relational Analysis (GRA) theory. Often classification accuracy gets affected by missing data values and to surpass this problem, a novel approach involving some typical classification algorithms is suggested in [11]. This method has its own disadvantages and the filling algorithms can be improved.

The present work deals with a crime record for Philadelphia region which contains missing values of some meaningful attributes present in the crime dataset. The proposed methodology follows an intuitive approach. Initially, a multiple regression model has been used that surpasses the persistence of missing values from the dataset. A modified dataset has been prepared leading to further analysis. A graph-based approach has been implemented for selecting significant features using the concept of rough set theory (RST) from the modified crime dataset. The proposed work is also compared with some other existing feature selection algorithms to determine the importance of missing value estimation.

The rest of the paper is organised as follows: Sect. 2 describes the proposed work in a skilful way. Section 3 shows the experimental results followed by conclusion in Sect. 4.

2   Proposed Methodology

This section demonstrates the underlined principles of the present methodology in an elaborate way. The objective of this work is to estimate the missing values present in the dataset and subsequently select the features to express the importance of attributes with missing values in crime analysis.

2.1   Data Collection and Preparation

Input data play an important role in the field of feature selection. The present methodology has collected data, documents, and various applications for public use from *OpenDataPhilly*. The collected data contain crime records in Philadelphia region from the year 2006 to 2016. Near about 100,000 crime records have been considered for the present work. It has been observed that 14 different features are present in the dataset which gives insight into the crimes occurring in Philadelphia. Out of these 14 features, some features can be derived from other features and so are removed from the dataset. Then the composite attribute *Dispatch_Date_Time* is decomposed into primitive attributes, namely, *Year*, *Month* and *Time_Of_Day*. Next, to get more relevant information from the attribute *Time_Of_Day*, higher level concepts like *Morning*, *Noon*, *Afternoon*, *Evening* and *Night* are generated from the low-level concepts of continuous values. Thus, we have a crime dataset $DS = (U, C, D)$, where U = universe of discourse that contains all crime records, C = set of conditional attributes that are prepared from the available dataset and D = decision attribute that indicates crime types happening in the country. Total 30 different crime types are highlighted in the dataset. The description of the prepared dataset is described in Table 1.

Table 1 List of features

Name of features	Description of features
Dc_Dist	District name
PSA	Police service area
Year	Year when crime incidents occurred
Month	Month of crime occurrence
Time of day	Time of dispatching the police to the crime scene
Location_Block	Location of the crime
DC_Key	Primary key that points to a specific record
UCR_General	Uniform Crime Report General code
Longitude	Longitude of the place
Latitude	Latitude of the place
Crime_Type	Type of crime

2.2   Estimation of Missing Values

The collected dataset often consists of missing values and predicting the missing values become troublesome and error-prone. Here, in crime dataset, two conditional attributes namely *Longitude* and *Latitude* contain missing values which are needed to be estimated for obtaining the exact location where the crime occurs. As the features from the data miners point of view are important, estimation of the missing values is necessary rather than simply ignoring them. There are many missing values in estimation techniques in [7, 12], which are biased and take information from small parts of the dataset. In this work, multiple regression model is used to remove the bias and receive information from all the records where no missing value occurs. Thus this method uses more information for missing value estimation.

In the dataset, out of 10 conditional attributes, 8 attributes have no missing value. To impute the missing values of *Longitude*, these 8 attributes and *Longitude* are taken and the dataset is divided into two partitions:

1. Set of tuples without missing values or *model generation* set and
2. Set of tuples with missing values or *missing value estimation* set.

Now using the *model generation* set, multiple regression model is generated with *Longitude* as the dependent variable which depends on other attributes. Then the tuples from *missing value estimation* set are fitted into the model to impute the missing values of *Longitude*. Similar treatment is made for *Latitude*. Thus the missing values are estimated for the attributes in the dataset.

2.3 Unsupervised Feature Selection Method

The present methodology proposes a graph-based method for getting the list of features which are mostly related to the class. First, the similarity coefficient between every pair of conditional attributes in C is calculated using the correlation coefficient as a similarity metric. Next, the similarity coefficient between every conditional attribute with the decision (class) attribute is computed using lower approximation of RST. Using both the similarity metrics, finally the important features will be selected.

(A) Similarity Measurement among Conditional Attributes: Similarity between every pair of attributes represented as vectors $x = (x_1, x_2, \ldots, x_{|U|})$, $y = (y_1, y_2, \ldots, y_{|U|}) \in C$ is computed using *Pearson Correlation Coefficient* using (1):

$$\rho(\mathbf{x}, \mathbf{y}) = \sum_{i=1}^{|U|} \frac{(x_i - m_x)(y_i - m_y)}{|U|\sigma_x\sigma_y}, \tag{1}$$

where m_x and m_y are mean of x and y, respectively, and σ_x and σ_y are respective standard deviations. Now, a weighted undirected graph $G = (V, E)$ is constructed with $V = $ set of vertices $= $ features in C and $E = \{(v_i, v_j)|v_i, v_j \in V \text{ and } \rho(v_i, v_j) > 0\}$.

We are not considering any relationship between two features which are independent to each other or one discourage the other.

(B) Dependency of Conditional Attributes on Decision Attribute: As the class value is given, the dependency of each feature in C on D can be measured to modify the obtained graph for selecting more relevant attributes for crime analysis. Here, the attribute dependency, a concept of RST, is used to measure the dependency of conditional attributes on decision attribute. To compute the attribute dependency of each attribute $A \in C$ on D, the steps applied are as follows:

1. Define an indiscernible relation $I(A)$ in (2) that computes a set of order pair of crime records which are indiscernible to each other with respect to feature A:

$$I(A) = \{(x, y) \in U \times U | A[x] = A[y]\}. \tag{2}$$

   This relation gives equivalent classes $[x]_A$ of objects in U. Thus the set U is partitioned into $U/A = \{[x]_A\}$.
   Similarly, based on decision attribute D, we get $U/D = \{[x]_D\}$, set of equivalent classes using (2).
2. For each class $X \in U/D$, lower approximation of X, i.e. $\underline{A}X$, is computed using (3) which proves that all the objects in $\underline{A}X$ positively belong to the target set X:

$$\underline{A}X = \{x|[x]_A \subseteq X\}. \tag{3}$$

Thus, dependency of D on A is obtained using (4):

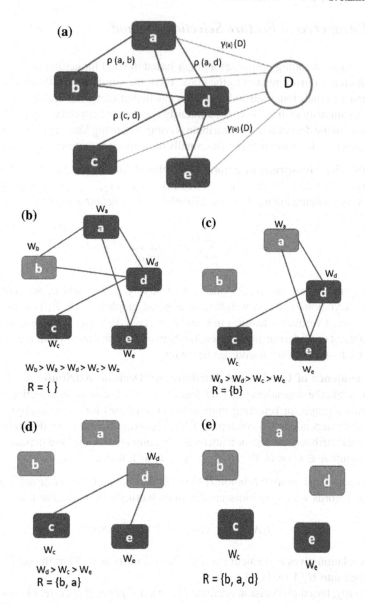

Fig. 1 Steps for feature selection according to proposed methodology

$$\gamma_A(D) = \frac{\bigcup\limits_{\forall x \in U/D} \underline{AX}}{|U|}. \tag{4}$$

This dependency value demonstrates how much similarity is there among the partitions of objects by attribute A and decision attribute D. As decision attribute gives the actual class of objects, higher γ value implies more significance of the attribute A.

(C) Feature Selection: For each attribute $a \in C$, there are some weighted edges in graph for $b \in C$ and a weighted edge to decision attribute D as shown in Fig. 1a. Now a weight factor for each attribute $a \in C$ is computed using (5)

$$W_a = \frac{1}{2}\left\{ \frac{\sum \rho(a,b)|(a,b) \in E}{degree(a)} + \gamma_{\{a\}}(D) \right\}. \tag{5}$$

Thus, a new graph as shown in Fig. 1b is obtained from the previous graph which is unweighted but a weight factor is assigned to each vertex of the graph, where weight between any two attributes $x, y \in C$ is denoted by $\rho(x, y)$ and $\gamma_{\{a\}}(D)$ is the weight of the edge between attribute A and D. Also from this graph vertex D is removed, as its importance on each conditional attribute is already measured using (5) for computing weight factor of each conditional attribute. Now the attribute (vertex) with the maximum weight is selected and its adjacent edges are removed from the graph. Then, the next highest weighted vertex is selected and so on, and finally the process terminates when a null graph remains. The feature selection in each iteration and modification of the graph are shown in Fig. 1b–e and final selected features are $R = \{b, a, d\}$. The selected features are considered as the important features for crime categorisation.

3   Experimental Results

The schemes illustrated in this work deal with Python with its several modules like numpy, scikit-learn, pandas, matplotlib, etc. Weka [13] software has been used for running the classifiers and getting their accuracy. The crime records of Philadelphia are obtained from *OpenDataPhilly* and preprocessed for crime analysis. As the dataset contains two features *Longitude* and *Latitude* with missing values, they are estimated and then the proposed graph-based feature selection method and other two popular supervised feature selection methods, namely, Correlation-Based Feature Selection (CFS) [14] and Consistency Subset Evaluation (CON) [15], are applied to select only the features relevant to crime information. It is observed that though CFS removes *Longitude* and *Latitude* as redundant features, CON and the proposed methods treat them as important features. Using these reduced datasets various classifiers are run and their accuracies are listed in Table 2. This table also shows the accuracies considering whole dataset. It is observed that the reduced dataset obtained by

Table 2 Classifier accuracy (in %) on dataset with various features

Feature selection methods (# of attributes used)

Classifiers	Unprocessed (10)	CFS (6)	CON (7)	Proposed method (7)
NaiveBayes	86.492	86.248	86.7	86.732
BayesNet	87.003	87.045	87.099	87.184
ClassifyViaClustering	20.767	25.738	26.697	28.26
FilteredClassifier	87.403	87.223	87.38	87.466
RandomTree	84.305	84.654	84.141	85.929

proposed method performs better for all classifiers compared to the dataset with all features which demonstrates that the irrelevant features are deleted by the method. At the same time, by selecting features like *Longitude* and *Latitude*, the method expresses that they are important for crime prediction as the accuracy obtained by the method outperforms than by CFS. This states the importance of missing value estimation and here it is estimated accurately.

4 Conclusion and Future Work

The features selected with the present work results in a better classification accuracy than some of the prominent feature selection methods. This concludes the fact that estimating missing values before applying feature selection techniques is important in case of large datasets. The method uses RST and correlation coefficient to select the conditional attributes more associated to decision attributes and at the same time eliminates the ambiguity and redundancy. As a future work, several crime reports from various countries will be extracted and experimental results will help to give an insight into the global crime scenario.

References

1. Acua, E., Rodriguez, C.: The treatment of missing values and its effect in the classifier accuracy. In: Classification, Clustering and Data Mining Applications. (July 2004) 639–647
2. Kalkan, H.: Online feature selection and classification with incomplete data. Volume 22. (2014) 1625–1636
3. Lou, Q., Obradovic, Z.: Margin-based feature selection in incomplete data. In: Proceedings of the Twenty-Sixth Association for the Advancement of Artificial Intelligence. (July 2012) 1040–1046
4. Meesad, P., Hengpraprohm, K.: Combination of knn-based feature selection and knn-based missing-value imputation of microarray data. In: The 3rd Intetnational Conference on Innovative Computing Information and Control (ICICIC'08). (August 2008)

5. Sun, Y., Braga-Neto, U., Dougherty, E.R.: Impact of missing value imputation on classification for dna microarray gene expression data model-based study. EURASIP Journal on Bioinformatics and Systems Biology (2009) 1–17

6. Doquire, G., Verleysen, M.: Feature selection with missing data using mutual information estimators. Neurocomputing **90** (August 2012) 3–11

7. Doquire, G., Verleysen, M.: Mutual information for feature selection with missing data. In: European Symposium on Artificial Neural Networks, Computational Intelligence and Machine Learning. (April 2011) 269–274

8. Saar-Tsechansky, M., Provost, F.: Handling missing values when applying classification models. Journal of Machine Learning Research **8** (July 2011) 1625–1657

9. Kraus, E.J., Dougherty, E.R.: Segmentation-free morphological character recognition. Proc. SPIE **2181** (1994) 14–23

10. Yao, C.S.C., Shen, L., Yu, X.: Improving classification accuracy using missing data filling algorithms for the criminal dataset. International Journal of Hybrid Information Technology **9**(4) (2016) 367–374

11. Sun, C., Yao, C., Li, X., Yu, X.: Detecting crime types using classification algorithms. Biotechnology-An Indian Journal **10**(24) (2014) 15452–15457

12. N.Poolsawad, C.Kambhampati, Cleland, J.G.F.: Feature selection approaches with missing values handling for data mining - a case study of heart failure dataset. Volume 5. (2011) 671–680

13. Witten, I.H., Frank, E.: Data Mining: Practical Machine Learning Tools and Techniques, Second Edition (Morgan Kaufmann Series in Data Management Systems). Morgan Kaufmann Publishers Inc., San Francisco, CA, USA (2005)

14. Dash, M., Liu, H.: Consistency-based search in feature selection. Artificial Intelligence **151**(12) (2003) 155–176

15. Hall, M.A.: Correlation-based feature selection for machine learning. Technical report, Waikato University, Department of Computer Science (1999)

Image Analysis for the Detection and Diagnosis of Hepatocellular Carcinoma from Abdominal CT Images

P. Sreeja and S. Hariharan

Abstract This paper describes the application of machine learning algorithms in detecting the liver lesion from abdominal CT images. Liver lesions are usually detected by different imaging modalities such as computed tomography (CT), ultrasonography (USG), or Magnetic resonance imaging (MRI). These lesions can be of benign or malignant in nature, both of which present similar appearance in the images. The segregation of the liver lesions into benign or malignant has been done with painful invasive techniques by the medical practitioners which create considerable difficulty to the patients. In this work the malignant liver lesion namely the hepatocellular carcinoma (HCC) is differentially identified from normal liver tissue with the help of supervised classification algorithms. The grainy nature of liver parenchyma is the prime factor in making the texture features for discriminating it from other tissues. Different textures features have been extracted from the Region of Interest (ROI) identified by an expert radiologist and are applied as an input to the classifiers for training and testing. The support vector machine (SVM) and naïve Bayes' classifier are supervised learning techniques. Thus our proposed Computer Aided Diagnostic (CAD) system can be easily operated by a medical practitioner in differentially detecting the HCC tissue from normal liver tissue with best classification accuracy which is more than 90%. Hence the painful invasive methods can be avoided to a great extent. The performance evaluation is done by calculating the accuracy, confusion matrix, sensitivity, and specificity. The receiver operating characteristics (ROC) also have been plotted.

Keywords Medical image · Features · Machine learning algorithm

P. Sreeja (✉) · S. Hariharan
Department of Electrical Engineering, College of Engineering Trivandrum,
Thiruvananthapuram, India
e-mail: sreeja@ceconline.edu

S. Hariharan
e-mail: harikerala2001@yahoo.com

© Springer Nature Singapore Pte. Ltd. 2018
Y.-C. Hu et al. (eds.), *Intelligent Communication and Computational Technologies*, Lecture Notes in Networks and Systems 19,
https://doi.org/10.1007/978-981-10-5523-2_11

1 Introduction

An image is an unstructured array of pixels. A feature is a piece of relevant information for solving a particular computational task. Feature extraction is an important preprocessing step in the image analysis. There are different objectives for feature extraction in image processing such as dimensionality reduction and classification [1]. In spatial domain analysis, texture features are extracted from the pixel statistics. The spatial feature extraction techniques are further classified as statistical, structural and spectral approach. Different textural, spectral, spatial, and shape features are reviewed by Wu et al. [2]. For images or regions with irregular size and shape, spatial features are more effective.

Different imaging modalities such as CT, MRI, and USG are usually adopted imaging modalities to study the nature of internal organs and find the reasons of its normal or abnormal condition. Among these imaging techniques, Computed Tomography (CT) becomes very popular.

CT provides better quality images with very short acquisition time. The simple image acquisition procedure promotes the wide spread application of CT regardless of its harmful radiation effect. CT images allow radiologists and other physicians to identify internal structures of human body and patients with the cancer history. Computer aided diagnostic systems can help to conduct the analysis of these images in a better and precise way. Specific CAD systems based on images have been developed for analysis of each organ and diseases. Such CAD systems promote early detection and diagnosis of the diseases which helps to provide better treatment at an early stage.

Liver is the largest and metabolic organ of human body which filters the blood from digestive tract and detoxifies drugs and chemicals. It has dual blood supply from hepatic artery and portal vein. The metabolic functions itself make the liver a potential candidate for malignant neoplasm. Also the dual blood supply favors the deposit of metastases arising from primary tumors in colon, breast, lung, pancreas, and stomach [3, 4]. The focal lesions occurring in the liver can be either benign or malignant. These focal liver lesions are identified with the help of different imaging techniques. These lesions present similar appearance in these images. This depicts a major difficulty in treatment planning. The patients were compelled to undergo surgery to differentiate between the benign or malignant nature of the lesion. To overcome this difficulty of painful invasive methods different CAD systems have been developed to distinguish benign and malignant from the image itself.

One of the classic works in the field of liver lesion classification was done by Chen et al. Detect before extract technique using fractal dimension estimation based neural network classifier was employed to detect hepatoma and hemangioma [5]. Various approaches have been proposed for the classification of liver tissue into normal, benign and malignant. All these works employed different spatial and frequency domain features. Jitendra Virmani presented characterization of primary and secondary malignant liver lesions from ultrasound images using SVM with good accuracy [6]. Lee et al. proposed an unsupervised segmentation scheme for the ultra sound liver images using M-band wavelet transform based fractal feature. In this work, the image is decomposed into sub images with the help of M-band wavelet transform. These sub images give the texture features of different orientation and scales. Fractal dimension was used for textural measurement. These feature vectors were applied to the k-means unsupervised clustering algorithm [7].

A CAD system based on textural features and neural network classifier to identify focal liver lesions from B mode ultrasound images was proposed by Deepti Mittal et al. Different segmented ROI are selected to find the optimal ROI and textural features of Gray level run length matrix (GLRLM), Textural energy measures (TEM), Gabor wavelet features, etc., were extracted to classify focal liver lesions [8].

A feature vector based on fractal dimension and M-band wavelet transform was applied to the classification of natural textured images by Wen-Li Lee et al. The result was validated with ultrasonic liver images with four different classifiers. In this work the authors presented a robust method for calculating the fractal dimension by effectively reducing the plateaus [9]. Zeumray Dokur and Tamer e Olmez proposed a neural network called Intersecting Spheres (In S) which has dynamic and incremental structures to segment US images using 2-D DCT of pixel intensities of ROI as features [10].

Constantine Kotropoulos et al. proposed support vector machine to segment liver lesions from ultra sound liver images. The SVM is trained using the local signal properties that discriminate the lesion area from normal liver tissue. In [11] the authors demonstrated the efficient segmentation of the lesion from US-B mode images using RBF-SVM. They have presented a comparative analysis with thresholding L2 mean filtered image. Vijayalekshmi and Subbiah Bharathi proposed Euclidean distance based classification of normal liver tissue and HCC from CT images using Legendre moments of local binary pattern transformed images [12].

From the literature review, it is evident that the development of different CAD system has been carried out for more than past 20 years. Because of the functional and structural nature, the liver becomes the prime candidate for getting malignancies and metastasis. Also the focal liver lesions are more prevalent in underdeveloped and developing countries which necessitate the detection of the diseases in the early stage at a lower cost. The advent of computing technology and embedded systems can together develop systems based on medical image analysis to provide a second opinion to the medical practitioner and hence to avoid painful invasive methods. In this work we have developed an autonomous system which extracts classical texture features, classifies the input image into normal or HCC, estimate accuracy and provides the performance characteristics for a clear interpretation.

2 Textural Features

Texture is basically the structural pattern of surfaces. It refers to the repetition of basic textural elements called Texels. A Texel contains several pixels whose placements could be periodic, quasiperiodic, or random. Natural textures are usually random. As per microscopic anatomy, the liver lobe is made up of hepatic lobules. The lobules are roughly hexagonal. This nature of the liver tissue makes it favorable for texture analysis.

Three principal approaches used to describe texture of a region are statistical, structural, and spectral. The statistical approach yield characterization of texture as smooth, coarse, grainy, etc. Structural techniques deal with arrangement of image primitives while spectral analysis is based on the properties of the Fourier spectrum.

In the statistical texture analysis, texture region is described by the first or higher order moments of the gray level histogram of the ROI. The various moments are explained as follows.

Mean:

$$\mu_n = \sum (Z_i - m)^n P(Z_i) \tag{1}$$

where Z_i is the intensity value of gray level histogram, $P(Z_i)$ is the corresponding histogram 'm' is the mean or average intensity.

$$m = \sum Z_i P(Z_i) \tag{2}$$

Variance: Variance is the second order moment which is having significant importance in the texture description. It is the measure of intensity contrast that can be used to establish descriptors of relative smoothness. Since the variance values tend to be large for gray scale images with intensity values ranging from 0 to 255, it is usually normalized in the interval of [0 1]. The square root of the variance is the standard deviation which is also usually used as a common texture measure.

The third moment is a measure of the skewness of the histogram while the forth moment is that of its relative flatness. The higher moments also provide further quantitative information of texture. Additional texture information based on histogram are uniformity and entropy. Entropy is a measure of randomness.

Gray level cooccurrence matrix (GLCM) provides the information regarding the distribution of intensity values and the relative positions of the pixels in the image. [15].

3 Classification

Image classification is perhaps the most important part of the digital image analysis. The main classification methods are supervised and unsupervised. In supervised learning a mapping from input to output is provided with the help of a supervisor. There is no such supervisor is associated with the unsupervised case. Different machine learning algorithms are existing which are utilized in the medical image processing applications. The detection and classification of the benign and malignant tissues of various organs using CAD systems are available in the literature.

Many numbers of classification algorithms are developed based on distance measure, probability, and statistical measures. Examples are k nearest neighbor, naïve Bayes, and support vector machine (SVM).

3.1 Naïve Bayes

Naive Bayes classifiers are a family of simple probabilistic classifiers based on Bayes' theorem. All naive Bayes' classifiers assume that the value of a particular feature is independent of the value of any other feature, i.e., the independence assumptions between the features. It is competitive in this domain with more advanced methods including support vector machines. It also finds application in automatic medical diagnosis. The accuracy of independence assumption is the main demerit of the algorithm.

3.2 Support Vector Machines

SVMs are nonprobabilistic supervised learning model used for classification and regression. In SVM the features are marked as points in the feature space. The points belonging to different classes are separated by a decision boundary which is a straight line for the case of linear binary classifiers. The features which determine the decision boundary is called as the support vectors. Nonlinear classification can also be done by SVM by using kernel trick that maps the features into high dimensional features space. Some common kernels are polynomial, Gaussian, radial basis, and hyperbolic tangent. SVM requires full labeling of the data [16].

4 CAD System for Classification of HCC

The proposed CAD system aims to detect hepatocellular carcinoma from the abdominal CT images. With the help of an experienced radiologist the HCC part is identified from the image. The diseased portion is cropped and saved from different CT images. Normal liver CT images of different persons are also collected. All these images were converted to gray scale and are saved in the same size and format.

These images formed the date base. Figure 1 shows different CT images of liver with hepatocellular carcinoma, normal liver, and the region of interest identified by an expert radiologist. The block diagram of the proposed system is shown in Fig. 2. The feature extractor module extracts various features from all the template images

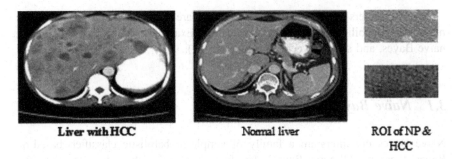

Liver with HCC   **Normal liver**   ROI of NP &
                                                             HCC

Fig. 1 CT images and region of Interest of proposed system

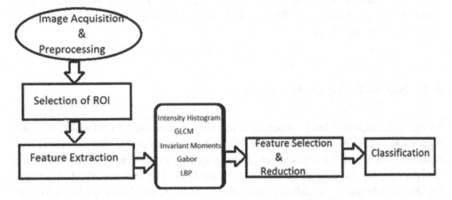

Fig. 2 Block schematic of proposed CAD system

of HCC and normal liver parenchyma (NP). In this work, GLCM features are selected for applying to classifier input. The classifier predicts whether the input ROI is HCC or normal.

4.1 Feature Extraction Module

Analysis with a large amount of data necessitates large amount of storage and computational power. Feature extraction process simplifies this problem by reducing the amount of data and describes it with sufficient accuracy. Texture plays a significant role in the accurate segmentation and classification in the image processing. In statistical texture analysis, texture features are computed from the statistical distribution of some combinations of intensities at specified positions relative to each other in the image. These features are grouped into first order, second order, or higher order statistics. From the Gray level cooccurrence matrix (GLCM) the second order statistical texture features were extracted [13].

The number of possible intensity levels in the image determines the size of GLCM. The gray-co matrix is calculated for different pixel distance and orientations. Some descriptors used for characterizing GLCM are maximum probability, correlation, contrast, energy, homogeneity, and entropy. In addition to these some more features such as dissimilarity, inverse difference moment, variance, mean, sum of average, auto correlation, skewness, kurtosis, etc., are also extracted [14]. All these features together contribute strong feature matrix. This feature matrix is input to the classifier.

4.2 Classifier

As explained in the third section support vector machine is used as the classifier. The features explained above (22 features) have been extracted and saved. These features of twenty different images of HCC and normal are fed to the input to the SVM. These feature vectors are used for training the SVM for the normal case and HCC. After completing the training a new image is fed to the SVM and the SVM predicts either normal or diseased.

These features were again applied to the naïve Bayes' classifier. The training and testing phase has done and the prediction has carried out.

4.3 Prediction

This block makes the prediction that the test image is diseased one or normal. Since there are lesser number of data set, ten-fold cross validation has been done. The confusion matrix is determined from which accuracy, sensitivity, and specificity are calculated.

5 Results and Discussions

In this work, 22 features were extracted from the given image data set. These features provide a strong feature vector. A set of features extracted from the ROI of each class (one from normal liver parenchyma(NP) and one from HCC) as been tabulated in Table 1.

The classification has conducted with the help of two versatile classifiers, viz., SVM and naïve Bayes'. Since the number of features is comparatively larger the accuracy of classification is as high as 95% with naïve Bayes' and 90% with SVM. Both classifiers work on different mathematical basis. But both of the classifiers yield approximately the same accuracy of classification. This implies that the

Table 1 Features extracted from normal liver parenchyma (NP) and HCC

Sl. No	Feature	NP	HCC
1.	Contrast	0.6258	0.44297
2.	Correlation	−0.0393	0.54577
3.	Energy	0.26012	0.19290
4.	Homogeneity	0.73006	0.79027
5.	Entropy	2.31472	2.71854
6.	Maximum Prob	0.39095	0.31569
7.	Dissimilarity	0.56114	0.42534
8.	IDM	0.72589	0.78909
9.	Mean	42.7128	29.1755
10.	Skew	0.02004	−0.0215
11.	Kurtosis	3.11128	2.38198
12.	Shade	0.18851	0.21664
13.	Prominence	1.31692	4.74730
14.	Variance	33.1428	18.3603
15.	Inertia	0.62581	0.44297
16.	IDN	0.93836	0.95294
17.	IDMN	0.99046	0.99321
18.	Auto corrln	32.9729	18.2087
19.	Corrp	−6325.59	−1751.19
20.	Sum average	11.4865	8.47178
21.	Sum entropy	1.1219	1.55131
22.	Difference entropy	0	0.79818

Table 2 Performance indices for SVM and naïve Bayes

Classifier	Confusion matrix	Accuracy (%)	Sensitivity	Specificity	PPV
SVM	$\begin{bmatrix} 8 & 0 \\ 2 & 10 \end{bmatrix}$	90	1	0.83	0.8
Naive Bayes'	$\begin{bmatrix} 9 & 1 \\ 0 & 10 \end{bmatrix}$	95	0 0.9	1	1

feature vector is strong enough for providing good classification job. Different classification measures are tabulated in Table 2.

The receiver operating characteristics (ROC) plots the variation of sensitivity (true positive rate) and specificity (false positive rate). ROC has been drawn (Fig. 3) which encloses the maximum area under curve showing the best classification performance of both the classifiers. The ideal nature of ROC is due to the less number of data set. The same features are used in weka 3.8 to train SVM. Figure 4 shows the ROC plot obtained from the weka.

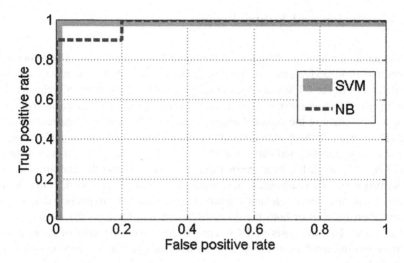

Fig. 3 ROC plotted using MATLAB

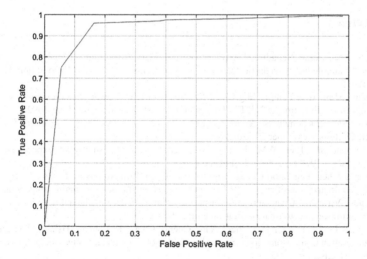

Fig. 4 ROC plotted from Weka

6 Conclusion and Future Scope

In this work, we have proposed a CAD system for the detection of hepatocellular carcinoma from abdominal CT images. Input to the CAD system was the ROI segmented from the CT images by an expert radiologist. The output is the decision whether the tissue is normal liver parenchyma or HCC. The image analysis was done with the help of the classical features extracted from the cooccurrence matrix. The classification is done with support vector machine and naive Bayes's classifier. The accuracy statistics indicates that the selected features perform well with the classifiers. The work has been carried out in MATLAB and the classification and performance characteristics are also evaluated in weka3.8. The work is intended to extend by finding more relevant features from spatial and frequency domain and combination features of both the domains which will improve the overall system performance. The diagnosis of HCC can be done with the help of these texture features and the machine learning algorithms which provides a second opinion to the radiologist.

Acknowledgements The authors acknowledge the expert radiologists from different hospitals from Trivandrum for providing labeled CT images of hepatocellular carcinoma and normal liver. The authors are grateful to College of Engineering, Trivandrum for providing MATLAB license and necessary literature resources.

References

1. Koel Dasa, Zoran Nenadicb, "Approximate information discriminant analysis: A computationally simple heteroscedastic feature extraction technique", Pattern Recognition 41 (2008) 1548–1557
2. Chung_Ming Wu, Yung-Chang Chen, Kai-Sheng Hsieh, "Texture Features for classification of ultrasonic liver images", IEEE Transactions on Medical Imaging, June 1992
3. Khan AN, Macdonald S, Amin Z, "Liver, metastasis", http://www.emedicine.com/radio/topic394.htm. November 21, 2003
4. Ackerman NB, Lien WM, Kondi ES, et al. "The blood supply of experimental liver metastases". 1969;66:1067–72. [PubMed]
5. E-Liang Chen, Pau-Choo Chung; Ching-Liang Chen, "An Automatic Diagnostic System for CT liver Image Classification", IEEE Transactions on Biomedical Engineering, June 1998
6. Jitendra Virmani, "Characterization of primary and secondary malignant liver lesions from B mode ultrasound", Journal of digit imaging, 1058 to 1070
7. Wen-Li Lee, Yung-Chang Chen, Ying-Cheng Chen, Kai-Sheng Hsieh, "Unsupervised segmentation of ultrasonic liver images by multiresolution fractal feature vector", Information Sciences, Elsevier, 2005
8. Deepti Mittal, Vinod Kumar, Suresh Chandra Saxena, Niranjan Khandelwal, Naveen Kalra, "Neural network based focal liver lesion diagnosis using ultrasound images", Computerized Medical Imaging and Graphics, 35, June 2011, 315–323, Elsevier
9. Wen-Li Lee, Kai-Sheng Hsieh, "A robust algorithm for the fractal dimension of images and its applications to the classification of natural images and ultrasonic liver images", Elsevier, Signal Processing 90(60), 1894–1904, (2010)

10. Zeumray Dokur, Tamer e olmez, "Segmentation of ultrasound images by using a hybrid neural network", Pattern Recognition Letters 23 (2002), 1825–1836, Elsevier
11. Constantine Kotropoulos, Ioannis Pitas, "Segmentation of ultrasonic images using Support Vector Machines", Pattern Recognition Letters 24 (2003) 715–727, Elsevier
12. B. Vijayalakshmi, and V. Subbiah Bharathi, " Classification of CT liver images using local binary pattern with Legendre moments", Current science, vol. 110, no. 4, 25, February, 2016
13. Belal K. Elfarra and Ibrahim S. Abuhaiba, "New Feature Extraction Method for Mammogram Computer Aided Diagnosis", International Journal of Signal Processing, Image Processing and Pattern Recognition Vol. 6, No. 1, February, 2013
14. P. Mohanaiah, P. Sathyanarayana, L. GuruKumar, "Image Texture Feature Extraction Using GLCM Approach", International Journal of Scientific and Research Publications, Volume 3, Issue 5, May 2013
15. Gonzalez and Woods, "Digital Image Processing", Pearson Education
16. Anil K Jain, "Fundamentals of Digital Image Processing", PHI

A Level-Set-Based Segmentation for the Detection of Megaloblastic Anemia in Red Blood Cells

N.S. Aruna and S. Hariharan

Abstract Anemia is a decrease in the amount of red blood cells or hemoglobin. It is observed that there are more than 400 types of anemia. Out of this megaloblastic anemia is one among them. Megaloblastic anemia is caused by deficiency of vitamin B12 and folic acid. Diagnosis of this type of anemia is a bit difficult process for the medical professionals. Digital image processing can help the physicians and researchers for the diagnosis of several blood-related diseases. In this paper an effort is made for the diagnosis of megaloblastic anemia precisely and accurately based on digital image processing principles.

Keywords Red blood cells · Megaloblastic anemia · Level-set segmentation Curve evolution

1 Introduction

Megaloblastic anemia is a blood disorder which can be identified by the appearance of very large red blood cells. Megaloblastic anemia can be included in a heterogeneous group of disorders that has the most common morphological characteristics like large cells which prevents the maturation of nucleus [1, 2]. Hyper-segmented neutrophils are the characteristics of megaloblastic anemia which can be seen on peripheral smears and giant bands that occur in bone marrow. Incomplete cell development and large immature red blood cells cause this type of anemia [3]. These anemic cells cannot function like a healthy red blood. As these cells are not developed as a healthy cell, they have a very short span of life. As there are different types of anemia, it is difficult for the health professionals to identify the type of anemia affected by the patient through a blood test.

N.S. Aruna (✉) · S. Hariharan
Department of EEE, College of Engineering Trivandrum, Thiruvananthapuram, Kerala, India
e-mail: arunasurendran2006@gmail.com

S. Hariharan
e-mail: harikerala2001@yahoo.com

© Springer Nature Singapore Pte. Ltd. 2018
Y.-C. Hu et al. (eds.), *Intelligent Communication and Computational Technologies*, Lecture Notes in Networks and Systems 19,
https://doi.org/10.1007/978-981-10-5523-2_12

119

Blood test is the only diagnosis method to determine the characteristics of anemia. It requires a detailed history of patient with a thorough clinical evaluation of blood test. Blood test gives clear idea about the abnormally large, disordered red blood cells which characterize megaloblastic anemia. The cause of megaloblastic anemia varies and may not exhibit any similar symptoms [4]. It is a difficult task to determine the true frequency of anemia in patients.

There are different existing tools in image processing to find the type of anemia present in blood smear images. Segmentation process extracts the boundary of the object from the background. Thus different segmentation processes are used to extract all details and information contained in the image. Level-set segmentation used by Chan-Vese and Shi [5, 6] is adopted in this paper as this method gives a proper idea about the regions and contours in the image. Shi method of segmentation is applied for getting exact boundary of the objects in the image that is to be segmented out to find the number of hyper-segmented neutrophils. The hyper-segmented neutrophils cannot be clearly seen in a microscopic view to count the number of neutrophils. These methods [5, 6] introduce an algorithm for the approximation of level-set-based curve evolution suitable for real-time implementation. Active contour and curve evolution techniques [7–11] can make it more flexible to identify nucleated regions of red blood cells in the image.

There has been extensive work done in the medical field of megaloblastic anemia but less in the field of processing of megaloblastic affected blood smear images. Requirement of a computer-aided system to detect this anemia is a very necessary product in this existing world. Different approaches in digital image processing [12–14] have been applied to the images of megaloblastic anemia and found that active contour method of segmentation is more suitable for the development of computer-aided system. We have applied active contour-based boundary detection of the hyper-segmented neutrophils avoiding the boundaries of other objects in the image. This brings the novelty of the method proposed in this work. The active contour method used to extract the boundary of the cells in the image helps the health professionals to decide the type of anemia.

The validation of the proposed system is performed on the basis of statistical analysis of different samples collected. The statistical analysis can be done by finding the true positive, true negative, false positive, and false negative methods [15]. After finding this, the accuracy of the system that we have proposed can also be evaluated.

The rest of the section in this paper discusses: the theory of proposed method used for image segmentation; then the proposed system, and the block schematic continued by the discussion of experimental results and a conclusion of the proposed system.

2 Theory

In this work, the image has been processed in different ways for automatic diagnosis of megaloblastic anemia. Segmentation is an important technique used in image processing to identify the objects in the image. The work is carried out using the observations, simulation results, and the experiments conducted; more interest is given to the segmentation portion and has been conducted in a robust and efficient manner on the image. The level-set segmentation is one of the easiest methods for segmentation in image processing. Real-time tracking of object boundaries is a great task in image processing. Active contour tracking based on region segmentation can be adopted for extracting a particular region.

Most of the techniques for finding the boundary of an object are by solving the partial differential equations. Shi method of segmentation is based on curve evolution. In level-set segmentation, the automatic topological changes are handled accurately with the numerical implementation of curve evolution. The implementation of curve evolution is used by solving certain partial differential equations which has become a burden and computational difficulties in real-time implementation. To avoid such computational errors and burden, Shi [6] has developed an algorithm for the approximation of level-set-based curve evolution which avoids the need for solving partial differential equations.

Consider the surface $R^l (l \geq 2)$ and a curve C is represented as zero-level set of a function ϕ [6]. Consider the following region-based segmentation problem and the curve evolution equation given below:

$$\frac{dC}{dt} = F\vec{N}, \tag{1}$$

where F is the data-dependent speed and N is normal of curve pointing outward. Equation (1) gives the curve depending on the speed field in normal direction. The speed field, F, is composed of two parts: an external speed derived from the image data and an internal speed that depends on the geometrical properties of C. Then we have to develop a curve C until it stops at point a (which is local) minima of the energy C^*, which is corresponding to a stationary point of Eq. (1), which is shown in Fig. 1 [6]. The curve C^* at stationary points must satisfy the following optimal condition in terms of the speed field F.

Fig. 1 Representation of optimal point curve

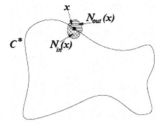

The list of neighboring grid points is represented as L_{in} and outside grid points as L_{out} for the object region Ω given by

$$L_{in} = \left\{ x. \middle| x \in \Omega \text{ and } \exists\, y \in N(x) \text{ such that } y \in \frac{D}{\Omega} \right\} \qquad (2)$$

$$L_{out} = \left\{ x. \middle| x \in \frac{D}{\Omega} \text{ and } \exists\, y \in N(x) \text{ such that } y \in \Omega \right\}, \qquad (3)$$

where

$$N(x.) = y \in D \middle| \Sigma_{l=1}^{l} |y_l - x_l| \middle| = 1. \qquad (4)$$

For a fixed curve C as shown in Fig. 2, L_{in} and L_{out}- are fixed points [6]. Here the point can be located using initial location of the curve. Background region on the discrete grid determined by the final location of the curve can be achieved using the relation between C, L_{in}, and L_{out}. To move the curve outward at the grid point, it just needs to switch the grid point x as shown in Fig. 2 from L_{out} to L_{in}. Similarly, it only needs to switch the grid point ϕ from L_{in} to L_{out} in order to move the curve inward. By applying such steps to all points in L_{in} and L_{out}, this can move C inward or outward one grid point everywhere along the curve with minimal computation. Continuing the switching operations, the method can achieve arbitrary object boundaries that are defined on the grid mentioned above. It should be noted that the switching process moves the curve each pixel for every execution, while the exact level-set method generates sub-pixel movement.

The switching process considering the pixel, approximating the sub-pixel level-set-based curve evolution process, results in the classification for the object and background region. Grid-level switch-based evolution with integer computation and level-set representation makes the system more robust and efficient. Thus Shi model is a faster algorithm that makes level-set segmentation easier for the user.

Fig. 2 Fixing the curve using active contour segmentation

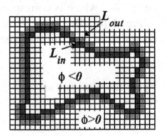

3 Proposed System

Proposed digital image processing system consists of different image analysis strategies, the block schematic diagram of which are shown in Fig. 3. The ultimate aim of image analysis in biomedical image processing is to collect clinical information from images which are used for diagnosis and treatment planning. In the present paper, preprocessing and segmentation techniques are performed which are implemented in clinical environments for retrieving data and information from images. Following are the steps in the proposed system:

1. Preprocessing makes the images more clear and hidden information can be made visible. The input image is preprocessed by filtering the image. This gives more clear vision of the objects in the image and helps the processing system to detect the edges of the objects.
2. Red blood cell segmentation and extraction are the process to distinguish between red blood cells and other cells in the blood smear image. The segmentation process is a high-level image processing technique. In image segmentation our aim is to segment or partition various regions of an image for visualizing the constituent parts clearly. Active contour method which is discussed in theory part is modified for segmenting the shape of neutrophils in red blood cells.
3. The segmentation of neutrophils can be counted by counting the number of closed boundaries present in the image.
4. In the processed images if number of segmented region is 5 or more, it shows the presence of megaloblastic anemia, otherwise there is presence of other type of anemia. The system will detect the presence of anemia by itself and display the output diagnosis result as 'presence of megaloblastic anemia'. If there

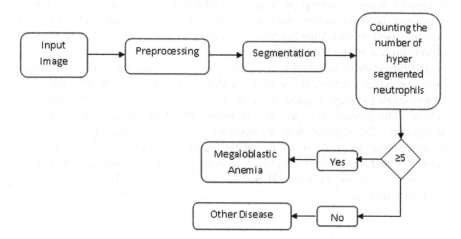

Fig. 3 Block schematic diagram of the proposed system

is absence of megaloblastic anemia, the system will display as 'absence of megaloblastic anemia, please check for other type of anemia'.

Successful results were obtained with the help of above-mentioned process under MATLAB platform by counting the number of segments in the hyper-segmented neutrophils.

4 Results and Discussion

The diagnosis has to be done with the blood samples which may or may not contain megaloblastic anemia. Our system will determine whether the person is mega-loblastic anemic patient or not. The images were acquired through digital camera used with scanning electron microscope (SEM). The image is preprocessed to sharpen the edges of image and to remove the noise. The image is then converted to grayscale image. The preprocessing steps involved in the system makes the image more compatible and further analysis of the image requires the noise content to be removed. One of the steps involved in preprocessing is to convert the image to grayscale.

The level-set segmentation is done to find the exact number of segmentation in the neutrophils of the cell. The basic idea of the level-set method is to represent the curve implicitly as the zero-level set of a function defined over a regular grid, which is then evolved according to the specified speed function based on active contour method.

The segmentation method used in this proposed work is an accurate method for the detection of megaloblastic anemia as shown in Fig. 4. This method uses dif-ferent functions in order to automatically detect interior contours, and to ensure the computation of boundaries and curves without considering a second initial curve. Shi model method is segmenting each neutrophil in the cell; hence we can get more idea about the segmentation. Shi model method of curve representation is an active contour based on which the boundary of neutrophils is detected avoiding all other objects present in the blood smear image.

The boundary of neutrophils is detected by initial contour drawn to the image. With the initial contour, the program coded in Matlab R2014b will detect the energy levels in image to detect the same levels over the curve and hence place the contour. The program is coded such that this method allows automatic change of topography. The segmentation process must be very accurate; otherwise, it will adversely affect the treatment of the patient. This method has the ability of detecting smooth boundaries, scaling adaptivity, automatic change of topology, and robust-ness with respect to the noise. Hence this method gives a highly accurate seg-mentation results.

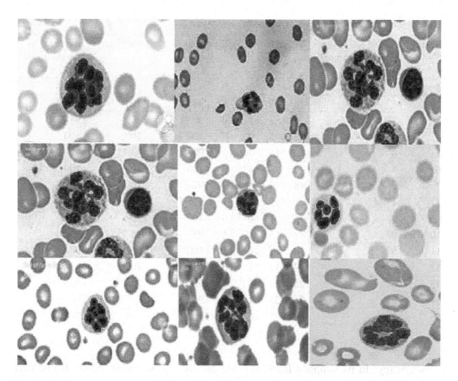

Fig. 4 Active contour segmentation performed on different images of blood smear affected by megaloblastic anemia

4.1 Statistical Analysis

The validation of the system is identified by using the statistical analysis of different sample images of blood. The result set can be evaluated with true positive (TP), true negative (TN), false positive (FP), and false negative (FN) using the following:

Sensitivity: It measures the ability of the test to detect the condition when the condition is present:

$$Sensitivity = (TP/(TP + FN)) \times 100. \tag{5}$$

Specificity: It measures the ability of the test to correctly exclude the disease when the disease is absent:

$$Specificity = (TN/(TN + FP)) \times 100. \tag{6}$$

Table 1 Performance evaluation of the proposed system

Sample	No. of Hyper-segmented neutrophils	TP	TN	FN	FP	Sensitivity (%)	Specificity (%)	Accuracy of the proposed system (%)
1	6	12	8	1	0	92.31	100	95.23
2	6	16	10	1	0	94.12	100	96.29
3	8	10	8	1	0	90.91	100	94.73
4	4	2	5	0	0	100	100	100
5	6	20	15	2	1	90.91	93.75	92.10
6	4	2	4	0	0	100	100	100
7	7	18	15	1	0	94.74	100	97.05
8	4	21	19	1	1	95.45	95	95.23
9	0	15	1	0	0	100	100	100
10	6	19	15	1	1	95	93.75	94.44
11	5	18	15	0	1	100	93.75	97.05
12	8	17	12	1	0	94.44	100	96.66
13	6	25	22	2	1	92.59	95.65	94
14	5	26	25	1	2	96.30	92.59	94.44
15	2	2	1	0	0	100	100	100

Accuracy: In its simplest form, it is the ability to distinguish between just two states of health or circumstances:

$$\text{Accuracy} = ((\text{TP} + \text{TN})/(\text{TP} + \text{TN} + \text{FN} + \text{FP})) \times 100. \qquad (7)$$

Table 1 shows the statistical analysis of 15 different blood samples to calculate the accuracy of the proposed system. The blood samples are collected from different hospitals for the processing of the samples. The proposed system has the ability to detect hyper-segmented nucleated red blood cells.

The accuracy plot is the plot between accuracy and different blood samples used, as shown in Fig. 4 and determines how accurately the proposed image processing system works in different blood samples provided. From the blood samples, the proposed algorithm-based system can identify where the patient is anemic or not. The system gives the output result after counting the segmented nucleated RBC. The accuracy is plotted against blood samples which are displayed in Table 1 and calculated using equation (7). The accuracy plot gives an idea how well the system accurately measures the blood sample (Fig. 5).

Fig. 5 Plot of accuracy against different blood image samples

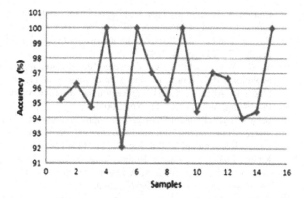

5 Conclusion

As there are 400 different types of anemia existing in present-day world, megaloblastic anemia being one among them require a special attention to detect the disease. Digital image processing techniques help in processing of such blood smear images. The blood smear images will have to undergo different preprocessing and segmentation procedures to exactly identify the disease. The level-set segmentation method is used to segment the hyper-segmented nucleated part of the red blood cells. Active contour-based boundary detection is applied to such images to detect the boundary of hyper-segmented neutrophils present in red blood cells. The method is so designed in Matlab in order to segment exactly the neutrophils. The level-set segmentation method has the property of evaluating the curve which will give clear segmentation as we are using 15 set of blood sample for diagnosis. As a part of quantitative analysis of SEM images of red blood cells calculated the accuracy of the method, the work helps in easier diagnosis of megaloblastic anemia.

Acknowledgements The authors would like to thank Centre for Engineering Research and Development under College of Engineering, Trivandrum, Kerala, India, for the financial support given for this project. And we would like to thank the doctors from different hospitals who provided advice and guidance for the detection of disease.

References

1. Meghana, Sricharan, Ujwala, "Megaloblastic Anemia", e-Journal of Dentistry, Vol. 4 Issue 2, pp. 608–620, June 2014.
2. Ulfar J., O.S. Roath, Charolette, "Nutritional Megaloblastic Anemia Associated with Sickle Cell States", Blood Journal, pp. 535–547, 1958.
3. Uma Khanduri, Archna Sharma, "Megaloblasticanemia: Prevelance and causative Factors" The National Medical Journal of India, Vol. 20, No. 4, pp 172–175, 2007.

4. Charles S. Hesdorffer, Dan L. Longo., "Drug induced Megaloblastic anemia", The New England journal of Medicine, vol. 373, pp. 1649–58, 2015.
5. T. Chan and L. Vese, "Active contours without edges," IEEE Trans. Image Process., vol. 10, pp. 266–277, February 2001.
6. Y. Shi, W.C. Karl, "A Real time algorithm for the approximation of level set based curve evolution", IEEE Trans. on Image Processing. Vol. 17, no. 05, pp. 645–656, 2008.
7. Shawn L., Allen T., "Localizing Region based Active contours", IEEE Transactions on Image Processing, Vol. 17, no. 11, Nov 2008.
8. Seema Chach, Poonam Sharma, "An improved region based active contour model for medical image segmentation", International Journal of Signal Processing, Image processing and Pattern Recognition, Vol. 8, Issue 1, pp. 115–124, 2015.
9. Kaihua Zhang, Lei Zhang, Huihui Song, Wengang Zhou, "Active Contours with selective local or global segmentation: A new formulation and level set method", Image and Vision computing, vol. 28, pp. 668–676, 2010.
10. Zhiheng Zhou, Kaiyi Liu, Xiaowen Ou, "Active contour energy used in object recognition method", IEEE Region 10 Conference, pp. 743–745, 2016.
11. K. Bikshalu, Srikanth, "Segmentation using active contour model and level set methodapplied to medical images", International Journal of Advanced Engineering and global technology, Vol-04, Issue-02, March 2016.
12. Rafael C. Gonzalez and Richard E. Woods, Digital Image Processing, Prentice-Hall, 2002.
13. A. K. Jain, Fundamentals of Digital Image Processing. Englewood Cliffs, NJ: Prentice-Hall, 1989.
14. Rafael C. Gonzalez and Richard E. Woods, "Digital Image Processing, using Matlab" Prentice-Hall, 2002.
15. Bland M. "An Introduction to Medical Statistics", 3rd Edn. Oxford: Oxford University Press; 2000.

Music Playlist Generation Using Facial Expression Analysis and Task Extraction

Arnaja Sen, Dhaval Popat, Hardik Shah, Priyanka Kuwor and Era Johri

Abstract In today's stressful environment of IT Industry, there is a truancy for the appropriate relaxation time for all working professionals. To keep a person stress-free, various technical or nontechnical stress releasing methods are now being adopted. We can classify people working on computers as administrators, programmers, etc. Each of them requires varied ways in order to ease them from the stressed environment. The work pressure and the vexation of any kind for a person can be depicted by their emotions. Facial expressions are the key to analyze the current psychology of the person. In this paper, we propose a user-intuitive smart music player. This player will capture the facial expressions of a person working on the computer and identify the current emotion. Intuitively, the music will be played for the user to relax them. The music player will take into account the foreground processes which the person is executing on the computer. Since various sorts of music are available to boost one's enthusiasm, taking into consideration the tasks executed on the system by the user and the current emotions they carry, an ideal playlist of songs will be created and played for the person. The person can browse the playlist and modify it to make the system more flexible. This music player will thus allow the working professionals to stay relaxed in spite of their workloads.

A. Sen (✉) · D. Popat · H. Shah · P. Kuwor · E. Johri
Department of Information Technology, K. J. Somaiya College of Engineering,
Mumbai, India
e-mail: arnaja.sen@somaiya.edu

D. Popat
e-mail: d.popat@somaiya.edu

H. Shah
e-mail: hns@somaiya.edu

P. Kuwor
e-mail: priyanka.kuwor@somaiya.edu

E. Johri
e-mail: erajohri@somaiya.edu

© Springer Nature Singapore Pte. Ltd. 2018
Y.-C. Hu et al. (eds.), *Intelligent Communication and Computational Technologies*, Lecture Notes in Networks and Systems 19,
https://doi.org/10.1007/978-981-10-5523-2_13

Keywords Facial expression analysis · Emotion recognition
Feature extraction · Viola–Jones face detection · Gabor filter · Adaboost
K-NN algorithm · Task extraction · Music classification · Playlist generation

1 Introduction

Music plays an imperative part throughout an individual's life. The inclination, identity, and sentiments of people can be perceived through their emotions. Music oftentimes communicates passionate qualities and characteristics of human identity, for example, happiness, sadness, aggressiveness, etc. Consequently, the kind of music a person prefers to listen at a particular time can reflect the state of the mind of a person right then and there. This marvel can be used in a few applications where human feelings assume a noteworthy part. Generally, people have a vast collection of songs in their computer. Manually separating the collection of songs and creating a suitable playlist based on an individual's emotions is an exceptionally monotonous task. Along these lines, users tend to play arbitrary music which may not suit their current passionate state. We have developed a music player which recognizes the user's mood and tasks performed by them at a particular instant, and accordingly manages the playlist for them. Since, the human face assumes a vital part for the extraction of an individual's conduct and passionate expressions, this software will capture the user's facial expressions and features to decide the present state of mind of the user. The facial expressions and emotions will be arranged into five unique categories which are anger, disgust, happiness, neutral, and sadness [1]. The images are captured through webcam. The captured image will be spared and passed on to the rendering stage. Simultaneously, the task being performed on the user's computer is extracted and classified into one of the three categories which are browsing, programming, and other tasks. Finally, a combination of the recognized emotion and the most recent task's category is provided to the neural networks for training and analysis which then provides the user with an ideal playlist. This playlist will help in improving the user's mood and maximize the efficiency of the task being performed. Furthermore, this will empower users to relieve their nerves with the assistance of fitting music. The current mood of the user may contrast after some time. Hence, the process will be reiterated after every 10 min to detect any variations in the emotions or task being performed. The system is constructed in a conventional way and comprises of five phases, which are, face detection, feature extraction, emotion recognition, task extraction and identification, and music playlist generation.

2 Literature Review

Music and its utilization in feeling control procedures, right up till today remains an uncertain question. Feelings can show up in many parts of human-to-human correspondence and regularly provide ancillary data about a message. Clinical and nonclinical reviews, all exhibit the successful utilization of music as a self-regulative device for feelings. The accompanying investigations of the current papers bolster the general understanding of this survey that music listening is most regularly utilized with a colossal arrangement of objectives and procedures for enthusiastic control purposes. The takeaways from the papers are as follows:

1. Face recognition and emotion detection can be carried out using the support vector machines (SVM) algorithm and facial action coding system [2] (FACS). Besides this, the detection of the various distinct feature points and different distances between the disparate feature face points are considered for recognizing facial expressions [2].
2. Identification of how the playlist can be played by users' inclination towards the diverse kinds of music of various periods [3]. Moreover, data cleansing and clustering concepts for music are considered. Further, various data visualization techniques and serendipity recommendation systems are likewise introduced in this paper [3].

3 Face Detection

Appropriate detection of the face and extraction of the relevant features is the most significant element of an emotion recognition system. Viola–Jones algorithm is selected for face detection as it has a very high detection rate, which makes it robust, and has an advantage of operating in real time.

The Viola–Jones algorithm [4] has four stages which are as follows:

1. Haar Feature Selection [4].
2. Creation of an Integral Image [4].
3. AdaBoost Training Algorithm [4].
4. Cascading Classifiers [4].

3.1 Haar Feature Selection

All the human faces have similar features. These similarities may be matched using Haar feature selection. Although, Haar feature Selection is a weak classifier, it has been employed since the Haar-based cascade works better as compared to other methods such as Local Binary Patterns in the standard platforms where there is more resource availability, and it has a better accuracy rate around 96.24% [5] as

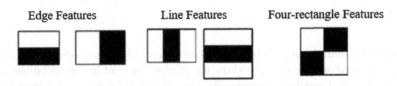

Fig. 1 Different features [4]

Fig. 2 Original image and integral image

Original Image

05	02	03	04	01
01	05	04	02	03
02	02	01	03	04
03	05	06	04	05
04	01	03	02	06

Integral Image

05	07	10	14	15
06	13	20	26	30
08	17	25	34	42
11	25	39	52	65
15	30	47	62	81

well. The possible types of features are two-, three-, and four-rectangle features [4]. These features are represented in Fig. 1.

The edge feature detects the contrast between two vertical or horizontal adjacent regions and the line feature detects the region of contrast present between two similar regions. The four-rectangle feature is beneficial in detection of similar regions placed diagonally.

$$\text{Feature value} = \sum (\text{pixels under black region}) - \sum (\text{pixels under white region}) \tag{1}$$

3.2 Creation of an Integral Image

Calculation of the areas under black [6] and white rectangles in real time is computationally expensive. Hence, the integral image concept is utilized. The calculation of an integral image can be done as follows [4]:

$$\text{Integral Image}(x, y) = \text{Integral Image}(x - 1, y) + s(x, y - 1) + \text{Image}(x, y), \tag{2}$$

where, $s(x, y)$ is the aggregate row sum (Fig. 2).

3.3 AdaBoost Training Algorithm

The working of the AdaBoost training algorithm is illustrated in Fig. 3.

In 24 * 24 pixel sub-window, there are approximately 162,000 features, many of which are not relevant. AdaBoost facilitates removal of these irrelevant features by

Fig. 3 AdaBoost training
algorithm [4]

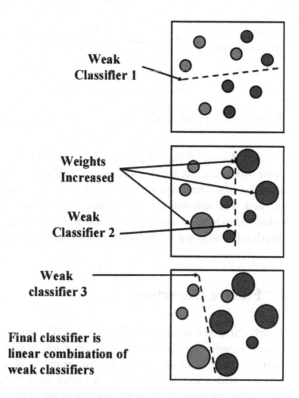

selecting the top features and trains the classifiers that use them. Using the Eq. (3), this algorithm forms a strong classifier [4] from various weak classifiers.

$$h(x) = \text{sign}\left(\sum_{j=1}^{M} \alpha_j h_j(x) \right), \tag{3}$$

where

- the threshold value θ_j, co-efficient α_j and the polarity s_j are found by training [4], and
-

$$h_j(x) = \begin{cases} -s_j & \text{if } f_j < \theta_j \\ s_j & \text{otherwise} \end{cases} \tag{4}$$

3.4 Cascading Classifiers

A cascade [4] of gradually more complex classifiers is employed in this case. Each classifier involves a pre-decided number of features which allows the differentiation between faces and non-faces. Hence, as the image goes through each stage in the

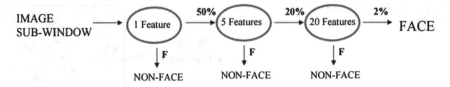

Fig. 4 Cascading process [4]

cascade, the probability of the image containing a face increases. A positive outcome from one classifier leads to the evaluation of the next classifier and this provides high detection rates. This process is continued till the last classifier is reached. The sub-window is rejected if a negative outcome occurs at any point. AdaBoost is used to train the classifiers in various stages in the cascade and then the threshold is adjusted to minimize false negatives (Fig. 4).

4 Feature Extraction

A collection of Gabor filters [7] having varied orientations is used to extract necessary features from an image. This filter has been adopted for the extraction of features, since it is more effective than geometric features-based filters, and works better in real-world environments. A Gabor filter is essentially a sinusoidal signal with given frequency and orientation, modulated by a Gaussian. Two-dimensional Gabor filters [8] in the discrete world are presented by the Eqs. (5) and (6).

$$G_c[i,j] = Be - \frac{(i^2 + j^2)}{2\sigma^2} \cos(2\pi f(i \cos\theta + j \sin\theta)) \tag{5}$$

$$G_s[i,j] = Ce - \frac{(i^2 + j^2)}{2\sigma^2} \sin(2\pi f(i \cos\theta + j \sin\theta)) \tag{6}$$

where,

- B and C are normalization factors,
- f is the frequency,
- θ enables us to find texture oriented in a particular orientation, and
- σ allows to change the dimensions of the input region being investigated.

Otsu method is used to compute the threshold level using which a binary image is obtained from a gray-level image [7]. This algorithm works under the assumption that the image whose threshold is considered comprises of only foreground and background pixels. The optimal threshold is then computed which separates those two classes, so that the variance between them is minimum.

The feature extraction process is shown in Fig. 5.

Fig. 5 Flowchart for feature
extraction [7]

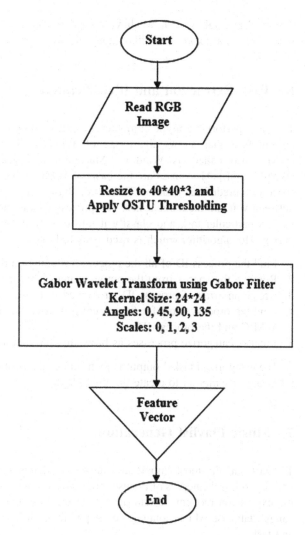

5 Emotion Recognition

The next task after the features have been extracted is the recognition of emotions based on these features. This task requires a training set consisting of images pertaining to different emotions and features that are distinctive to a particular expression. The training set consists of two databases: The Japanese Female Facial Expressions (JAFFE) database and YALE database. The JAFFE database for our study is a collection of around 213 images with 7 facial expressions in all to classify expressions like neutral, sadness, surprise, happiness, fear, anger, and disgust [9].

The YALE database that we have considered is a collection of 165 images to recognize facial expressions like happy, neutral, sad, sleepy and surprised [10].

6 Task Extraction and Identification

The tasks performed on the computer are extracted by tracking the processes using an interface that allows the usage of Windows Management Instrumentation (WMI) also called as Windows Management Instrumentation Command-line (WMIC) tool [11]. Aliases are the basis of WMIC which act upon the WMI name space in a predefined manner. Therefore, aliases are friendly syntax intermediaries between you and the name space. The process ID and the process name are used to track a particular task and classify it into browsing, programming, or other categories. The algorithm which is used is as follows:

1. Find the process ID of all the processes running on the computer using WMIC.
2. Remove the spaces from the output string obtained in step number 1.
3. Create an array of process ID's retrieved from step number 2.
4. Find the process name for five recent processes running on the computer using WMIC and the process ID.
5. Finally, categorize processes as browsing, programming, and other tasks.

The categorized tasks' output is then sent to the neural network [12] along with the analyzed emotion to create the final playlist.

7 Music Playlist Generation

The last and the most important component of this system is playing the songs which soothes the nerves of the user and refreshes them. Based on a combination of the expression recognized and the task detected, a playlist is generated from our songs' database which contains songs pertaining to each combination of emotion and task.

In the songs database, the songs are classified into various genres [13] using k-nearest neighbors (k-NN) [14] after performing an analysis of the lyrics present in the song. The k-NN algorithm provides class membership to each song where each class corresponds to a different combination of emotion and task. Fifteen such classes are created and the songs are added to them. K-NN algorithm [14] has been employed since it does not require much training and can handle noisy training data very well.

The generated playlist contains three songs at a time. After a duration of 10 minutes, the tasks are tracked again and the emotion recognition process is repeated to generate a playlist of three songs. This process continues until the user closes the application (Fig 6).

Fig. 6 Playlist generation method

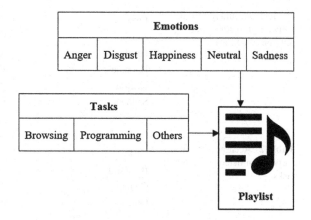

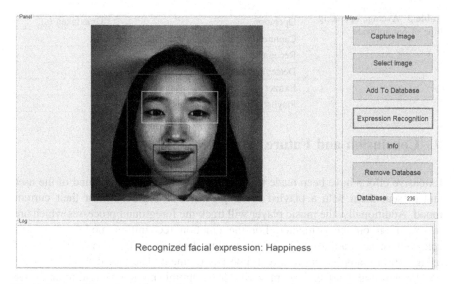

Fig. 7 Intermediate output of our software

8 Results

'X-Beats—A Smart Music Player' is created as a desktop application utilizing MATLAB as a development tool. MATLAB version 2016a is used for the implementation of this software. After testing the system for a duration of over 2 months in a real-world environment, we have attained the following performance measures for our software (Fig. 7):

Average accuracy of emotion recognition: 83%

Average accuracy of task detection: 96% (Tables 1, 2, and 3)

Table 1 Accuracy of our software in recognizing different emotions

Emotion	Accuracy (%)
Anger	82
Disgust	86
Happiness	92
Neutral	75
Sadness	78

Table 2 Accuracy of our software in detecting different tasks

Task	Accuracy (%)
Browsing	96
Coding	93
Others	98

Table 3 Average operating time of our software

Operation	Time taken (s)
Capture the image and detect face	2
Recognize the emotion	4
Detect and categorize task	2
Extract songs from the database	4
Playlist generation process	3

9 Conclusion and Future Work

Extensive efforts have been made to naturally identify the state of mind of the user and present them with a playlist of songs which is reasonable for their current mood. Additionally, the music player will track the foreground processes which are in execution on the computer. For the emotion recognition module, we have achieved an accuracy of 83% for the testing data. The software has been tested using various sample data in a real-time environment. The task detection module has been tested with an accuracy of 96%. Optimizing the algorithm used for implementation reduces the overall operating time of the system.

The music genre classification has been beneficial in reducing the searching time of songs, and thereby increasing the overall efficiency of the system. Furthermore, this has proved to be advantageous in swiftly generating the music playlist. The major strengths of the system are complete automation of the software as well as user independence.

In the future, we intend to increase the accuracy of the application by training the neural network with no tolerance for the error rate signal. Moreover, we plan to enhance our software by developing a mechanism that will be helpful in the music therapy treatment, and also assist the music therapists in treating patients suffering from disorders like anxiety, acute depression, and trauma.

References

1. P. Ekman and W. V. Friesen: Facial Action Coding System: A Technique for the Measurement of Facial Movement. In: Palo Alto, California, USA: Consulting Psychologists Press, 1978.
2. F. ABDAT, C. MAAOUI and A. PRUSKI: Human-computer interaction using emotion recognition from facial expression. In: UKSim 5th European Symposium on Computer Modeling and Simulation, 2011, pp. 196–201.
3. Chim Chwee WONG, Emy Salfarina ALIAS and Junichi KISHIGAMI: Playlist environmental analysis for the serendipity-based data mining. In: 2013 International Conference on Informatics, Electronics and Vision, May 2013.
4. P. Viola and M. J. Jones: Robust real-time object detection. In: International Journal of Computer Vision, Vol. 57, No. 2, pp. 137–154, 2004.
5. Souhail Guennouni, Ali Ahaitouf and Anass Mansouri: A Comparative Study of Multiple Object Detection Using Haar-Like Feature Selection and Local Binary Patterns in Several Platforms. In: Modelling and Simulation in Engineering Journal, pp. 365–374, 2015.
6. Jerome Cuevas, Alvin Chua, Edwin Sybingco, Elmi Abu Bakar: Identification of river hydromorphological features using Viola-Jones Algorithm. In: 2016 IEEE Region 10 Conference (TENCON), 2016.
7. Rakesh Kumar, Rajesh Kumar and Seema: Gabor Wavelet Based Features Extraction for RGB Objects Recognition Using Fuzzy Classifier. In: International Journal of Application or Innovation in Engineering & Management, Vol. 2, Issue 8, August 2013, pp. 122–127.
8. A.G. Ramakrishnan, S. Kumar Raja and H.V. Raghu Ram: Neural network-based segmentation of textures using Gabor features. In: Proc. 12th IEEE Workshop on Neural Networks for Signal Processing, pp. 365–374, 2002.
9. Michael J. Lyons, Shigeru Akemastu, Miyuki Kamachi, Jiro Gyoba: Coding Facial Expressions with Gabor Wavelets. In: 3rd IEEE International Conference on Automatic Face and Gesture Recognition, pp. 200–205, 1998.
10. P. N. Bellhumer, J. Hespanha, and D. Kriegman: Eigenfaces vs. fisherfaces: Recognition using class specific linear projection. In: IEEE Transactions on Pattern Analysis and Machine Intelligence, Special Issue on Face Recognition, 17(7):711–720, 1997.
11. Hui Peng and Yao Wang: WMIC-based technology server network management software design. In: Second Pacific-Asia Conference on Circuits, Communications and System, August 2010.
12. Derrick H. Nguyen and Bernard Widrow: Neural networks for self-learning control systems. In: IEEE Control Systems Magazine, Vol. 10, Issue 3, April 1990.
13. T. Li and M. Ogihara: Music genre classification with taxonomy. In: Proc. of IEEE Intern. Conf. on Acoustics, Speech and Signal Processing, pages 197–200, Philadelphia, USA, 2005.
14. Qingfeng Liu, Ajit Puthenputhussery and Chengjun Liu: Novel general KNN classifier and general nearest mean classifier for visual classification. In: IEEE International Conference on Image Processing, September 2015.

Digital Forensic Enabled Image Authentication Using Least Significant Bit (LSB) with Tamper Localization Based Hash Function

Ujjal Kumar Das, Shefalika Ghosh Samaddar
and Pankaj Kumar Keserwani

Abstract Addition of hash component in a message is always for verification of authentication of message without integrity. In order to retain integrity certain schemes such as Merkle–Damgard scheme or Rabin scheme based on the block cipher can be used. However, the block cipher schemes in messages do not get applicable on image which are encoding message in a visual format. Therefore, block cipher based schemes are not able to provide tamper localization in case of image authentication. Tamper detection is needed in case of images as it is not visualized without proper verification. Moreover, localization is not possible by hash verification making the authentication processes only detectable but not identifiable. The paper presents a generation process of three least significant bits of hash function. The image insertion is not visualized due to human eye limitation. The method proposed is able to embed a secret image (gray scale) inside a corner image. Experimental results show an improvement in peak signal-to-noise ratio (PSNR) values of the proposed technique over the other techniques available, such as hash-based LSB(2-3-3) image steganography in spatial domain. This paper proposes an image authentication process not only for robustness, security, and tamper detection but also successfully able to identify tamper by a process of tamper localization. The parameters compared include the classical mechanism along with newly introduced entropy measurement of images. A performance analysis is able to show the result claimed in the paper. The paper uses a least significant bit (LSB) based hash function, which is a blind technique as image-related information is never sent to receiver separately. In fact, the proposed

U.K. Das (✉)
Srikrishna College, Bagula, Nadia, West Bengal, India
e-mail: ujjalmnnit@gmail.com

S.G. Samaddar · P.K. Keserwani
Department of Computer Science & Engineering,
National Institute of Technology Sikkim, Ravangla, Sikkim, India
e-mail: shefalika99@yahoo.com

P.K. Keserwani
e-mail: pankaj.keserwani@gmail.com

© Springer Nature Singapore Pte. Ltd. 2018
Y.-C. Hu et al. (eds.), *Intelligent Communication and Computational Technologies*, Lecture Notes in Networks and Systems 19,
https://doi.org/10.1007/978-981-10-5523-2_14

141

approach is a clever design as the generated hash function will be embedded in the image itself. The proposed mechanism is able to offer good imperceptibility between original image and embedded image with hash bits. Even a minute tampering can be detected with proper localization; the localization of tampering is having its wide applicability in case of forensic evidence of malicious image morphing.

Keywords Image authentication · Image hashing
Least significant bit (LSB) based technique · Integrity · Tamper detection

1 Introduction

The data produced by multimedia devices have characteristics that are different from the data due to textual messages or data having character representation of any language. The multimedia devices, such as scanners, digital camera, smartphone, multimedia cell phones can not only capture the image but also able to communicate, share, embed, filter, and morph the images using various tools and techniques. Such tools and techniques (e.g., Adobe Photoshop, Wings3D, 123D, Digital Studio, visual studio, Maya, Corel Draw, Cinema 4D, K-3D, Sketchfab, Modo, Clara.io, Lightwave Autocad GMSH, etc.) produce multimedia products in different format and also capable of introducing various synchronization of other media over images. The images in communication demand image integrity, image authentication as well as entity authentication. A cryptographic hash function is generally used to form an image object which generates the integrity of the image. Though an ideal mathematical model for hash function is given by Random Oracle Model (ROM), there are inherent problems in the model as these strings of 0 and 1 of fixed length are independent of the message or image. The hash functions take a message of arbitrary length and create a message digest of fixed length out of a variable-size message. There are various designs such as Merkle–Damgard scheme, message digest schemes (MD2, MD4, MD5 SHA-1, SHA-224, SHA-256, SHA-384, SHA-512), RACE Integrity Priorities Evaluation Message Digest (RIPMED-160), HAVAL, Rabin Scheme, Davies Meyer Scheme, Matyas–Meyer–OSeas Scheme, Miyaguchi–Preneel Scheme Whirlpool [1]. The image authentication with integrity remains a challenging problem. Such tamper detection mechanisms are in place using various integrity checks and authentication but tamper localization remains a problem again. For finding a particular object from large databases of high-dimensional form, a technique of locally sensitive hashing (LSH) [1] is used which churns out a high probability equal to $1 - \delta$, being the nearest neighbor for any query point. However, such type of query search and retrieval are not forensically valid as the probable object has to come out with a discrete value 1 of probability and therefore the requirement of image authentication with tamper detection and localization in a viable digital forensic is not ensured.

The image authentication technique as proposed must address the core issues of security, tamper detection, tamper localization, and robustness. To carry out the work, Sect. 2 provides the limited literature survey in the context of the problem. Section 3 presents the proposed approach to solve problem and justifies it against the standard measure such as histogram, entropy, etc. Section 4 analyses the experimental result on the basis of results obtained, and it is concluded that the proposed approach can provide an image authentication technique with purpose of applicability in digital forensic. Section 5 finally concluded the paper with future direction of work along with the peer researchers.

2 Literature Survey and Background

A content-based image copy detection method was suggested by Kim [2] which was partially successful in deleting copies with rotation. Method was modified by Wu et al. [3] using a dynamic centre point detection mechanism for the elliptical track for the images for dealing with different kind of shifting process. The paper satisfies the basic requirement of copy detection of digital media such as image but does not recover the tamper or manipulation for digital evidence building in digital forensic. Wang and Zhang [4] surveyed a number of recent developments in perceptual image hashing with focus on a two-stage structure and some geometric distortion resilient techniques, which are not suitable for images that undergo geometric modification, enhancement, or compression coding (lossy and lossless). Their properties must be satisfied by perceptual image hash [5]. The algorithm based on the wavelet decomposition as proposed by Venkatesan [6] is a technique in which subbands of the image are obtained using wavelet decomposition. An error correcting decode generates the final hash value. A tolerant image authentication technique in JPEG compression was presented by Lin and Chang [7]. The technique proved useful in original image searching in a database containing different distortions of the same image. Ahmed et al. [8] tried to address the core issues (though conflicting in nature) of tamper detection, security, and robustness. The technique proved successful against JPEG compression, low-pass- and high-pass filtering. Another content-based hash function for image authentication was devised by Schneider and Chang [9]. The paper proposes the procedure of tampering localization and content restoration. However, restored image is usually not considered as digital evidence. Lu et al. [10] indicated that media hashing technique are not resistant against geometrical attacks and therefore, proposed a geometry invariant image hashing scheme. Tang [11] worked for tamper deletion using nonnegative matrix factorization (NMF) in robust image hashing. The technique can be utilized in digital forensics. Hassan et al. [12] used feature combination in Kernel space for distance-based image hashing where a genetic algorithm is used for optimization of tasks. Forensic hash is short signature usually attached to an image before transmission [13] as an additional information. The forensic hash is generated based on visual words perception. The short signature in a multimedia document is generated using a framework for multimedia forensic called FASHION (Forensic hash for information assurance) [13]. The multimedia forensic assumes a new analytical role through the use of robust image hashing. The FASHION framework is based on a number of image hashing works of other authors [14]. Perceptual image hashing has also been used in reduced reference image quality assessment [15]. Even an earlier image authentication method tried to distinguish in normal JPEG compression from malicious tampering in an image [16]. The direction has been taken into consideration in a paper by Lu and Liao [17] for an incidental distortion resistant scheme using structural digital signature for image authentication.

Image authentication techniques are typically either digital signature based or watermark based. The approaches can be classified into quantization based [18], feature point based [19], and relation based [16]. Fridrich and Goljan contributed toward implementation of a robust hash function for digital watermarking that can be used in image authentication [20]. Similar robust image fingerprinting system for multimedia content is suggested by Seo et al. [21] using the mechanism of Radon transform. Yong and Chen obtain robust image hashing that is based on set partitioning in hierarchical trees (SPIHT) algorithm [22]. Similar effort for generating a robust and secure image hashing has been located in later years. Swaminathan's [23] technique is based on Fourier Transform features and controlled randomization. Further improvement in secure image hashing scheme using Radon transform was seen in 2009 [24].

Secure hashing schemes are also being perfected over the years keeping in mind the applicability such as image forgery detection. Deepa and Nagajothi have defined their hashing scheme for image authentication using Zernike moments and local features with histogram features [25]. Earliar, Monga, and Mihcak obtained a similar technique based on nonnegative matrix factorization (NMF) [26]. A number of DCT-based image hash was available. The algorithm proposed by Kailasanathan and others was particularly interesting as it was compression tolerant [27]. A comparison has been made with other paper in this area [19]. Lu and Hsu devised an image hashing scheme that can not only resist geometric attack but also detect copy and problem content authentication [28]. Perceptual image hashing through feature points that are visually significant. It has been suggested by Monga and Evans [29].

An area of image authentication that has been introduced quite late in this millennium is digital image forensics. The forensic problem may occur in counterfeiting of origin as well as of the content. The solution in these two cases is addressed by providing the

(1) Identification of imaging device.
(2) and detection of traces (data and metadata) of the forgery.

There are a number of forensic tools (both open source and proprietary) that are used in solution or in sequence (offline and online) to find out forensic information and valid evidence that can be authenticated by those who are authorized to do so.

Thus, digital image forensic requirement is different from the earlier applications as it requires to detect.

- Authentic information
- Tamper information (intentional/malicious and unintentional/accidental/ technical)
- Tamper localization
- Evidence criteria and evidence preservation at the court of law.

Image forensic can, loosely, categorized into.

- Digital image forensic (collection of data and metadata using various analytical tools)

- Source device identification such as ballistic analysis to identify the device that captured the image and vice versa.

- Tempering detection.

Image processing for forensics make extensive use of techniques of authentication such as digital watermarking and stenography. Blind image forensic techniques can capture traces left by different processing steps in the image acquisition and storage phases. The traces may be camera fingerprints that can be used for origin authentication.

The methods of forgery detection accordingly are applied if it is a single image or a composite. Researchers from Columbia University in 2004 published dataset of authentic and spliced image block [30]. There are mechanisms of tempering detection independent of the type of forgery.

The area of counter counter-forensics is studied from the view point of first two categories of the forgery tampering. Kirchner and Fridrich [31] targeted the median filter track. Compared to image forgery detection, video forgery detection is cumbersome when considering the properties of the video. Video forgery detection can be broadly classified into active- and passive-based approaches [32] (Fig. 1).

Having so many categories demand many techniques and the area emerges as a vast field of research. Even tampering has been made easy due to application of various editing tools such as Adobe Photoshop, Illustrator, GNU Gimp, Premier, Vegas, etc. Therefore, requirement of digital video forgery has a serious repercussion on digital rights management. Image authentication in wavelet domain has given rise to many techniques, especially based on structural digital signature and semi-fragile fingerprinting. A semi-fragile authentication scheme based on block mean quantization in the wavelet domain is investigated [33] which is based on a structural digital signature process and semi-fragile watermarking/fingerprinting algorithm.

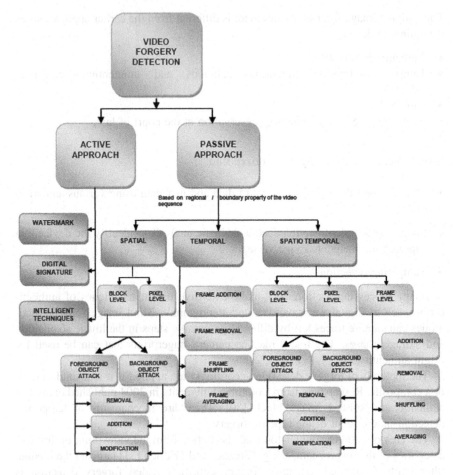

Fig. 1 Video forgery (tampering) classification [32]

3 Proposed Approach of the LSB in the Paper

Proposed hash function is based on LSB (Least Significant Bit) replacement technique. This hash scheme uses pixel-wise approach in spatial domain. A host (cover) image I of size r × c is changed to size of 255 × 255 and then three LSBs are set to zero. Embedding and extracting procedures of generated hash bits are applied. Block diagrams for embedding and extracting hash bits are shown in Fig. 2.

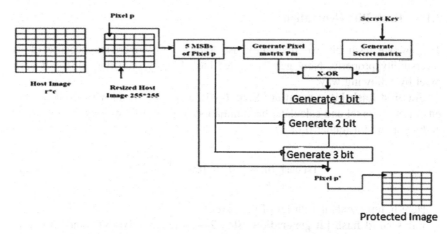

Fig. 2 Hash bit generation and embedding procedure for proposed approach

3.1 Hash Bit Generation and Embedding Process

Following pseudocode/algorithm is proposed in the paper.

3.1.1 Preprocessing

Before applying the proposed hash function generation an input image must be preprocessed in following steps:

```
1-  Take a grayscale Image I of size r x c, such that r, c>0 as input
2-  Change the size of the Grayscale Image I into a fixed size, such
    that 1≥r, c≤255.
3-  Generate Secret random matrix Sm, by using a user defined Secret Key
    srt_key.
4-  In an Image I (for every image pixel), do
        for i= 1 to r
            for j= 1 to c
                a)  Represent ith pixel p in 8-bit binary format from
                    decimal format.
                b)  Set first 3 LSB of pixel p to zero.
                c)  Applying bit wise X-OR on bit of binary pixel and
                    generated secret matrix Sm, and resultant image is
                    called Im.
                Im= bitxor(Sm(i), I(i))
            EndFor
        EndFor
```

3.1.2 Hash Bits Generation

To generate hash bit, we frequently use logical XOR operation because of its reversibility property. Now generate three hash bits by only five MSBs of each pixel by following way:

For first hash bit generation: **Step 1**—Take successive bit-wise XOR of five bits of every pixel and calculate the summation of them followed by modulo 2 for getting single bit (either 0 or 1).

$$1\text{st hash bit} = \left(\sum_{i=1}^{4} b_i \oplus b_{i+1} \right) mod\, 2 \tag{1}$$

Here b_i represents the ith bit of the pixel.

For second hash bit generation: **Step 2**—Generate a five bits random matrix R_m of size r x c whose values ranges from 0 to 31 using another secret key. Do bit-wise XOR between corresponding pixels of Rm and Im.

$$2\text{nd hash bit} = \left(\sum_{i=1}^{5} b_i(I_m) \oplus b_i(R_m) \right) mod\, 2 \tag{2}$$

After doing bit-wise XOR take the summation of the obtained result followed by modulo 2. Because of modulo 2 the final result will be single bit either 1 or 0.

For third hash bit generation: **Step 3**—Calculate the complement I_c of each pixels of Im by following way

$$I_c = 31 - I \tag{3}$$

Since the maximum intensity value for five bits will be 31, we take the difference from that.

Take the bit-wise XOR with its original pixel. Calculate the summation of all bits followed by modulo 2 for getting single hash bit either 0 or 1.

$$3\text{rd hash bit} = \left(\sum_{i=1}^{5} b_i(I_m) \oplus b_i(R_m) \right) mod\, 2 \tag{4}$$

Step 4—Embed all generated three bits into first three LSBs of its corresponding pixel. Now the resultant image will be called protected image.

3.2 Tamper Localization Process

It takes the following steps:

```
1 - Input Tampered Image T of size r x c, such that r=255 and c=255.
2 - Input a Secret Key srt_key used in insertion process to generate
random matrix.
3 - Generate Secret Matrix Sm,pseudo-random function on randomly
selected Secret Key srt_key generates Sm.
4 - Generate 255 x 255 Tampered Pixel Localization Image TPL_img to
locate tampered pixels. Initially every pixel p of TPL_img will be zero
i.e. black.
5 - In Tampered Image T for every Pixel p do
   For i= 1 to r
        For j= 1 to c
        a) Pixel p in 8-bit binary format gets changed to decimal
           format.
        b) Store 3 LSB of pixel p in a vector, LSB_ORG.
        c) Now,3 LSB of pixel p are set to zero.
        EndFor
   EndFor
6 - Generate three bits by using eq. 1, 2, 3 and 4. These are called
recalculated hash bits. Now compare them by the corresponding three LSBs
of vector LSB_ORG.
7 - If any mismatch is found then mark the corresponding pixel of TPL by
white pixels.
```

4 Experimental Result and Analysis

Proposed approach has been implemented in MATLAB 2015, windows 7 (operating system) with processor Intel core 2 duo. Experimental results show different types of attacks such as addition attack (addition of some object or text), deletion attack (deletion of some object or text), content removal attack, etc. Different grayscale images of different sizes show the effectiveness of the proposed approach. Since hash bits are embedded in the image itself, the PSNR between original image and hash bits embedded image in order to calculate the imperceptibility is used. The experimental results show that this scheme is able to achieve PSNR value equal to 44.09 and efficiency up to 100%. Figure 3a, b shows the original image and results after embedding the hash bits in the image respectively. Figure 3c, d shows the tampered image and tamper localization results after applying hash bits extraction and comparison procedure. Section 4.1 show different types of result on grayscale images and Sect. 4.2 show result.

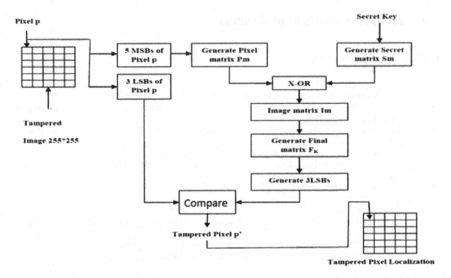

Fig. 3 Tamper localization procedure of proposed approach

Fig. 4 Tamper localization result for Lena image

4.1 Experimentation with Lena Image

When the proposed algorithm is applied, the following visual difference of content may be noticed. It is difficult to find the difference with content due to human eye limitation.

In Fig. 4, alteration is performed by adding a rose flower on the hat of Lena with a tampering of 720 pixels. Then, by applying hash bit comparison procedure, 656 pixels are detected. Thus, efficiency is equal to 91.1% and PSNR value equal to 44.09 dB.

4.1.1 Number Plate Image

Using the same algorithm on number plate image (Fig. 5).

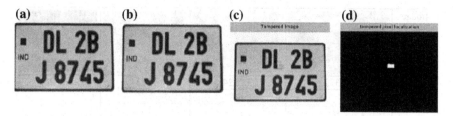

Fig. 5 Tamper localization result for number plate image

Table 1 Observations for various images

Host image	PSNR value	Altered pixel	Detected pixel	% of accuracy	Detection time (sec)
Baboon	41.14 dB	1515	1421	93.79	147.3
Lena	40.98 dB	1739	1656	95.22	140.5
Camera man	41.11 dB	818	766	93.64	150.3
Number plate	40.79	859	841	97.7	148.8

Fig. 6 Plot between altered and detected pixels

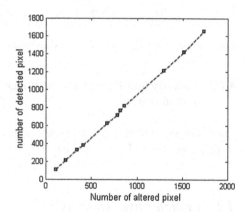

Parameters of various observations can be perceived by Table 1 which shows the effectiveness of the proposed approach. Images have sufficient PSNR value even after embedding the hash bits are presented. We are able detect almost 90% plus altered pixels.

Table 1 demonstrates the plot between number of altered pixel and detected pixels. Straight line shows the effectiveness of the proposed approach. A hash-based significant bit (2-3-3) stage steganography in spatial domain is presented by Manjula and the Danti [34] (Fig. 6).

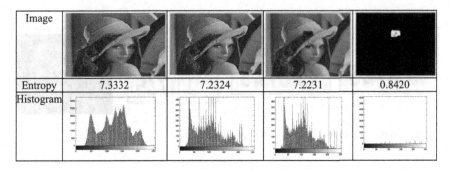

Fig. 7 Entropy and Histogram of used Lena images

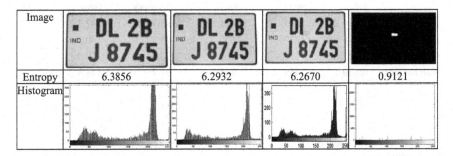

Fig. 8 Entropy and Histogram of used *Gray Scale* images

4.1.2 Entropy and Histogram for Images and Their Entropy Calculated

The following figures presents a table with image, entropy, and histogram for the images considered (Figs. 7 and 8).

4.2 Comparative Analysis

LSB insertion method was not very novel technique, but optimum bit selection to make the image(s) visually indifferent insertion requires investigation. Human vision and its limitation play a major role in such embedding, tampering, localization of tampering, watermarking, fingerprinting, and above all authentication levels of certain degree which is determined at times by one or two threshold values. The attributes of measure that is generally considered in case of original image and inserted bit(s) image(s) of mean square error (MSE), peak signal-to-noise ratio (PSNR) and normalized absolute error (NAE), and structural similarity index (SSIN). Formula used for calculation(s) are as follows:

$$PSNR = 10\,log_{10}L^2/MSE$$

L is the peak signal level for an image.

$$MSE = \frac{1}{HN} \sum_{i=1}^{H} \sum_{j=1}^{W} \left(P(i,j)*S(i,j)^2 \right),$$

where H = Height, W = Width, P(i, j) = Original image, S(i, j) = Corresponding inserted image.

Comparing the PSNR value of Lena image with the value given by Dasgupta et al. [35], there is improvement is PSNR of Lena 128.jpg is 37.6828 whereas our proposed algorithm gives 40.98. However, the PSNR value obtained is lesser from the values obtained by Manjula and the Danti [34] which is 42.4208. Other results are more or less comparable through the percentage of accuracy of temper detection and localization remains high (in all the cases above 93%). The result may be considered as providing positive direction towards digital forensics enabled image authentication. The LSB-based hash function may further be investigated by changing the techniques of insertion as well as hash bit generation.

4.3 Analysis

In proposed approach, three hash bits are generated for each pixel. Hence if the size of the image is m x n, the length of the hash bits L for any image I will be

$$L = (m \times n) \times 3 \ (5)$$

Single hash bit for a pixel can guarantee for authentication with 0.5 probabilities because a pixel can be either 0 or 1. If we increase the number of authentication hash bits per pixel, it will definitely increase the tamper detection rate but the imperceptibility of embedded image will be compromised. That is why there is a tradeoff between number of hash bits and imperceptibility of images after embedding. Three bits per pixel for hash bit have been selected which can alter the intensity up to 0–7. Three (3) bits can generate maximum 7 and minimum 0 intensity. Hence, the proposed approach generates an acceptable PSNR.

5 Conclusion

The paper proposes LSB-based algorithm for hash bit generation to authenticate images. Proposed approach is utterly based on spatial domain. Changes made in every pixel of the images can be captured in to spatial domain. Proposed hash bits generation algorithm is able to increase PSNR value. It has low complexity. Proposed scheme achieves PSNR value up to 44.09 dB since embedded hash bits alter only three LSBs per pixel. Thus, this proposed algorithm gives a way to combat temporization attacks and proves its applicability in image forensics.

Acknowledgements The authors express gratitude to Cloud Computing Laboratory of National Institute of Technology, Sikkim India.

References

1. Forouzan, B. A., & Mukhopadhyay, D.,: Cryptography and Network Security (Sie). McGraw-Hill Education (2011)
2. Kim, Changick: Content-based image copy detection. Signal Processing: Image Communication 18, no. 3, 169–184 (2003)
3. Wu, Y. T., & Shih, F. Y.,: Digital watermarking based on chaotic map and reference register. Pattern Recognition, 40(12), (2007) 3753–3763
4. S. Wang and X. Zhang: Recent Development of Perceptual Image Hashing. In: J. Shanghai Univ. (English ed.), vol. 11, no. 4, pp. 323–331 (2007)
5. Monga V.: Perceptually based methods for robust image hashing (Ph. D. thesis). Electrical Engineering, Electrical and Computer Engineering, The University of Texas at Austin (2005)
6. Venkatesan R, Koon SM, Jakubowski MH, Moulin P.: Robust image hashing. In: Image Processing. Proceedings. 2000 International Conference on 2000 (Vol. 3, pp. 664–666). IEEE (2000)
7. Lin CY, Chang SF.: Robust image authentication method surviving JPEG lossy compression. In: Photonics West'98 Electronic Imaging (pp. 296–307). International Society for Optics and Photonics. Dec 23 (1997)
8. F. Ahmed, M.Y. Siyal, and V.U. Abbas.: A Secure and Robust Hash-Based Scheme for Image Authentication. In: Signal Processing, vol. 90, no. 5, pp. 1456–1470 (2010)
9. M. Schneider, S.-F. Chang: A robust content based digital signature for image authentication, in: Proceedings of the IEEE International Conference on Image Processing, vol. 3, Lausanne, Switzerland, pp. 227–230 (1996)
10. C.S. Lu, C.Y. Hsu, S.W. Sun, and P.C. Chang.: Robust Mesh-Based Hashing for Copy Detection and Tracing of Images. In: Proc. IEEE Int'l Conf. Multimedia and Expo, vol. 1, pp. 731–734 (2004)
11. Z. Tang, S. Wang, X. Zhang, W. Wei, and S. Su.: Robust Image Hashing for Tamper Detection Using Non-Negative Matrix Factorization. In: J. Ubiquitous Convergence and Technology, vol. 2, no. 1, pp. 18–26 (2008)
12. E. Hassan, S. Chaudhury, and M. Gopal.: Feature Combination in Kernel Space for Distance Based Image Hashing. In: IEEE Trans. Multimedia, vol. 14, no. 4, pp. 1179–1195, Aug. (2012)
13. W. Lu and M. Wu.: Multimedia Forensic Hash Based on Visual Words. In: Proc. IEEE Int'l Conf. Image Processing, pp. 989–992 (2010)
14. Roy, Sujoy, and Qibin Sun.: Robust hash for detecting and localizing image tampering. In: IEEE International Conference on Image Processing. Vol. 6. IEEE (2007)

15. X. Lv and Z.J. Wang.: Reduced-Reference Image Quality Assessment Based on Perceptual Image Hashing. In: Proc. IEEE Int'l Conf. Image Processing, pp. 4361–4364 (2009)
16. C.-Y. Lin, S.-F. Chang: A robust image authentication method distinguishing JPEG compression from malicious manipulation. In: IEEE Trans. Circuits Syst. Video Technol. 11 (2) 153–168 (2001)
17. C.-S. Lu, H.-Y.M. Liao.: Structural digital signature for image authentication: an incidental distortion resistant scheme. In: IEEE Trans. Multimedia 5 (2) 161–173 (2003)
18. Kundur, Deepa, and Dimitrios Hatzinakos.: Digital watermarking for telltale tamper proofing and authentication. In: *Proceedings of the IEEE* 87.7: 1167–1180 (1999)
19. Bhattacharjee, Sushil, and Martin Kutter.: Compression tolerant image authentication. In: *Image Processing, 1998. ICIP 98. Proceedings. 1998 International Conference on.* Vol. 1. IEEE (1998)
20. J. Fridrich, M. Goljan.: Robust hash functions for digital water- marking. In: Proceedings of the IEEE International Conference on Information Technology: Coding and Computing, Las Vegas, NV, USA, pp. 178–183 (2000)
21. J.S. Seo, J. Haitsma, T. Kalker, C.D. Yoo.: A robust image fingerprinting system using the Radon transform.: In Signal Process.: Image Commun. 19 (4) 325–339 (2004)
22. S.-H. Yang, C.-F. Chen, Robust image hashing based on SPIHT, in: Proceedings of the IEEE International Conference on Information Technology: Research and Education, pp. 110–114 (2005)
23. A. Swaminathan, Y. Mao, M. Wu, Robust and secure image hashing, IEEE Trans. Inf. Forensics Secur. 1 (2) 215–230 (2006)
24. Y. Ou, K.H. Rhee, A key-dependent secure image hashing scheme by using Radon transform, in: Proceedings of the IEEE International Symposium on Intelligent Signal Processing and Communication Systems, Kanazawa, Japan, pp. 595–598 (2009)
25. Deepa, Nagajothi.: a secure hashing scheme for image authentication using zernike moments and local features with histogram features. In: American International Journal of Research in Science, Technology, Engineering & Mathematics (2014)
26. V. Monga, M.K. Mhcak.: Robust and secure image hashing via non- negative matrix factorizations. In: IEEE Trans. Inf. Forensics Secur. 2 (3) 376–390 (2007)
27. C. Kailasanathan, R.S. Naini, P. Ogunbona.: Compression tolerant DCT based image hash, in: Proceedings of the IEEE International Conference on Distributed Computing Systems Workshops, pp. 562–567 (2003)
28. C.-S. Lu, C.-Y. Hsu: Geometric distortion-resilient image hashing scheme and its applications on copy detection and authentication. In: Multimedia Syst. 11 (2) 159–173 (2005)
29. V. Monga, B.L. Evans, Perceptual image hashing via feature points: Performance evaluation and tradeoffs, IEEE Trans. Image Process. 15 (11) 3452–3465 (2006)
30. Ng, T. T., Chang, S. F., & Sun, Q: A data set of authentic and spliced image blocks. In: Columbia University, ADVENT Technical Report, 203–2004 (2004)
31. M. Kirchner and J. Fridrich. On detection of median filtering in digital images. In: Media Forensics and Security II, part of the IS&T-SPIE Electronic Imaging Symposium, San Jose, CA, USA, January 18–20, Proceedings, Nasir D. Memon, Jana Dittmann, Adnan M. Alattar, and Edward J. Delp, Eds., vol. 7541 of SPIE Proceedings, p. 754110, SPIE (2010)
32. Sowmya, K. N., & Chennamma, H. R.: A survey on video forgery detection. In: International Journal of Computer Engineering and Applications, 9(2), 17–27. (2015)
33. Zhu, Y., Li, C. T., & Zhao, H. J.: Structural digital signature and semi-fragile fingerprinting for image authentication in wavelet domain. In: Information Assurance and Security, 2007. IAS 2007. Third International Symposium on (pp. 478–483). IEEE. (2007, August)
34. Manjula, G. R., & Danti, A.: A novel hash based least significant bit (2-3-3) image steganography in spatial domain. In: arXiv preprint arXiv: 1503.03674 (2015)
35. Dasgupta, K., Mandal, J. K., & Dutta, P.: Hash based least significant bit technique for video steganography (HLSB). In: International Journal of Security, Privacy and Trust Management (IJSPTM), 1(2), 1–11 (2012)

Automatic Cotton Leaf Disease Diagnosis and Controlling Using Raspberry Pi and IoT

Asmita Sarangdhar Adhao and Vijaya Rahul Pawar

Abstract The cotton crop is one of the maximum crucial cash crops among India. Plant diseases are generally due to pest insect and pathogens and every year reduce the productiveness to huge scale if not managed within time. When farmers observe that crop is infected, they detect the disease by naked eyes and decide the pesticide as solution by their previous experience. It is not the accurate way and instead of increasing crop production, they become responsible for the loss of production due to use of wrong pesticide in large quantity. So it is very essential to use automatic system for cotton leaf disease detection. This paper proposed such a system for diagnosis as well as controlling of disease infection on cotton leaves together with soil quality monitoring. The present system uses a support vector machine classifier for recognition of five cotton leaf diseases, i.e., Bacterial Blight, Alternaria, Gray Mildew, Cereospra, and Fusarium wilt. After disease detection, the name of a disease with its remedies is given to the farmers using Android app. The Android app is also used to display the soil parameters values inclusive of humidity, moisture, and temperature along with the water degree in a tank. By using Android app, farmers can ON/OFF the relay to control the motor and sprinkler in order to achieve crop suitable environment or to spray pesticides. All this leaf disease detection system and sensors for soil quality monitoring are interfaced with the usage of Raspberry Pi, which makes it impartial and price powerful system. This system shows the overall accuracy of 83.26% for disease detection.

Keywords Cotton leaf disease · Resizing · Median filter · Color transform
Gabor filter · Support Vector Machine (SVM) · Raspberry pi · Android app
Sensors

A.S. Adhao (✉) · V.R. Pawar
Department of Electronics and Tele-Communication, Bharati Vidyapeeth College
of Engineering for Women, Savitribai Phule University, Dhankawadi, Pune, India
e-mail: adhao.asmita87@gmail.com

V.R. Pawar
e-mail: pawar.vijaya40@gmail.com

© Springer Nature Singapore Pte. Ltd. 2018   157
Y.-C. Hu et al. (eds.), *Intelligent Communication and Computational
Technologies*, Lecture Notes in Networks and Systems 19,
https://doi.org/10.1007/978-981-10-5523-2_15

1 Introduction

Cotton is the most essential cash vegetation among the most essential standout in India and influences Indian economy from numerous points of view [1]. A massive number of the populace relies on cotton crops either for its cultivation or for the motive of processing. It is found that the advancement in farming is sluggish these days because of the attack of disease. Numerous farmers recognize disease by their past experience or some take assistance from specialists. However, the specialists typically detect the symptoms of disease with naked eyes. So there may be the possibility of an inaccurate diagnosis of diseases having huge closeness in their signs. If any mistakes occurred during the analysis of the disease, then in some cases it may prompt to wrong controlling and overuse of pesticides [1].

That is why it is far essential to move toward the new strategies for automatic diagnosis and controlling of disease. A number of assortments of pesticides are available to manipulate disease and increase production; however finding the most reasonable and powerful pesticide to control the infected disease is troublesome and required specialists help which is tedious and costly. Attack of disease on the cotton crop is particularly reflected from signs on leaves. So there may be need of an automatic, accurate, and cost-effective machine vision system for detection of diseases from cotton leaves and to recommend the best possible pesticide as a solution. This paper focuses on detection of most commonly occurring disease on cotton leaves and their controlling using IoT-based Android app. The diseases are classified using SVM classifier algorithm written in Python code in Raspberry Pi. In this paper, four unique sensors, that is, temperature, moisture, humidity, and water sensors are used and interfaced with Raspberry Pi for soil quality monitoring. Two Android apps are used; one for displaying soil parameters and other app for displaying disease details along with the handling of external devices which include sprinkler or motor On/Off. The app also helps to handle the motion of the complete system from one place to another to check soil parameters at exclusive places. So this device is beneficial in huge farm for accurate detection and controlling of disease.

1.1 Cotton Leaf Diseases

The attack of disease on crop is generally identified from symptoms such as appearance of spots or change in the color of the leaves. Symptoms are considered as the main indicator of the presence of disease on leaves and are given below

Bacterial Blight. Bacterial blight is bacterial disease mainly caused as a result of the bacteria "Xanthomonas Campestris pv. Malvacearum" [1]. The indications of Bacterial curse begin as dark green, water splashed precise spot of 1–5 mm on a leaf with ruby to cocoa fringe. At the beginning, those precise leaf spots look as

Fig. 1 **a** Bacterial Blight disease leaf **b** Alternaria infected leaf **c** Cerespora infected leaf **d** *Gray mildew* infected leaf **e** Fusarium Wilt infected leaf [1]

water-absorbed ranges which, after a while changes from dim chestnut to dark shade [2]. The spots at the sore area of leaves can likewise spread over the essential veins of leaf and in later petioles and stems get corrupted and ominous tumble off of the leaf develops [3]. Figure 1a depicts symptoms of bacterial blight disorder.

Alternaria. It is a fungal disease specially caused due to A. Alterneta or Alternaria macrospora [3]. This disease is more intense on the lower some portion of leaf when contrasted with the upper part and might get confused with the symptoms of bacterial blight disease as the signs are closely similar [4]. At the start, brown, gray-brown to tan-colored small round spots appears on leaves and vary from 1–10 mm in size which later on emerges as dry, useless with gray facilities which crack and fall out [4]. Sometimes, old spots integrate together and create abnormal dead regions. Figure 1b depicts the symptoms of Alternaria disorder.

Cerespora. Cercospora arises by the Cercospora Gossypina [3]. The contaminated leaf has red spot blemishes on the leaf which extends in breadth to around 2 cm. The spots are round or sporadic fit as a fiddle with yellowish chestnut, purple, dim cocoa or blackish outskirts with white focuses [4]. The precise leaf spot shows up because of the confinement of the sore zone by fine veins of the leaves. The disease influences more seasoned leaf of developed plants. Figure 1c depicts Cerespora disease symptoms.

Gray Mildew. It is fungal disease typically because of the Ramularia Areola Atk. in blemish stage and Mycosphaerella areola in flawless stage [3]. In the preliminary level, the disease shows up as triangular, square, or sporadically spherical whitish spots of 3–4 mm size on leaves [4]. This disease generally seems as irregular angular and pale translucent spots of diameter 1–10 mm on older leaves of matured plant [5]. The light to yellowish infected regions in shade are experienced on the upside surface. As disease infection increases, the small spots merge collectively forming bigger spots and the leaf tissues turn yellowish chestnut while whitish chilly increase appears on a decreased surface however occasionally on the top floor [5]. Figure 1d depicts the symptoms of gray mildew disorder.

Fusarium wilt. It is the contagious infection for the most part, brought about by Fusarium oxysporum [3]. The organism can assault cotton seedlings, however the infection generally shows up when the plants are matured [4]. The influenced plants first get to be distinctly darker green and hindered [1]. The yellowing of the leaves and loss of foliage appears. At first, the side effects show up on lower leaf around the season of first blossom and the leaf edges wither, turn yellow initially took after by chestnut, moving internal. Figure 1e depicts Fusarium Wilt disease symptoms.

2 Literature Survey

Ashourloo et al. [6] proposed the strategies for Wheat Leaf Rust infection diagnosis in addition to withstand the preparation test size and impact of infection indications consequences on the techniques. This paper looks at the general execution of PLSR, v-SVR, and GPR with the PRI and NBNDVI. The combinations of malady indications at every ailment seriousness level outcomes in complex spectra which declined the exactnesses of PRI and NBNDVI while they don't impact on the performance of PLSR, v-SVR, and GPR. The GPR's execution utilizing short training dataset brings about higher exactness than different techniques.

Rothe et al. [1] proposed design acknowledgment method for the identification of three cotton leaf disorders such as Myrothecium, Alternaria, and Bacterial Blight. Segmentation is done by using active counter model and Hu's moments are extracted to train adaptive neuro-fuzzy inference system. Finally disease is identified by BPNN.

Gulhane and Kolekar [7] utilized principle component analysis (PCA) and nearest neighborhood classifier (KNN) for determination of disease on cotton leaves. After implementing PCA/KNN multivariable techniques, statistical analysis of the data related to the Green (G) channel of RGB image is done. Here, Green channel is thought about for capacity arrangement since turmoil or inadequacies of components are reflected well by utilizing green channel.

Rothe and Kshirsagar [2] proposed the concept of automated feature extraction in which the RGB image is first captured and transferred to gray scale. The noise is removed by using LPF and Gaussian filter. K-means clustering followed by graph cut energy minimization operation is used for segmentation purpose. For color feature extraction, the segmented RGB image converted to YCbCr and DCT of YCbCr is obtained and zigzag scanning is used for achieving color layout descriptor of input image.

Revathi and Hemalatha [8] proposed two techniques to recognize the lesion region of the disease. First edge identification procedure is utilized for segmentation and afterward image examination and identification of disease is done utilizing the developed HPCCDD Algorithm. This paper discusses RGB color based methods in which, the caught pictures are preprocessed first and afterward color image segmentation is completed to get lesion spots. The edge features are extracted to identify the disease spots using Sobel and Canny filter.

3 System Architecture

This system is used to detect and control the disease on cotton leaves. It can also be used to monitor the soil quality parameters such as temperature, moisture, and humidity. By using this system farmer can automatically ON/OFF the external devices such as motor, sprinkler assembly according to need with Android app. Figure 2 represents the block diagram of present system.

3.1 Block Diagram of Present System

This system shown in Fig. 2 consists of Raspberry Pi- model B which is the main part of the system used for interfacing purpose. Initially, the input image is selected and given to the Raspberry Pi. The programming for leaf disease detection is done in Python code. According to selected image, disease is detected and the name of disease and its remedies are displayed on the Android app. After knowing the

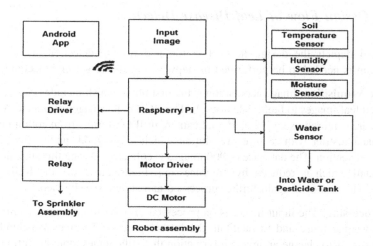

Fig. 2 Block diagram of cotton leaf disease detection and controlling system

(a)   (b)   (c)

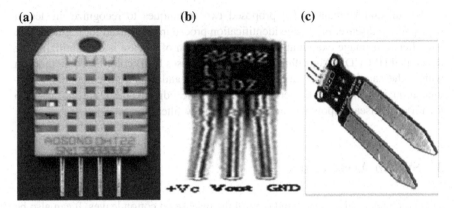

Fig. 3 Different sensors used for soil quality measurement **a** DHT22 humidity sensor **b** LM35 temperature sensor **c** Moisture sensor

infected disease, farmers can take necessary decision whether to turn ON/OFF the sprinkler or motor assembly. ON/OFF of the external devices is done automatically by using app in order to spray pesticides or fertilizers by mixing it in water. Relay driver and single pole double throw relay is used to control the ON/OFF of external devices. Farmer can also check the soil condition and water level in the tank with the help of sensors. These sensors include LM 35 temperature sensor, DHT-22 humidity Sensor, water sensor, and moisture sensor as shown in Fig. 3. All these sensors are interfaced with Raspberry Pi. The motor driver and DC motor are used for the movement of overall system. The moving system helps to monitor soil condition at different places.

3.2 Design Flow of Leaf Disease Detection

Figure 4 depicts the design flow of cotton leaf disease detection. The disease detection is needed to be performed in stepwise manner to get high accuracy.

Image Acquisition. Image acquisition is the first phase which includes capturing of infected leaf images to build database [9]. The RGB color images of infected cotton leaves are captured using Nikon digital camera in JPEG format from cotton farm in Buldhana district with required resolution which gives good quality images for disease detection. The database of 900 images is collected. During this step, noise is generated which is reduced by preprocessing. Figure 5a shows the RGB input image. This image is again further processed and shown step by step.

Preprocessing. The input image is preprocessed in order to enhance by removing the unwanted noise and to facilitate the further process. The reason behind pre-processing is to change an image information that stifle undesirable distortions and

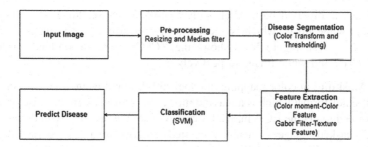

Fig. 4 Design flow of cotton leaf disease detection

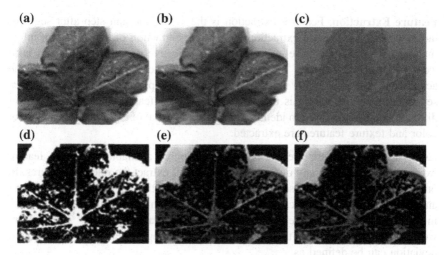

Fig. 5 Image processing results **a** Input RGB image **b** Preprocessed image **c** YCbCr color transform image **d** Logical *black* and *white* image **e** Masked image **f** *Gray*-level image

improves some image features important for further processing. In this system, resizing and median filter techniques are used. First the RGB input image is selected and then resized into 250 × 250 px and then median filter is applied. As compared to other techniques median filter is more accurate and during the process of noise removal it helps to preserve edges. Figure 5b shows the preprocessed image.

Segmentation. Segmentation is an important step after preprocessing for leaf diseases detection which is used to extract the lesion region from images [9]. Segmentation separates the infected region from the healthy region and makes further process easy. In this system, RGB to YCbCr color transformation and bi-level thresholding is used to extract the region of interest from the image. First the preprocessed RGB color space image is transformed to YCbCr color space and then bi-level thresholding is applied. In bi-level thresholding, upper and lower threshold is defined for Y, Cb, Cr color plane and pixel between these two ranges are considered as diseased part. After applying bi-level thresholding, we get logical black and white image which is needed to be converted to gray image to extract region of interest. Figure 5c, d shows the YCbCr color transformed image and logical black and white image respectively.

Color Masking. In color mapping, the logical black and white image is converted to RGB color format which is a masked image. In color masking bit-wise operation is performed to get the RGB masked image. We are interested only in diseased part so this masked image is again converted to gray image to extract diseased part. Figure 5e shows the RGB masked image while Fig. 5f shows the gray image. In this gray image white color shows diseased part which is our Region of Interest (ROI).

Feature Extraction. Features extraction is the next important step after segmentation to perform. Feature extraction aims at the extraction of the important information that portrays diverse classes [5]. The objective of feature extraction is to extract a set of features representing each character, which augments the acknowledgment rate with minimal number of components. The diseased area separated by segmentation is the area of interest for feature extraction to extract different features that used to identify the disease. A total of eight features including color and texture features are extracted.

Color Features. There are various methodologies available for color feature extraction. We can use any one method which will extract the relevant features. In our case, color moment is used for color feature extraction. In color moment, mean and standard deviation are calculated. Therefore, an image is characterized by six moments as two color moments for each three color channels (i.e. 3 * 2 = 6). Here we define Pij pixel having i-th color channel at j-th image pixel. Mean and standard deviation can be defined as

$$Mean = E_i = \sum_{j=1}^{N} P_{ij} \tag{1}$$

$$Std.deviation = \sqrt{\frac{1}{N} \sum_{j=1}^{N} \left(P_{ij} - E_i \right)^2} \tag{2}$$

Texture Features. Two texture features are extracted using 2D Gabor filter. The Gabor texture feature include mean and standard deviation of the magnitude of Gabor wavelet transform coefficient. Mean gives the average value and standard deviation decides the deviation of data from center of Gaussian filters. In present system, Gabor filter with different parameters such as frequency, angle in 16 orientation with 10 different frequencies is considered.

Classification. After feature extraction, classifier is used to classify the disease based on extracted features. It is the last step, in which classifier is used to recognize the kind of the cotton leaf diseases. Classification deals with matching the given data vectors with one of the trained data vectors of different classes which are already trained. In machine learning, there are various types of classifiers available for classification. In our case, support vector machine classifier is used. To get high accuracy, SVM is trained and tested with different kernels and it is known that Gaussian kernel gives high accuracy. In this system, SVM classifier with nonlinear Gaussian kernel is designed for classification of the diseases on cotton leaves. This classifier finds the nonlinear relationship between input vectors and response variables by finding the best hyper plane.

4 Results and Discussions

As discussed, the primary aim of this work is to identify and control diseases on cotton leaves. The secondary aim is to monitor the soil quality. For disease detection, we used the database of 900 cotton leaves images. This database includes cotton leaf images of five different diseased. Out of the total collected database, 629 images are trained and 271 images are used for testing. Table 1 shows the accuracy of correctly classified disease while Table 2 shows comparison of disease detection rate of present and existing system. Figure. 6a shows the graphical representation of Table 1 while Fig. 6b depicts the implementation of overall system showing raspberry Pi, Relay, and sensors. Figure 7a shows the results of Android app displaying cotton leaf disease name with its remedies. It also shows forward, reverse, left, and right movement and stop options along with motor ON/OFF to ON or OFF the relay. Figure 7b shows the result of Android app displaying different sensors reading such as temperature, liquid level moisture, and humidity, respectively. This system gives the accuracy of 83.26% for cotton leaf disease detection and it is calculated from Eq. 3. The individual accuracy shown in Table 1 is also calculated by the same formula.

$$\text{Overall Accuracy} = \frac{TP + TN}{TP + TN + FP + FN} \tag{3}$$

where TP—True Positive, TN—True Negative, FP—False positive, FN—False Negative

Table 1 Accuracy of cotton leaf disease detection

Name of disease	Correctly classified	Incorrectly classified	Accuracy (%)
Bacterial Blight	67	11	85.89
Alternaria	55	10	84.61
Cerespora	39	08	82.97
Fusarium wilt	31	06	83.78
Gray mildew	28	06	82.35
Healthy leaf	8	02	80

Table 2 Comparision of different classification techniques based on disease detection rate

Paper	Classification technique	Accuracy (%)
Muthukannan et al. [10]	Feed forward neural network(RBF)	71.18
Ehsanirad and Sharath Kumar [11]	Gray-level co-occurrence matrix	78.46
Present approach	Support vector machine	83.26

(a)   **(b)**

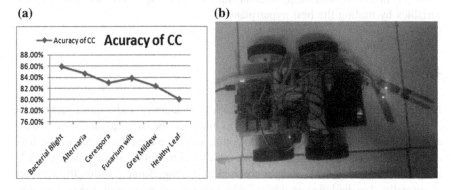

Fig. 6 a Graphical representation of Table 1 **b** Overall system implementation

(a)   **(b)**

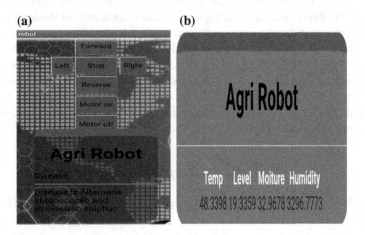

Fig. 7 a Android app displaying cotton leaf disease information **b** Android app displaying different sensors reading

5 Conclusion

This paper presents the support vector machine classifier for detection of five cotton leaf disease. To control the disease, the name of detected disease and pesticides is suggested to the farmer within a short time. The Android app is developed to display disease and sensor information along with the ON/OFF of the relay. The app also handles the movement of the whole system from one place to another. So by using this system, the farmers can automatically detect the disease and know the remedies to control that disease. The farmer can move the system from one place to another place to check the soil condition at different locations with the help of sensor and can change soil condition by turning motor ON/OFF with relay. All these processes are done by using Android app which saves human hard work in large field area. The use of Raspberry pi makes this system cost-effective and independent. This system gives accuracy of 83.26% for disease detection and proves its effectiveness to the farmers for cotton leaf disease detection and controlling by improving the crop production.

References

1. Rothe, P.R., Kshirsagar, R.V.: Cotton Leaf Disease Identification using Pattern Recognition Techniques. In: International Conference on Pervasive Computing.pp. 1–6. IEEE (2015). doi:10.1109/PERVASIVE.2015.7086983.
2. Rothe, P. R., Dr. Kshirsagar, R. V.: Automated Extraction of Digital Images Features of three kinds of Cotton Leaf Diseases. In: International Conference on Electronics, Communication and Computational Engineering. IEEE (2014). doi:10.1109/ICECCE.2014.7086637.
3. Plant village Cotton. https://www.plantvillage.org/en/topics/cotton.
4. Texas plant disease handbook. http://plantdiseasehandbook.tamu.edu/industryspecialty/fiber-oil-specialty/cotton/.
5. Revathi, P., Hemalatha M.: Advance Computing Enrichment Evaluation of Cotton Leaf Spot Disease Detection Using Image Edge detection. In: IEEE. Coimbatore (2012). doi:10.1109/ICCCNT.2012.6395903.
6. Davoud Ashourloo, Hossein Aghighi, Ali Akbar Matkan, Mohammad Reza Mobasheri and Amir Moeini Rad, "An Investigation Into Machine Learning Regression Techniques for the Leaf Rust Disease Detection Using Hyperspectral Measurement" *IEEE Journal Of Selected Topics In Applied Earth Observations And Remote Sensing*, pp. 1–7, May 26, 2016. Doi:10.1109/JSTARS.2016.2575360.
7. Gulhane, V.A., Kolekar, M.H.: Diagnosis Of Diseases On Cotton Leaves Using Principal Component Analysis Classifier. In: Annual IEEE India Conference. IEEE (2014). doi:10.1109/INDICON.2014.7030442.
8. Revathi, P., Hemalatha M.: Classification Of Cotton Leaf Spot Disease Using Image Processing Edge Detection Technique. In: International Conference on Emerging Trends in Science, Engineering and Technology. IEEE, pp. 169–173. (2012). doi:10.1109/INCOSET.2012.6513900.
9. Es-saady, Y., El Massi, I., El Yassa, M., Mammass, D., Benazoun, A.: Automatic recognition of plant leaves diseases based on serial combination of two SVM classifiers. In: 2nd International Conference on Electrical and Information Technologies. IEEE (2016). doi:10.1109/EITech.2016.7519661.
10. Muthukannan, K., Latha, P., Selvi, R. P., and Nisha, P.: Classification Of Diseased Plant Leaves Using Neural Network Algorithms. In:Asian Research Publishing Network(ARPN) Journal of Engineering and Applied Sciences. (2015).
11. Ehsanirad, A., Sharath Kumar Y. H.: Leaf recognition for plant classification using GLCM and PCA methods. In: Oriental Journal of Computer Science & Technology, pp. 31–36. (2010).

Part III
Networks and Mobile Communications

Symmetric Key Based Authentication Mechanism for Secure Communication in MANETs

Sachin Malhotra and Munesh C. Trivedi

Abstract In MANETs, secure communication is very difficult to achieve because its nature-like communication medium is open wireless which can be easily accessed by anyone who comes in the radio range of the communicating devices, nodes are physically vulnerable, less-efficient (processing power, memory) communicating devices, the absence of central authority, free from the constraint of topological structure of network, etc. In last decade, applications of MANETs are increasing rapidly. Most of the application demands the secure communication of information. In this work, symmetric key based authentication mechanism has been proposed to ensure the secure communication between the communicating parties. The proposed model secures the network from the well known and frequently occurred attacks (impersonation, modifies routing information, black hole). In this work, two levels of authentication have been used, first level for hop-to-hop authentication (MD5 algorithm has been used for authentication code generation) and second level for end-to-end authentication (SHA1 algorithm has for authentication code generation). For the simulation purpose, AODV protocol is used for checking the effectiveness of the proposed model and PDR; AE2ED and TP are used to measure the performance of the proposed, AODV and existing models in the absence and presence of the malicious nodes. NS2.35 on Ubuntu 12.04 LTS with 4 GB RAM is used for simulation. Simulation results show the advantage of proposed model over the original AODV and some related models published recently. A result showing our model implements the security scheme with less overhead.

Keywords MANETs · AODV · Attacks · Secure communication · Message authentication

S. Malhotra (✉)
Department of Information Technology, IMS, Ghaziabad, India
e-mail: sachin_malhotra123@yahoo.com

M.C. Trivedi
Department of Computer Science & Engineering, ABES Engineering College,
Ghaziabad, India
e-mail: munesh.trivedi@gmail.com

© Springer Nature Singapore Pte. Ltd. 2018
Y.-C. Hu et al. (eds.), *Intelligent Communication and Computational
Technologies*, Lecture Notes in Networks and Systems 19,
https://doi.org/10.1007/978-981-10-5523-2_16

1 Introduction

Suppose that we want to establish the communication connection between two floors of any organization building or between two buildings in the same campus using wireless short-range communicating electronic devices. In the organization each employee has mobile devices and some fixed devices such as printer, computer, etc. We can connect these devices using fixed wired network or by infrastructure access points, but this restricts the mob ability of the devices. Another option is used in base station based network, i.e. cellular network which allows the large communication range; but the limitation is that, these cellular networks are costly and time-consuming deploying networks. Alternatively, at last we need a network that should be fast and cheap in deployment, provide sufficient range for communication, easily scalable, support mob ability, etc., and these features are provided by only one network, i.e., mobile ad hoc network (MANET) [1, 2]. Apart from these types of applications, MANETs are also used in defense, emergency relief operations, environment monitoring, VANETs, WSNs, etc.

With the increasing number of applications, the demand for secure communication is increasing. Due to the fundament characteristics of MANET [3], secure communication is very difficult [4, 5] to achieve, because its nature-like communication medium is open wireless which can easily be accessed by anyone who are coming in the radio range of the communicating devices, nodes are physically vulnerable, less-efficient (processing power, memory) communicating devices, the absence of central authority, free from the constraint of topological structure of network, etc.

Communicating devices used in MANET have limited memory and processing power, because of this asymmetric key based security scheme is not the good solution for implementing security scheme in MANETs. With considering this limitation in mind, in this work symmetric key based authentication mechanism has proposed to ensure the secure communication between the communicating parties. The proposed model secures the network from the well-known and frequently occurred attacks (impersonation, modifies routing information, black hole). In this work, two levels of authentication have been used, first level for hop-to-hop authentication [6, 7] (MD5 algorithm has used for authentication code generation) and second level for end-to-end authentication (SHA1 algorithm has for authentication code generation). Digest_1 is the size of 128 bits generated by the MD5 algorithm and Digest_2 is the size of 160 bit generated by SHA 1 algorithm.

The remainder of this paper is organized as follows: Sect. 2 introduces related work and motivation to do this research. Section 3 describes the proposed methodology. Section 4 explains simulation result and discussion, and finally conclusion is defined in Sect. 5.

2   Related Work and Motivation

In this section, research work done in the field of secure communication in MANETs is discussed. Especially, we cover the mechanisms that have used symmetric key based security schemes. Arya K. V. and Rajput S. S. [8] have proposed the model to secure AODV routing protocol using nested MAC. In this model, the author has also used the concept of key pre-distribution (distribution of symmetric key at the time of network deployment) to overcome the drawback of methods [9] that distributes the keys at run time (when communication connection establishing between the sender and receiver). This method [8] significantly prevents the networks from many attacks (impersonation, modifies routing information, black hole). The limitation of this method is that it works efficiently when the attacker is outsider and work little bit inefficient when attacker is insider, i.e., our genuine node is compromised by the attacker.

Similar concept was used by Rajput S. S. [10] to protect MANETs against frequently occurred attacks. In this particular paper, ZRP routing protocol is used to major the performance of the proposed model. These two papers played the key role in the motivation to do the research in this field.

3   Proposed Security Mechanism

In this model, symmetric key based authentication technique is used for securing the network from various attacks. In this model, two levels of authentication are used to test the integrity of the message. The first level of authentication is used for hop-to-hop authentication and second level of authentication is used for end-to-end authentication. At first level, MD5 algorithm is used to generate 128 bits digest for checking the integrity of the message at each node of the route. SHA 1 is used to generate the 160 bits digest for checking the integrity of the message at the intended receiver of the message.

For speedup, the algorithm key table of size 15 keys (K0–K14) is stored at each node at the time of deployment. These keys are used for generating MAC (MD5) at first level that helps us to check the integrity of the message at the intermediate nodes. For generating MAC code at second level, SHA 1 uses second key generated by the random number generator. Random number generator function uses random four-digit number as seed to generate the second key.

The advantage of making two levels of authentication are to speed up the algorithm compare to Method [1], Method [2], and increase the security level. This mechanism helps us to protect our network from internal as well as external attacks.

The working model of the proposed work is given in Fig. 1. Here we are not discussing the behavior of the attacks in details. Here we are using the same behavior of the attacker as discussed in Method [1] and Method [2]. In working model as shown in Fig. 1, taking four nodes in the route one is sender node S, one

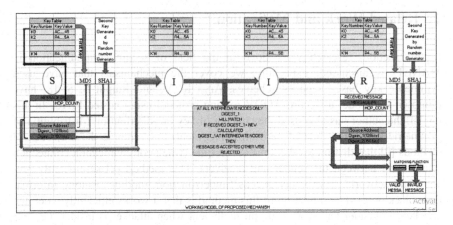

Fig. 1 Working model of proposed algorithm

is receiver node R, and two intermediate nodes I. Each node has a key table used for the authentication of the message at first level. For better understanding, whole working model is divided into three parts: Process at the Sender, Process at the receiver, and Process at the intermediate nodes.

At the sender side: Sender S first generates the message, then first key from the key table for generating the Digest_1 using MD5 is selected according to the value of hop count field in the header of the message. (First_key = Key number (Hop_count mode 15) then random number function call to generate Second_key for creating the second-level authentication code using SHA 1 algorithm then whole message (message + random 4 digit seed + Digest_1 + Digest_2) send to the next node in the route.

At the intermediate nodes, first check the integrity of the message by generating only the Digest_1 of message using the key (First_key = Key number (Hop_count mode 15) and if new Digest_1 equals to the received Digest_1, then message is treated as valid message and then again Digest_1 is created using next key in the key table and message is forwarded to the next hop in the route. If new Digest_1, does not equal the Received Digest_1 then message is treated as invalid message and message is discarded.

At the receiver side: new Digest_1 and Digest_2 is created by using First_key (First_key = Key number (Hop_count mode 15) from the key table and Second_key (generating by random number generator using same seed that has been used by the sender). If both new digest matches with received digest, then only message is received as valid message otherwise discard the invalid message. For more understanding, pseudocode for the proposed model is given in the Algorithm 1

ALGORITHM 1: ALGORITHM TO IMPLEMENT   PROPOSED SECURITY MECHANISM

Abbreviations:

H_Count: Hop Count

Digist_1: 128 bits Message Authentication code generate by MD5 Algorithm

Digist_2: 160 bits Message Authentication code generate by SHA 1 Algorithm

First_Key: first key select from the key table uses by MD5

> First_key = Key number (H_Count mode 15) For ex. K0, K1, --- K14

Second_Key: generating by random number generator using 4 dist seed number and

> It uses by SHA 1 Algorithm

Seed: 4 digit random value

At Sender Node (S):

Step1: Generate message **M**

Step2: Select **Seed**

Step3: Select **First_Key = K0** (because **H_count** = 0 At sender) from the Key table

Step4: Generate **Second_Key**

Step5: Calculate **Digest_1** = MD5 (M, First_Key)

Step6: Calculate **Digest_2** = SHA 1 (M, Second_Key).

Step7: Send (M+Seed+Digest_1+Digest_2)

At Intermediate Nodes (I):

Step1: Receive (M+Seed+Digest_1+Digest_2)

Step2: Select **First_Key** = K(H_Count mode 15) From the Key table

Step3: Calculate **Digest_1** = MD5 (M, First_Key)

> **IF (New Digest_1 = = Received Digets_1)**
>
> **THEN:**
>
> > H_Count = H_Count + 1;
> >
> > First_Key = K(H_Count mode 15)
> >
> > Calculate **Digest_1** = MD5 (M, First_Key)
> >
> > **Send (M+Seed+Digest_1+Digest_2)**
>
> **ELSE**
>
> Discard **M** (M is invalid);

At Receiver Node (R):

Step1: Receive (M+Seed+Digest_1+Digest_2)

Step2: Select **First_Key** = K(H_Count mode 15) From the Key table

Step3: Generate **Second_Key**

Step4: Calculate **Digest_1 = MD5** (M, First_Key)

Step5: Calculate **Digest_2 = SHA 1** (M, Second_Key).

> **IF** (New Digest_1 = = Received Digets_1
>
> > **&&** New Digest_2 = = Received Digets_2)
>
> **THEN**
>
> > **M** is valid and accepts
>
> **ELSE**
>
> > **Discard M** (M is invalid);

---

4 Simulation and Result Analysis

To implement proposed model NS2.35 [11] has been used. Simulation parameters are given in Table 1.

In the work, proposed model is compared with original AODV, previous model was proposed by K. V. Arya and S. S. Rajput [8] (named as Method [1]) and another model was proposed by S. S. Rajput et al. [10] (named as Method [2]). Original Method [2] is proposed for the ZRP routing protocol, but in this work we have changes in this model to make it compatible with the AODV routing protocol. Performances of all the models are measured in the presence and absence of the malicious nodes in terms of AE2ED, TP, and PDR.

First we simulate, analyze, and compare the performance all four models, i.e., AODV, Method [1], Method [2], and Proposed in terms of different pause time V/s all four performance parameters (AE2ED, PDR, and Average TP). Simulation results are shown in Figs. 2, 3 and 4 shows that proposed model work better than Method [1] and Method [2] and almost similar than original AODV [12] in the absence of malicious nodes. This is because, in the proposed model at intermediate node only one digest is generating and verifying (Digest_1) while in previous model digest generated two times (used NMAC). It means proposed mechanism increases negligible over head in term of computational complexity.

Comparison of proposed model, AODV, Method [1], and Method [2] in terms of average AE2ED, PDR, Average TP with increasing number of malicious nodes (No_of_Mlcius_Nodes) shown in Figs. 5, 6 and 7. For this fixed pause

Table 1 Simulation parameters

Simulator	Ns2 (v-2.35)
Simulation time	150 s
Performance Parameters	TP (Through Put), PDR, AE2ED (Average End To End Delay)
Area size	800 m × 600 m
Transmission range	100 m
Number of nodes	10–150
Previous models	Method [1], Method [2]
Protocol	AODV
Transmission range	250 m
Maximum speed	0–20 m/s
Application traffic	CBR
Packet size	512 bytes
Traffic rate	4 packet/s
Node mobility model	Random Way-Point Model
Pause time	10, 20, 60, 100–140 s
Mac method	802.15.4

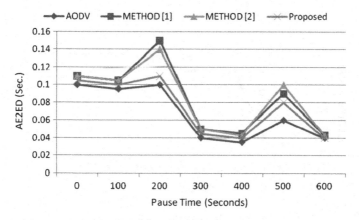

Fig. 2 Pause_Time versus AE2E Delay

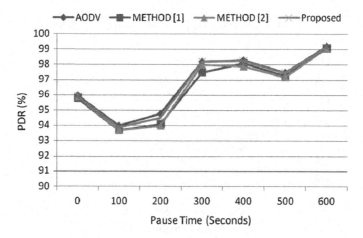

Fig. 3 Pause_Time versus PDR

time = 300 s and number of connections = 10 have been used. Results show that proposed model performing outstanding as comparing to the other three models in the presence of malicious nodes. Out of all malicious nodes in each simulation, 50% nodes are taken as inside attackers and remaining 50% as outside attacker. Simulation results also show that our model perform better in the presence of inside attacker; it means our model significantly overcomes the drawback of Method [1] and Method [2].

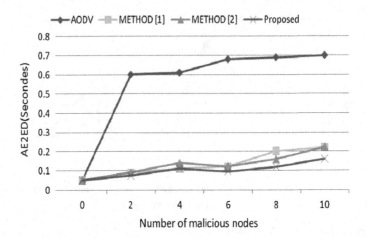

Fig. 4 Pause_Time versus Average_TP

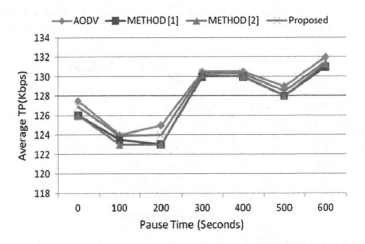

Fig. 5 AE2E Delay versus No_of_Mlcius_Nodes

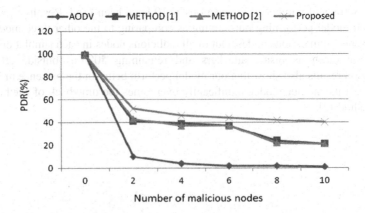

Fig. 6 PDR versus No_of_Mlcius_Nodes

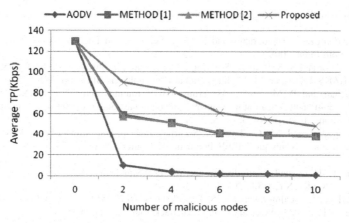

Fig. 7 Average_TP versus No_of_Mlcius_Nodes

5 Conclusion

In this paper, symmetric key based authentication mechanism has been proposed to secure the AODV against various attacks (impersonation, modifies routing information, black hole). Simulation results show that proposed model work better than Method [1] and Method [2] and almost similar than original AODV in the absence of malicious nodes. It means proposed mechanism increases negligibly over head in terms of computational complexity. Results also show that proposed model perform in an outstanding way compared to the other three models in the presence of malicious nodes. Out of all malicious nodes in each simulation, 50% nodes are taken as inside attackers, and remaining 50% as outside attacker. Simulation results also show that our model performs better in the presence of inside attacker; it means our model significantly overcomes the drawback of Method [1] and Method [2].

References

1. N. Sharma, A Gupta, SS Rajput and V. Yadav, " Congestion Control Technique in MANET: A Survey" 2nd IEEE International Conference on CICT, pp. 280–282, Feb. 2016.
2. S. S. Rajput, V. Kumar and S. K. Paul, "Comparative Analysis of Random Early Detection (RED) and Virtual Output Queue (VOQ) in Differential Service Networks" IEEE International Conference on SPIN, pp. 281–285, Feb 2014.
3. S. S. Rajput, V. Kumar and K. Dubey, "Comparative Analysis of AODV and AODV-DOR routing protocol in MANET" International Journal of Computer Application, vol. 63, no. 22, pp. 19–24, Feb 2013.
4. D. Djenouri et al. "A survey of security issues in mobile ad hoc networks" IEEE Communications Surveys & Tutorials. Fourth Quarter 2005.
5. M. Charvalho, "Security in Mobile Ad hoc Networks" Published by the IEEE Computer S ociety, pp:72–75, 2008.

6. William Stalling, Cryptography and Network Security, 4th Ed. Pearson Education, India, 2006.
7. B. A. Forouzan, Cryptography and Network Security, 2nd Ed., Tata McGraw-Hill Higher Education, India, 2008.
8. K. V. Arya and S. S. Rajput, "Securing AODV routing protocol in MANET using NMAC with HBKS Technique," IEEE International Conference on SPIN, pp. 281–285, Feb 2014.
9. P. Sachan and P. M. Khilar, "Securing AODV routing protocol in MANET based on cryptographic authentication mechanism," International Journal of Network Security and Its Applications (IJNSA), vol. 3, no. 5, 2011.
10. S. S. Rajput and M. C. Trivedi, "Securing ZRP routing protocol in MANET using Authentication Technique," IEEE International Conference on CICN, pp. 872–877, Nov. 2014.
11. E. H. T. Issariyakul, "Introduction to Network Simulator NS2," Springer Science and Business Media, NY, USA, 2009.
12. C. Perkins, E. Beldingroyer, and S. Das, "AODV RFC 3561," Internet Engineering Task Force (IETF), 2003. Available at http://datatracker.ietf.org/doc/rfc3561/.

Performance Comparison of Rectangular and Circular Micro-Strip Antenna at 2.4 GHz for Wireless Applications Using IE3D

Mahesh Gadag, Shreedhar Joshi and Nikit Gadag

Abstract This paper compares the performance parameters of rectangular and circular micro-strip antenna for 2.4 GHz RFID wireless applications using IE3D simulation software. The main parameters of comparison are return loss, gain, and area. The main objective of the work is to obtain the reduction in the size of circular antenna compared to rectangular for same frequency of operation. Impedance matching between the feeder line and radiating patch is done using planar micro-strip transmission lines. Quarter wave feeding technique is used for both the antennas to provide best matching between antenna and the source. Simulation is done using IE3D simulation software and parameters like bandwidth, return loss, and gain are measured. Results are optimized using optimizing techniques and results are compared with practically measured results. Finally antennas are fabricated and tested using in-house facilities.

Keywords Rectangular micro-strip antenna (RMSA) · Circular micro-strip antenna (CMSA) · Return loss · Bandwidth · Impedance · IE3D · Quarter wave feed

M. Gadag (✉)
Department of ECE, HIT, Nidasoshi, India
e-mail: mahgadag@gmail.com

S. Joshi
Department of ECE, SDMCET, Dharwad, India
e-mail: shreedhar.joshi@gmail.com

N. Gadag
Department of Electrical Engineering and Information Technology,
Otto Van Gureeke University, Magdeburg, Germany
e-mail: nik93cool@gmail.com

© Springer Nature Singapore Pte. Ltd. 2018
Y.-C. Hu et al. (eds.), *Intelligent Communication and Computational Technologies*, Lecture Notes in Networks and Systems 19,
https://doi.org/10.1007/978-981-10-5523-2_17

1 Introduction

Rectangular and circular micro-strip antennas are narrowband, low profile, light-weight, and planar antennas. These antennas can be fabricated by using suitable dielectric substrate, such as glass epoxy, FR-4, and RT-Duroid [1–6]. For rectangular micro-strip antenna, the order of modes can be changed by changing the relative dimensions of the width and length of patch. For circular patch there is only one degree of freedom to control its radius of patch which changes the value of resonant frequency. RMSA can be analyzed using rectangular coordinates whereas circular patch uses cylindrical coordinates. Various parameters of rectangular and circular micro-strip antennas [7–9] like, dual characteristics, circular polarizations, dual frequency operation, frequency agility, broad bandwidth, feed line flexibility, beam scanning can be easily obtained from these patch antennas.

A basic rectangular and circular micro-strip patch (Fig. 1a and b) consists of a radiating patch on one side and ground planes on the other side separated by suitable dielectric substrate as in Fig. 1.

1.1 Feeding Mechanisms

A. *Micro-strip quarter wave line feed*

The micro-strip antenna of impedance Z_A can be matched to a transmission line of characteristic impedance Z_0 by using a Quarter wave transmission line of characteristic impedance Z_1 by below equation.

$$Z_1 = \sqrt{Z_0 Z_A} \tag{1}$$

Patch antennas with quarter wave feed are as shown in Fig. 2.
Characteristics of MSA
The advantages and disadvantages of the MSA are given below

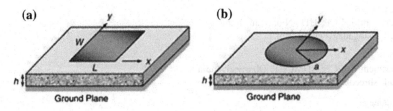

Fig. 1 Comparison of rectangular and circular microstrip antenna

Fig. 2 Quarter wave feed for rectangular and circular microstrip antenna

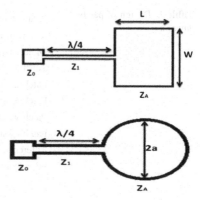

B. *Advantages*

- The size can be made compact for use in personal mobile communication.
- They can be design for dual- and triple-frequency operations.
- The printed-circuit technology can be used for mass production.
- They are easier to integrate with other MICs on the same substrate.
- They allow both linear and circular polarization.
- They are lightweight and have a small volume and a low-profile planar configuration.
- They can be made conformal to the host surface.

C. *Disadvantages*

- Narrow Bandwidth.
- Lower Gain.
- Low Power-handling capability.

D. *Applications*

- Aircraft and ship antennas for navigation and blind landing system.
- Satellite communication—DTH Service and vehicle based antenna.
- Mobile Radio—Pagers and hand telephones.
- Remote Sensing.

2 Design of Patch Antenna

2.1 Design Equations of Rectangular Micro-strip Patch [3]

See Table 1.

1. Width of patch antenna

Table 1 Design of patch antenna

Parameter name	Designed values
Resonant frequency, f_r	2.4 GHz
Dielectric constant, ϵ_r	4.4
Substrate height, h	1.58 mm
Patch thickness, t	0.017 mm
Patch length, L	29.50 mm
Patch width, W	38.03 mm
Patch radius, a	17.508 mm
Reference impedance, Zc	50 Ω
Feed length, $\lambda/4$	3.125 cm

$$W = \frac{1}{2f_r\sqrt{\mu_0 \varepsilon_0}} \sqrt{\frac{2}{\varepsilon_0 + 1}} \tag{2}$$

2. Effective dielectric constant

$$\varepsilon_{eff} = \frac{\varepsilon_r + 1}{2} + \frac{\varepsilon_r - 1}{2}\left[1 + 12\frac{h}{w}\right]^{-1/2} \tag{3}$$

3. Length due to fringing effect

$$\Delta L = 0.412 * h \frac{(\varepsilon_{eff} + 0.3)}{(\varepsilon_{eff} - 0.258)} \frac{\left(\frac{W}{h} + 0.264\right)}{\left(\frac{W}{h} + 0.8\right)} \tag{4}$$

4. Length of the patch

$$L = \frac{1}{2f_r\sqrt{\varepsilon_{eff}}\sqrt{\mu_0 \varepsilon_0}} - 2\Delta L \tag{5}$$

5. Effective length of the patch

$$L_{eff} = L + 2\Delta L \tag{6}$$

6. Patch input impedance

$$Z_A = 90 * \frac{\varepsilon_r^2}{\varepsilon_r - 1} \frac{(L)^2}{(W)^2} \tag{7}$$

2.2 Design Equation of Basic Circular Patch

1. Radius of Patch

$$a = \frac{F}{\sqrt{\left\{1 + \frac{2h}{\pi \varepsilon_r F}\left[\ln\left(\frac{\pi F}{2h}\right) + 1.7726\right]\right\}}} \tag{8}$$

and

$$F = \frac{87.91 * 10^9}{f_r \sqrt{\varepsilon_r}} \tag{9}$$

3 Simulated Results

3.1 Simulated Patches

See Figs. 3 and 4.

Fig. 3 Circular micro-strip antenna

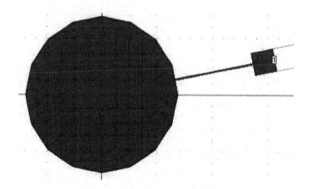

Fig. 4 Rectangular micro strip antenna

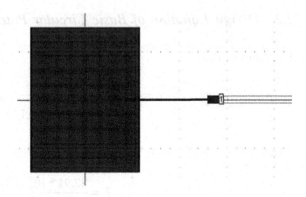

3.2 Return Loss Curves

See Figs. 5 and 6.

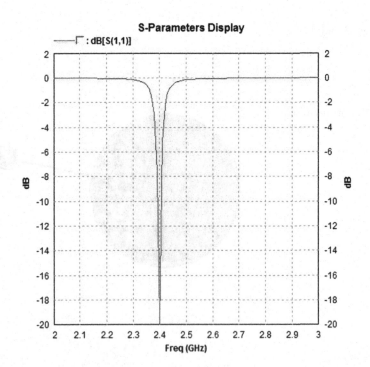

Fig. 5 Return loss curve for rectangular quarter wave feed

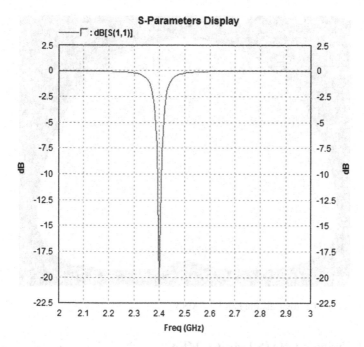

Fig. 6 Return loss curve for circular quarter wave line feed

3.3 Current Distributions

See Figs. 7 and 8.

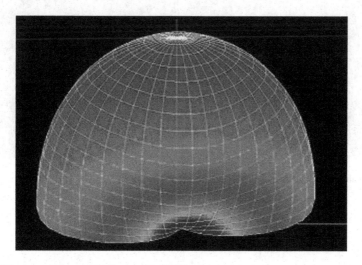

Fig. 7 3D view of current distribution for CMSA

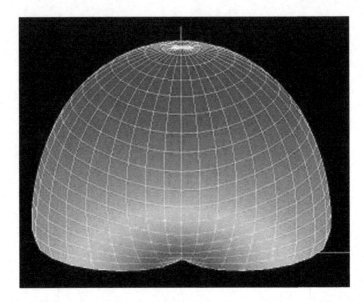

Fig. 8 3D view of current distribution for RMSA

3.4 *Fabricated RMSA and CMSA*

See Fig. 9.

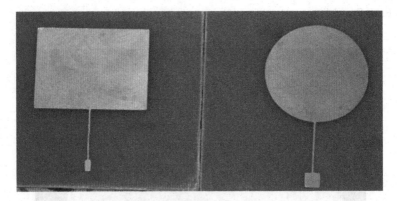

Fig. 9 Fabricated RMSA and CMSA with quarter wave feed

Table 2 Comparison of simulated results

Parameter	Circular micro strip antenna	Rectangular micro strip antenna
Resonant frequency (GHz)	2.4	2.4
Gain (dB)	8	6
Return loss curve (dB)	−20	−20.5
VSWR	1.226	1.204
Area (cm^2)	9.6299 cm^2	11.139 cm^2
Bandwidth (MHz)	50	60

Table 3 Comparison of measured Results

Parameter	Circular micro strip antenna	Rectangular micro strip antenna
Resonant frequency (GHz)	2.4	2.4
Gain (dB)	8	5
VSWR	1.32	1.30

4 Comparative Study

See Tables 2 and 3.

5 Conclusion

A comparative study of rectangular and circular patches at 2.4 GHz for wireless application is done by simulating, fabricating, and testing. Return loss curves of rectangular and circular micro-strip antenna with quarter wave feed resonates at 2.4 GHz with return loss of 20.5 dB for each. Current distribution of circular micro-strip shows more radiation resulting in more gain and less bandwidth compared to rectangular micro-strip antennas. It was found that circular patch occupies lesser area with lesser return loss, higher gain, and bandwidth compared with rectangular patch which occupies little larger area with higher return loss and bandwidth. Thus, both patches are suited with array structures for wireless applications like radars, WLANS, etc.

References

1. T. I. Huque, K. Hossain, S. Islam, and A. Chowdhury, "Design and Performance Analysis of Microstrip Array Antennas With Optimum Parameters For X-band Applications," *International Journal of Advanced Computer Science and Applications*, Vol. 2, No. 4, pp. 81–87, 2011.
2. R. Garg, P. Bhartia, I. J. Bahl, and P. Ittipiboon, *Microstrip Antenna Design Handbook*, Artech House, Boston. London, 2001.

3. S. Sudharani, P. Rakesh Kumar,"Microstrip Circular Patch Array Antenna for Electronic Toll Collection," *International Journal of Research in Engineering and Technology (IJRET)*, Vol 2, Jan 2013.
4. M. M. Olaimat and N. I. Dib, " A Study of 15-75o-90o Angles Triangular Patch Antenna," *Progress In Electromagnetics Research (PIER)*, Vol. 21, pp. 1–9, 2011.
5. M. M. Olaimat and N. I. Dib, "Improved Formulae for the Resonant Frequencies of Triangular Microstrip Patch Antennas," *International Journal of Electronics,* Vol. 98, No. 3, pp. 407–424, March 2011.
6. Olaimat, M., "Design and analysis of triangular microstrip patch antennas for wireless communication systems," Master Thesis, Jordan University of Science and Technology, 2010.
7. Prof. Mahesh M. Gadag, Mr. Dundesh S. Kamshetty, Mr. Suresh L. Yogi, Mr. Vinayak C. D "Design and Comparative Study of Different Feeding Mechanisms for Microstrip Antenna for Wireless Communication", *International Conference on Computational Intelligence & Computing Research (IEEEICCIC), 978-1-61284-693-4/11, pp-638-641, December 15–18, 2011.*
8. Prof. Mahesh M. Gadag, Ms. Lubna F Shaikh, Ms. Aahaladika, Mr. Kundan Kumar, "Rectangular Microstrip Antenna using Air-Coupled Parasitic Patches for Bandwidth Improvement", *International Conference on Electronics and Communication Engineering,* 978-93-81693-56-8, Vizag, 2012, pp. 45–49.
9. C. A. Balanis, "Antenna Theory, Analysis and Design," John Wiley & Sons, New York, 2008, pp 723–725.

Optimal Power Flow Problem Solution Using Multi-objective Grey Wolf Optimizer Algorithm

Ladumor Dilip, Rajnikant Bhesdadiya, Indrajit Trivedi
and Pradeep Jangir

Abstract The optimal power flow is highly constrained, nonlinear, and non-convex optimization problem of power system. This paper proposed multi-objective grey wolf optimizer (MOGWO) algorithm to evaluate optimal power flow (OPF) problems. Three objective functions such as emission, fuel cost, and active power loss are preferred as single-objective OPF problems. Proposed MOGWO algorithm was used to find pareto-optimal solution for two different multi-objective cases like Minimization of Fuel cost with Emission value and Minimization of Fuel cost with Active Power loss. The proposed MOGWO algorithm was tested on standard IEEE-30 bus test system with above two multi-objective functions to determine the efficiency of proposed algorithm. The outcomes of MOGWO algorithm were related with well-known NSGA-II (Non-dominated Sorting Genetic Algorithm) which was reported in the literature. MOGWO algorithm is given fast convergence and best pareto-optimal front setting with compared to NSGA-II.

Keywords Active power loss · Emission value · Fuel cost
IEEE-30 bus system · MOGWO · OPF

L. Dilip (✉) · R. Bhesdadiya · P. Jangir
Lukhdhirji Engineering College, Morbi, Gujarat, India
e-mail: Ladumordilip56@gmail.com

R. Bhesdadiya
e-mail: rhblec@gmail.com

P. Jangir
e-mail: pkjmtech@gmail.com

I. Trivedi
GEC Gandhinagar, Gandhinagar, Gujarat, India
e-mail: forumtrivedi@gmail.com

© Springer Nature Singapore Pte. Ltd. 2018
Y.-C. Hu et al. (eds.), *Intelligent Communication and Computational
Technologies*, Lecture Notes in Networks and Systems 19,
https://doi.org/10.1007/978-981-10-5523-2_18

1 Introduction

Optimal Power Flow problems [1] are highly non-convex and nonlinear optimization problem due to having some equality and inequality constraints to satisfy desired objective functions. There are large numbers of optimization techniques for the solution of the optimal power flow problems. Earlier some classical (deterministic) methods like Gradient Method, Lagrangian method, newton Method, Steepest Decent Method, Quadratic Programming Method, Penalty and Barrier Method, GRG Method, etc. used for the solution of OPF problem [2]. But the main difficulties are all these methods are deterministic and slow rate of convergence. These deterministic methods are not guaranteed for the global solution and for finding out the local solution for the OPF problem. However, some of these methods have extremely good convergence characteristics and mainly used in the industry for the solution of different optimization problems.

Due to the recent development in computational intelligent tools, the shortcomings of the above deterministic methods are avoided. Since last two decades, some meta-heuristic techniques such as Particle Swarm Optimization (PSO) [3], Artificial Immune Algorithm (AIA), Genetic Algorithm (GA) [4], Ant Colony Algorithm (ACO) [5], Harmony search Algorithm (HSA) [6] are used for the solution of different nonlinear and non-convex optimization problems. The advantages of these methods are it gives the guarantee for the Global solution.

In this paper, three objective functions such as fuel cost, emission and real power loss are taken as single-objective functions. Single-objective optimization techniques will give best optimal solution for each individual objective function but not guaranteed for rest of the objective functions. It is cleared that when single-objective techniques are used for fuel cost minimization, it gives minimum fuel cost but other two objectives emission and active power losses may or may not be minimized. If more than one objective function is considered simultaneously, then it is called multi-objective techniques and it will give compromising solution for each selected objective functions. Hence to overcome this difficulty multi-objective GWO approaches are used for solution of two different multi-objective functions such as Minimize Fuel cost with Emission value and Minimize Fuel cost with Active Power loss is considered simultaneously.

2 OPF Formulation

To obtain the optimal setting of control variables, some equality and inequality constraints must be followed. The mathematical formulations of OPF problem has been given by the following [2].

$$Min\ K(X_S, X_C) \tag{1}$$

$$g(X_S, X_C) = 0 \tag{2}$$

$$w\ (X_S, X_C) \le 0, \tag{3}$$

where, K is main objective function, X_S are dependent variables and X_C are control variables. g and w are equality and inequality constraints respectively.

2.1 Independent Variables

The variables which can be controlled to satisfy desired objective functions and some of the control variables (X_C) to solve load flow studies are as follows:

P_{gen} Generation of active power at PV bus expect slack bus.
V_g PV bus voltage magnitude
T Transformer Tap setting
Q_C Shunt VAR compensators

So X_C is expressed by following equation:

$$X_C = \left[P_{gen1} \ldots P_{genNG}, V_{g1} \ldots V_{gNG}, Q_{c1} \ldots Q_{cN}, T_1 \ldots T_{NT} \right] \tag{4}$$

2.2 Dependent Variables

The variables that represent electrical state of the system is called dependent variables (X_S) and expressed as follows:

P_{gen} Real power of slack bus
V_L Voltage magnitude of load bus (PQ)
Q_g Reactive power of all generator buses
S_l Line loading

Hence, X_C is described as per follows:

$$X_S = \left[P_{gen1}, V_{L1} \ldots V_{LNL}, Q_{gen_1} \ldots Q_{gen_{NG}}, S_{l_1} \ldots S_{l_{nl}} \right] \tag{5}$$

2.3 Equality Constraint

The generation of active power and reactive power are equality constraints and given by the following equations:

$$P_{gen^{ith}} - P_{demi^{th}} - V_{i^{th}} \sum_{j=1}^{N_b} V_{j^{th}} \left[g_{ij^{th}} \cos{(\theta_{ij})} + b_{ij^{th}} \sin{(\theta_{ij})} \right] = 0 \qquad (6)$$

$$Q_{gen^{ith}} - Q_{demi^{th}} - V_{i^{th}} \sum_{j=1}^{N_b} V_{j^{th}} \left[g_{ij^{th}} \sin{(\theta_{ij})} + b_{ij^{th}} \cos{(\theta_{ij})} \right] = 0 \qquad (7)$$

where NB indicate number of buses, P_{dem} demand of active power, P_{gen} generation of real power, Q_{gen} is the reactive power generation, g_{ij} is conductance, and b_{ij} is susceptance between ith and jth bus respectively.

2.4 Equality Constraint

The generation of voltage, active power and reactive power, transformer tap changer and Shunt VAR compensator limits are the inequality constraints and given by the following equation:

$$V_{gen^{ith}}^{MIN} \leq V_{gen^{ith}} \leq V_{gen^{ith}}^{MAX}, \; i = 1, \ldots, N_G \qquad (8)$$

$$P_{gen^{ith}}^{MIN} \leq P_{gen^{ith}} \leq P_{gen^{ith}}^{MAX}, \; i = 1, \ldots, N_G \qquad (9)$$

$$Q_{gen_{ith}}^{MIN} \leq Q_{gen_{ith}} \leq Q_{gen_{ith}}^{MAX}, \; i = 1, \ldots, NG \qquad (10)$$

$$T_{i^{th}}^{MIN} \leq T_{i^{th}} \leq T_{i^{th}}^{MAX}, \; i = 1, \ldots, N_T \qquad (11)$$

$$Q_{C_i}^{MIN} \leq Q_{GC_i} \leq Q_{C_i}^{MAX}, \; i = 1, \ldots, NG \qquad (12)$$

3 Main Objective Functions

3.1 Minimize Fuel Cost

The primary objective of power system is to provide uninterruptable power supply to the consumer at minimum cost. So fuel cost minimization is the most common objective function of the OPF problem and objective function is given by following formula:

$$J = \sum_{i=1}^{N_G} a_i + b_i P_{geni^{th}} + c_i P^2_{geni^{th}} (\$/h) \tag{13}$$

where, N_G represent generators, $P_{geni^{th}}$ represent active power generation of ith bus and a_i, b_i and c_i are the generator coefficient.

3.2 Minimize Emission

By increasing the awareness of people to the environment pollution, it is essential to meet the demand not only economically but environment criteria are also considered. Total emission $EM(P_{geni^{th}})$ (tonnes/h) which is caused atmospheric pollution is represented by the following equation

$$EM(P_{gen}) = \sum_{i=1}^{N} 10^{-2} \left(\alpha_i + \beta_i P_{geni^{th}} + \gamma_i P^2_{geni^{th}} \right) \tag{14}$$

where, α_i, β_i and γ_i are coefficients of emission

3.3 Minimize Active Power Loss

Transmission power loss is one of the major problems of power system; so it will be as minimum as possible. The objective function of minimization of active power loss is given by the following equation:

$$J = \sum_{i=1}^{N_B} P_L = \sum_{i=1}^{N_B} P_{geni^{th}} - \sum_{i=1}^{N_B} P_{demi^{th}} \tag{15}$$

4 Multi-objective GWO

Seyedali Mirjalili et al. proposed multi-objective GWO algorithm in 2015. This algorithm was inspired from social headship and hunting performance of grey wolves. Mostly grey wolves prefer to live in troop and average size of troop is around 5–12. To formulate the mathematical model of proposed GWO algorithm based on social hierarchy of grey wolves, the fittest solution- Alpha (α), beta (β) and delta (δ) are adopted as second and third best solution. Rest of the solution or agents are considered as Omega (ω) and they follow above three wolves to find global optimum solution. Alpha has higher dominance in pack and they are decision makers. They may be male or female. Alpha has higher dominance in pack and

Fig. 1 Social hierarchy of
grey wolves

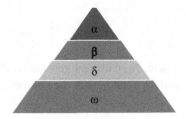

they are decision maker for hunting, Wakeup time, Place of sleep and so on. Other wolves are follower of alpha wolves, it is not mandatory that alpha should be the strongest member of pack but it has ability to manage pack properly. The social hierarchy of grey wolves shown in the Fig. 1 the dominance power decreased sequentially [7].

4.1 Encircling of Prey

The mathematical model for encircling of prey is given by following equations:

$$\vec{X} = \left| \vec{A} \cdot \vec{P}_{(t)} - \vec{W}_{(t)} \right| \tag{16}$$

$$\vec{W}_{(t+1)} = \vec{P}_{(t)} - \vec{B} \cdot \vec{X}_t \tag{17}$$

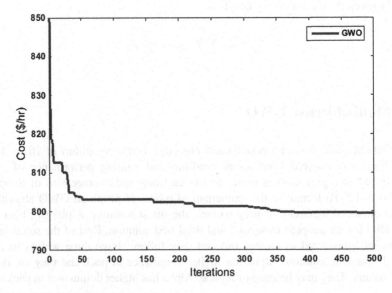

Fig. 2 Individual solution of fuel cost

where, $\vec{A}$ and $\vec{B}$ are the vectors, $\vec{P}$ and $\vec{W}$ are the position vectors of prey and wolves respectively.

The coefficient vectors are calculated by following mathematical equations:

$$\vec{B} = 2\vec{b} \cdot \vec{r}_1 - \vec{b} \tag{18}$$

$$\vec{A} = 2\vec{r}_2 \tag{19}$$

Two additional components inserted in GWO algorithm to developed multi-objective grey wolf algorithm. First component is archive and leader selection strategy is second component. Former component is nothing but simple memory unit which protects or rescues the non-dominated solution obtained so far. The size of archive is propositional to number of non-dominated solution obtained. So when archive becomes full, new solution is stored in least crowded segments to obtain optimal solution. Leader selection components choose best solution from the least crowded archive to decide alpha, beta and delta wolves for hunting process and roulette wheel method is used for selection.

Almost all characteristics of GWO is inherited in MOGWO but the main difference between MOGWO and GWO is that MOGWO searches around the set of archive where GWO searches only for first three finest solutions. The complexity of proposed MOGWO algorithm is same as other well-known multi-objective optimization algorithms but it is better compared to other algorithm like NSGA-II [8] and SPEA [9].

5   Results

To determine the effectiveness of proposed algorithm, it was applied on IEEE-30 bus system and data of IEEE-30 bus [10] system are as per: Six buses namely bus 1, 2, 5, 8, 11, and 13 taken as a Generator or PV buses. Four Transformer Tap Changer at the bus 11, 12, 15, and 16. Nine Shunt VAR Compensators at bus 10, 12, 15, 17, 20, 21, 23, 24, and 29. All the control parameters are shown in Table 1. PG1 to PG6 represents the generator power, VG1 to VG6 represents the generator voltage limits, T11, T12, T15, and T16 represents transformer tap setting at the bus 11, 12, 15, and 16 respectively. QC10, QC 12, QC 15, QC 17, QC 20, QC 21, QC 23, QC 24, and QC 29 represents the Shunt VAR compensators at the bus 10, 12, 15, 17, 20, 21, 23, 24, and 29 respectively.

Table 1 Optimal setting of control variables

Case-1			Case-2		
Control variable	MOGWO	NSGA-II	Control variable	MOGWO	NSGA-II
PG1	90.99889	159.473	PG1	53.48344	134.5544
PG2	54.96999	57.459	PG2	53.67177	46.2891
PG3	36.39849	20.049	PG3	36.31428	32.936
PG4	34.99809	19.892	PG4	34.99082	30.1163
PG5	21.73458	16.383	PG5	29.25878	18.735
PG6	29.07727	20.536	PG6	24.51069	26.5392
Fuel cost ($/h)	**833.8528**	859.849	Fuel cost ($/h)	**847.9695**	823.8875
Emision (tonne/h)	**0.245126**	0.3214	Power loss (MW)	**4.588649**	5.7699

5.1 Case 1: Minimize Fuel Cost with Emission

In case 1, emission and fuel cost are taken simultaneously as multi-objective functions and it is solved by proposed MOGWO algorithm. The proposed MOGWO algorithm is given best compromise solution for both the objective functions compared to NSGA-II [8, 11] and it was derived by fuzzy membership function. The MATLAB code of NSGA-II taken from [12]. Table 1 shows the optimal setting of control variables and best compromise solution for both fuel cost and emission value. Pareto-optimal front obtained by MOGWO is shown in Fig. 3.

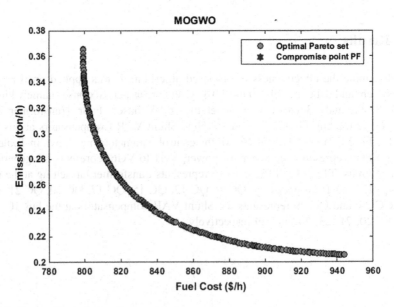

Fig. 3 Compromise solution for case-1

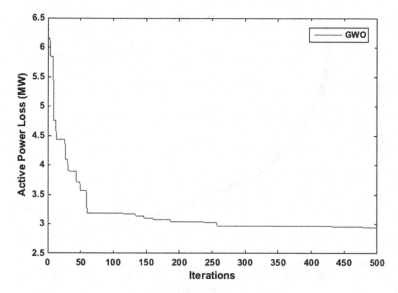

Fig. 4 Individual solution of active power loss

So, proposed MOGWO algorithm has given compromising solution for both fuel cost and emission instead of individual solution which gives only optimal value of fuel cost. Figure 2 shows the individual solution of fuel cost.

5.2   Case 2: Minimize Fuel Cost with Active Power Loss

In case 2, both minimization of fuel cost and minimization of active power loss are together taken as a multi-objective function. Proposed MOWGO algorithm gives best compromising solution for this two objective function compared to NSGA-II. Pareto-optimal front is shown in Fig. 5 and the optimal setting of control variables derived using MOGWO [11] is shown in Table 1. Single objective gives only optimal solution of active power loss not guaranteed for fuel cost and it shows in Fig. 4.

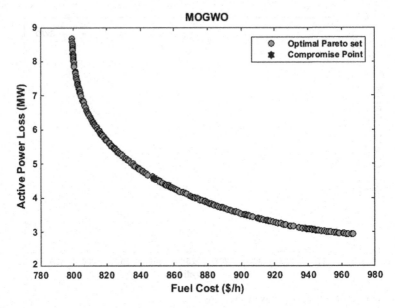

Fig. 5 Compromise solution for case-2

6 Conclusion

This paper introduced MOGWO algorithm for the solution of two different multi-objective functions such as to Minimize fuel cost with Emission and Minimize fuel cost with Active power loss. To determine the efficiency of suggested MOGWO algorithm, it was tried on IEEE-30 bus system with above two multi-objective functions. From the results, it was clear that proposed MOGWO algorithm has given best pareto-optimal solution compared to other algorithm (NSGA-II), reported in the literature. Using proposed algorithm, we will get pareto-optimal front for more than two objective functions.

References

1. H. Dommel and W. Tinny, "Optimal Power Flow Solutions," *IEEE Trans Power Appar Syst*, vol. PAS-87, no. 10, pp. 1866–1876, 1968.
2. H. R. E. H. Bouchekara, "Optimal power flow using black-hole-based optimization approach," *Appl. Soft Comput. J.*, 2014.
3. B. E. Turkay and R. I. Cabadag, "Optimal Power Flow Solution Using Particle Swarm Optimization Algorithm," no. July, pp. 1418–1424, 2013.
4. M. S. Osman, M. A. Abo-Sinna, and A. A. Mousa, "A solution to the optimal power flow using genetic algorithm," *Appl. Math. Comput.*, vol. 155, no. 2, pp. 391–405, 2004.

5. M. F. Mustafar, I. Musirin, M. R. Kalil, and M. K. Idris, "Ant Colony Optimization (ACO) based technique for voltage control and loss minimization using transformer tap setting," *2007 5th Student Conf. Res. Dev. SCORED*, no. December, 2007.

6. M. M. Bhaskar and S. Maheswarapu, "A Hybrid Harmony Search Algorithm Approach for Optimal Power Flow," *Telkomnika*, vol. 9, no. 2, pp. 211–216, 2011.

7. S. Mirjalili, S. Saremi, and S. Mohammad, "Multi-objective grey wolf optimizer : A novel algorithm for multi-criterion optimization," *Expert Syst. Appl.*, vol. 47, pp. 106–119, 2016.

8. N. Srinivas and K. Deb, "Multiobjective Optimization Using Nondominated Sorting in Genetic Algorithms," *Evol. Comput.*, vol. 2, no. 3, pp. 221–248, 1995.

9. M. Ghasemi, S. Ghavidel, E. Akbari, and A. Azizi, "Solving non-linear, non-smooth and non-convex optimal power fl ow problems using chaotic invasive weed optimization algorithms based on chaos," *Energy*, 2014.

10. I. Data and O. F. The, "IEEE-30 Bus System Data," pp. 493–495, 2003.

11. A. R. Bhowmik and A. K. Chakraborty, "Electrical Power and Energy Systems Solution of optimal power flow using nondominated sorting multi objective gravitational search algorithm," *Int. J. Electr. POWER ENERGY Syst.*, vol. 62, pp. 323–334, 2014.

12. A. SESHADRI, "a Fast Elitist Multiobjective Genetic Algorithm: Nsga-Ii," pp. 1–4.

Bisection Logic Based Symmetric Algorithm for the Design of Fault Tolerance Survivable Wireless Communication Networks

B. Nethravathi and V.N. Kamalesh

Abstract In current technological development on the design of wireless communication network, it is observed that one of the critical issues is dependable network. The dependable network encompasses the properties like availability, reliability and survivability. The wireless networks are more complicated with problems such as allocation of frequencies, service quality and node or link failures. Hence, survivability is one of the significant factors to be considered during the design process. Survivability is attained by achieving fault tolerance. Survivability can be achieved by having high node connectivity (k) in the topological design of network. This chapter presents a symmetric method based on bisection logic (BSA) to generate a k-node connected network structure for maximal connectivity number k and hence highly survivable network structures.

Keywords Wireless network · Fault tolerance · Survivability
k-node connectivity

1 Introduction

The primary vision of a wireless and mobile communication network is to provide services to the end clients irrespective of the location, movement of the end user and terminal access type used. The rapid growth of wireless networks and increasing demand of mobile communication attracts researchers same. End users

B. Nethravathi (✉)
Department of Information Science and Engineering, JSSATE,
Affiliated to Visvesvaraya Technological University, Belgaum, Karnataka, India
e-mail: nethravathi.sai@gmail.com

V.N. Kamalesh
Department of Computer Science and Engineering,
Sreenidhi Institute of Science and Technology, Autonomous Under UGC, Hyderabad,
Telengana, India
e-mail: kamalesh.v.n@gmail.com

© Springer Nature Singapore Pte. Ltd. 2018
Y.-C. Hu et al. (eds.), *Intelligent Communication and Computational
Technologies*, Lecture Notes in Networks and Systems 19,
https://doi.org/10.1007/978-981-10-5523-2_19

insist on reliable services. The major objective is to achieve end-end performance and dependability in the network. The dependable networks are valued on parameters like reliability, survivability and availability. The estimation of network failure probability is the reliability of the network structure. The estimation of network ready time for usage without any interference is the availability of the network structure. Survivability is the ability of a network to execute its intended services upon infrastructure component failure. The wireless networks lack credibility by the factors of survivability, reliability and availability. The factors affecting these enhancements are diverse types of links like Zigbee, RF, Bluetooth, etc.; further the factors like lack of regulations, high CAPX, user mobility and OPEX. Also due to diversities in technology like CDMA, GSM, UMTS, 3G LTE environments degrade the performance [1].

Wireless communication networks are more vulnerable to failures, which can be a node failure or link failure. Even a single node failure could cause serious damage to the network services. Thus dependability issue has to be compulsorily addressed in the wireless communication network design problem for maximum fault tolerance, which in turn guarantees high performance [2].

The network engineers and computer scientists have unequivocally accepted that the mathematical deterministic measure for network survivability and fault tolerance is the k-node connectivity of the underlined network graph structure [3]. Higher the value of the k more is the survivability of the network structure and hence the dependability of the network structure. In a complex wireless network structure generating a fault-tolerant survivable communication network following the principles of network design is a challenging and open research problem. Here bisection logic based symmetric algorithm (BSA) is proposed to generate a k-node connected network structure for maximal connectivity [4] number k.

2 Related Literature

This research work considers two major related literature works,

1. Fault isolation strategies.
2. Generation of fault tolerance survivable computer communication network.

Fault identification and fault isolation technique identifies and determines the exact fault location in a communication network. Fault recognition and fault Isolation is an integral part in the design of computer network structures. Over the decades, a good number of techniques have evolved for the same. A detailed analysis of techniques for the same can be found in the article "Analysis of Techniques for Isolation of Faults in Survivable Computer Networks" [5].

Few algorithms and methods for designing fault tolerance k-node connected communication networks are available in the literature. The oldest one is the link deficit algorithm put forward by Stegliz et al. [6].

S. Latha et al. [7, 8] have also proposed techniques using Kbit gray code and Harrary circular graphs.

S. K. Srivatsa et al., have used graph theoretical concepts and have proposed incremental method [9] and bipartite method [10]. The detailed literature review and recent progress on the design of k-node connected fault tolerance survivable networks structure is found in the article "Topological Design of Computer Communication Network Structures: A Comprehensive Review" [11]. However, it is clear from the above comprehensive review, that there is no single method or technique which addresses most of the design principle issues. In this research paper we are proposing bisection logic based symmetric algorithm (BSA).

3 Proposed Method

This section presents the proposed BSA method for the design of fault tolerance survivable communication networks. To start with, the geographical positions of node concentrators are given. The nodes are indexed by using first n natural numbers in sequence. Establish link between node 1 and node 2, node 2 and node 3, so on and in the end between node n and node 1, which results in 2 connected network. Bisect the node set of the network into two sets P and Q such that P = {1, 2, 3, ..., (n/2)},

$$Q = \{(n/2)+1, (n/2)+2, (n/2)+3, \ldots, n\}$$

Establish links between the node set P and the node set Q to get 3-node connected network in the following manner. Connect 1st node of set P with 1st node of set Q, connect the second node of set P with the second node of set Q, continuing the same, finally connect the last node of P with the last node of set Q.

Further establish links between the node set P and the node set Q to get 4-node connected network in the following way. Connect the first node of set P with 2nd node of set Q, connect the second node of set P with the third node of set Q, continuing the same, finally connect the last node of P with the first node of set Q circularly.

Continuing in the above manner for the required node connectivity k ≤ (n/2) + 1, by shifting the parallel edges to the next node circularly if exist.

Algorithm: BSA

Input: (i) n, number of node concentrator.
               where n=2e, e is an integer.
     (ii) k, the required node connectivity,
               where $k \leq (n/2)+1$.
Output: k-node connected fault tolerance survivable network structure N.
Method:
Step 1: Geographical location of the n distant node
        concentrators is given.
Step 2: Index the nodes by using first n natural numbers in
        sequence.
Step 3: // while (k=2)
        For i=1 to n-1
            Connect node i to i+1
        End for
        Connect node n to node 1
        Output: 2-node connected network structure.
Step 4: Split the node set in to two sets L1 and L2
        Such that,
            P={1, 2, 3,......,(n/2)}
            Q= {(n/2)+1, (n/2)+2, (n/2)+3,..........,n}
Step 5: //while (k=3)
        For i=1 to (n/2)
            Connect edges from P[i] → Q[n/2+i]
        End for
        Output: 3- node connected network structure.
Step 6: while (k=4 to (n/2)+1)
        Repeat step 5 by incrementing index of set Q by 1
        till the last node n and continue circularly.
        Output: k-node connected network structure.
   Algorithm end

4 Illustration

In this section the proposed method is illustrated phase by phase for 12 number of node concentrators (n) and node connectivity 5(k).

The geometrical positions of 12 node concentrators are given as shown in Fig. 1. The nodes are numbered using first n natural number (Fig. 2).

For easy understandability and demonstration let us place the node concentrators on a single line and establish connection as said above. Figure 3 shows the same 2-node connected network structure.

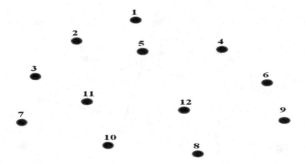

Fig. 1 Geographical position of 12 nodes

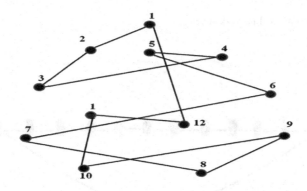

Fig. 2 2-node connected network structure

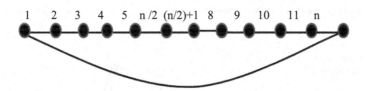

Fig. 3 2-node connected network structure

In the next step we get 3-connected network structure as shown in Fig. 4.

Figure 5 shows the same 3-node connected network structure when the node concentrators are on the single line.

Figure 5 shows the same 3-node connected network structure when the node concentrators are on the single line.

In the next step we get 4-node connected network structure as shown in Fig. 6.

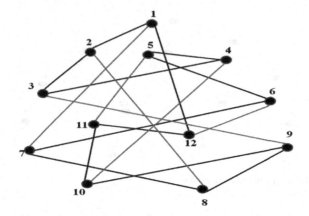

Fig. 4 3-node connected network structure

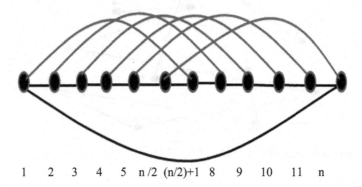

1   2   3   4   5   n /2 (n/2)+1 8   9   10   11   n

Fig. 5 3-node connected network structure

Fig. 6 4-node connected
network structure

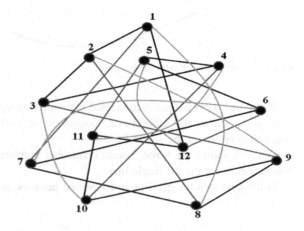

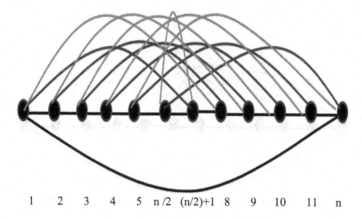

1   2   3   4   5   n /2   (n/2)+1   8   9   10   11   n

Fig. 7 4-node connected network structure

Fig. 8 5-node connected
network structure

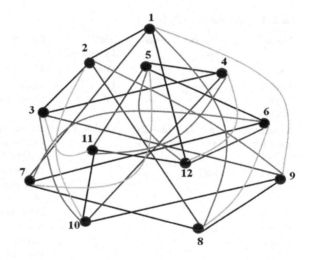

Figure 7 shows the same 4-node connected network structure when the node concentrators are on the single line.

In the next step we get 5-node connected network structure as shown in Fig. 8. The number of links used for n = 12 and k = 5 is 30 (Fig. 9).

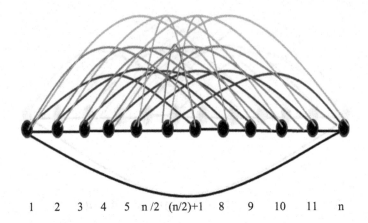

1   2   3   4   5   n/2 (n/2)+1   8   9   10   11   n

Fig. 9 5-node connected network structure

Table 1 Comparative analysis

Method	Number of links used	Time Complexity	Remarks
BSA	kn/2	O(n)	Maximum connectivity is (n/2) + 1
Bipartite graph method	k(n − k)	O $(n^{k/2} + n^2)$	Method is link optimal, but maximum connectivity is n/2
Incremental approach	(n − 1) + (n − 2) + · · · + (n − k)	O$(n^k + n^2)$	Maximum connectivity is (n − 1), but method needs more number of links compared to BSA
Latha's method	kn/2		Works on condition that number of node exactly equal to 2^k, e.g., 2, 4, 8, 16, 32, etc., but BSA works for all even numbers
Steiglitz's method	kn/2		It is a generic method, needs more computation, not mathematically modeled

5   Results and Discussions

The proposed method works only for even number of node concentrators. The number of links used to generate the required k-node connected network structure is kn/2; hence the proposed method is link optimal compared to incremental approach. This method does not involve repeated searching of nodes concentrators due to conflicts. Hence demands less computation efforts compared to methods due to S. K. Srivatsa et al. and S. Latha et al. Further the geographical location of the node concentrators does not play any specific role. The time complexity of the proposed method is linear, which is better compared to bipartite method and incremental method.

The comparative analysis is given in Table 1.

6　Conclusion

One of the basic principles of wireless communication network design is the symmetry of the communication network. The symmetric property imposes regularity on the network graph structure. High symmetric network structure is preferred as it is advantageous in the construction and simulation of network algorithm, which in turn helps in network management. The BSA algorithm proposed here attains symmetry as the network nodes are divided equally. The algorithm is link optimal and generic. The time complexity of this algorithm is O(n). Further the only constrain is that the method works only for even number of node concentrators.

References

1. Andrew S. Tanenbaum: Computer Networks, Prentice Hall, Englewood Cliffs, (2001).
2. K. Rajalakshmi & Krishna gopal, (2014), A Survey on Cost Effective Survivable Network Design in Wireless Access Network, Int. J. Com. Sci. & Eng. Survey, Vol. 5, No. 1, DOI: 10.5121/ijcses.2014.5102.
3. Junming Xu: Topological Structure and Analysis of Interconnection Networks, Kluwer Academic Publishers (2002).
4. Douglas B. West: Introduction to Graph Theory, Pearson Education, Inc 2nd Edition, (2001).
5. Nethravathi, B., Kamalesh, V. N., Nidhi H Kulakarni, and Apsara, M. B. (2016). Analysis of Techniques for Isolation of Faults in Survivable Computer Networks, Accepted for publication and presentation at IEEE conference- ICEECCOT 2016 on 9–10th Dec 2016.
6. K. Steiglitz, P. Weiner and D.J. Kleitman. (1969). The design of minimum cost survivable network. IEEE Trans. Circuit Theory (1969) 455–460.
7. S. Latha and S. K. Srivatsa. (2007). Topological design of a k-connected network. WSEAS transactions on communications vol. 6 [4], pp. 657–662.
8. S. Latha and S.K. Srivatsa. (2007). On some aspects of Design of cheapest survivable networks. International journal of computer science and network securities, Vol. 7, No. 11, 210–211.
9. Kamalesh, V. N. and Srivatsa, S. K. (2009). On the design of minimum cost survivable network topologies. Nat. Conf. on communication at IIT, Guwahati, India, pp. 394–397.
10. Kamalesh, V. N. and Srivatsa, S. K. (2009). Topological design of minimum cost survivable computer communication networks: A bipartite graph method. Int. J. Computer Science and Information Security, Vol. 3, pp. 148–152.
11. Nethravathi, B. & Kamalesh, V. N. (2016). Topological Design of Computer Communication Network Structures: A Comprehensive Review. Ind. J. Sci. and Tech., vol 9(7), DOI:10.17485/ijst/2016/v9i7/85211.

A Bibliometric Analysis of Recent Research on Machine Learning for Cyber Security

Pooja R. Makawana and Rutvij H. Jhaveri

Abstract In today's world a huge amount of information is shared around the globe using internet. While connected to cyber word, cybersecurity is an increasing problem. Nowadays, various machine learning techniques are used to deal with cybersecurity threats. To enlighten the researchers about recent trends in this research area, we have analyzed 149 research papers from January 2015 to December 2016 and present a graphical and organized view of the referred research works. We observe that machine learning for cybersecurity has a great potential in carrying out further research. We have carried out bibliometric analysis by categorizing the referred papers using the method of implementation, article type, publishers and article efficiency. This analysis will provide insights for researchers, students, publishers and experts to study current research trends in the area of machine learning for cybersecurity.

Keywords Machine learning · Cybersecurity · Trend analysis · Graphical interpretation

1 Introduction

Cyber-attacks are malicious activities intended to disrupt, deny, degrade or destroy vital information and services residing in the host computer. Conventional security software requires a lot of human effort to identify threats, to extract characteristics from the threats, and encode the characteristics into software to detect the threats. This labor-intensive process can be more efficient by applying machine learning

P.R. Makawana (✉) · R.H. Jhaveri
Department of Information Technology, Shri S'ad Vidya
Mandal Institute of Technology, Bharuch, India
e-mail: poojakatariya1408@gmail.com

R.H. Jhaveri
e-mail: rhj_svmit@yahoo.com

© Springer Nature Singapore Pte. Ltd. 2018   213
Y.-C. Hu et al. (eds.), *Intelligent Communication and Computational
Technologies*, Lecture Notes in Networks and Systems 19,
https://doi.org/10.1007/978-981-10-5523-2_20

algorithms. As a result, a number of researchers have devised various machine learning algorithms to detect attacks more efficiently and reliably [1].

Trend analysis is an aspect of technical analysis to predict movement of particular research area in future, based on past data. Since cybersecurity is an emerging field, our objective behind this trend analysis is to see what and how much of work has been done. In this paper we propose a bibliometric analysis on Machine learning for cybersecurity showing the current trend of year 2015 till now. We have analyzed 149 research papers of different well known journals and reputed publishers.

The reminder of this paper is organized as follows: in Sect. 2 we include graphical representation of this trend analysis and Sect. 3 consists of the final conclusion of paper.

2   Graph Interpretation

2.1   Percentage of Papers Published by Countries

Machine learning is an emerging field. There are large number authors from different countries participate in the research work of machine learning for cybersecurity. For this trend analysis we have analyzed 149 papers from different publishers, i.e., ACM, Elsevier, IEEE, Springer, Taylor and Francis and Wiley from January 2015 to December 2016. Figure 1 shows that 30% of authors in our data set are from UK, 18% of authors are from Germany, 7% of authors are from Australia and Israel followed by other countries.

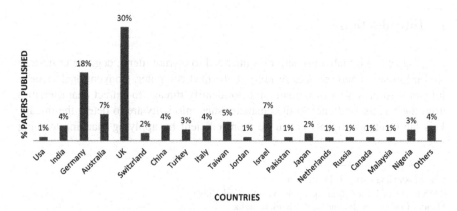

Fig. 1 Percentage of papers published by countries

2.2 Percentage of Papers Publisher

High quality literature as resource is necessary to get direction towards accurate analysis. As shown in the Fig. 2, IEEE is the most well recognized publisher having 58% of Papers followed by ACM (17%), Elsevier (13%), Springer (9%). Other publishers Wiley, Hindawi and Taylor and Francis are having 3% of published papers.

2.3 Citations per Publisher

This analysis shows number of citations acquired by different publishers. Quality of paper increases by the increasing number of citations. As shown in Fig. 3, IEEE has the maximum number of citations (229) from January 2015 to December 2016, followed by Elsevier (114), Springer (44), ACM (32) and Taylor & Francis (6).

2.4 Conference Papers Versus Journals

Conference papers are the papers which are written with the goal of being accepted to a conference. The place where the researchers can present their research to community is called conference. The article that is published in an issue of journal is journal papers. Figure 4 shows that maximum of our dataset consists of Conference papers (57%), followed by Journal papers (43%).

Fig. 2 Percentage of publishers

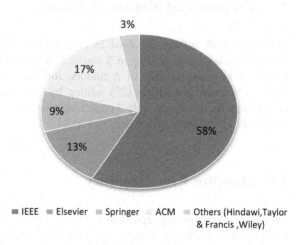

■ IEEE ■ Elsevier ■ Springer ACM ■ Others (Hindawi,Taylor & Francis ,Wiley)

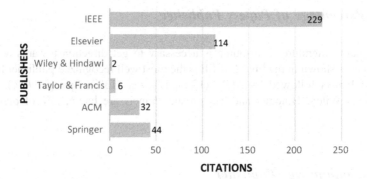

Fig. 3 Citations per publishers

Fig. 4 Conference papers
versus journal paper

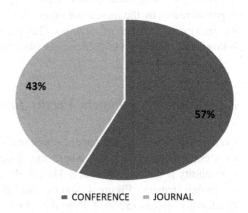

2.5 Frequency of Number of Authors

The number of authors involved in covering the whole topic in co-authorized manner is analyzed here. Figure 5 shows that maximum of the papers (41) were published by co-authorship of 3 authors, followed by papers (33) written by 4 authors, followed by papers (29) written by 2 authors and followed by papers (17) written by 6 authors. Only 2 papers have a total 10 number of authors and one paper has 13 numbers of authors.

2.6 Objective Wise Frequency

Objectivewise frequency represents the most trending objective from January 2015 to December 2016. As shown in Fig. 6, detection of malicious activities have maximum of the chart covering (37%) each. Defense against adversary is having

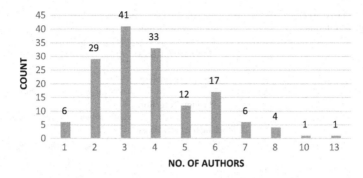

Fig. 5 Frequency of number of authors

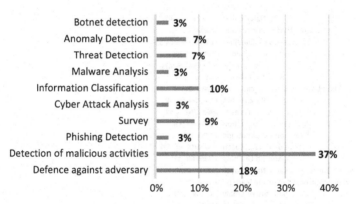

Fig. 6 Objective wise frequency

18% of chart covering followed by Information classification (10%), Survey (9%), Anomaly Detection (7%), and Threat Detection (7%). Least addressed objective here are botnet detection, malware analysis cyber-attack analysis and phishing detection.

2.7 Objective Wise Percentage of Citations

This analysis shows the percentage of citations per objective. As shown in Fig. 7, Survey (32%), Detection of malicious activity (29%), Malware Analysis (22%) were the most trending objective from January 2015 to September 2016 and were followed by Threat Detection (3%), Anomaly detection (2%) and Cyber-Attack Analysis (1%).

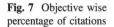

Fig. 7 Objective wise
percentage of citations

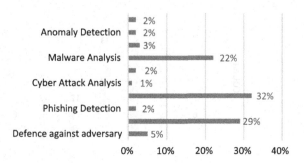

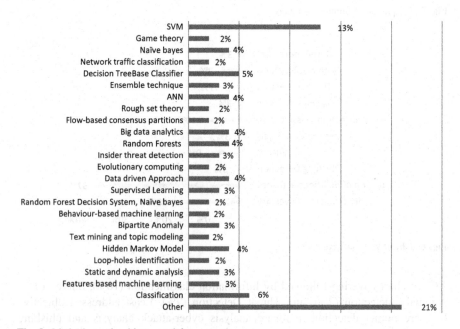

Fig. 8 Methods or algorithms used for research

2.8 Methods or Algorithms Used for Research

The methods or algorithms used for research for our data set include classification, clustering and regression. As shown in Fig. 8 the most acknowledged algorithm is SVM (Support Vector Machine) with 13% of chart covering. Classification and decision tree base classifier 6% and 5% of papers, respectively.

2.9 Paper Efficiency/Ranking

Efficiency of the paper shows how resourceful the paper is. For this analysis, we first need to normalize the efficiency evaluation for all the papers of our dataset from January 2015 to December 2016 needs to normalize the efficiency. For all articles in our dataset, the equation shown below is proposed to standardize the ranking/efficiency [2].

$$E = C/(24 - PM)$$

where, **E**—Efficiency of the Paper, **C**—Citation of the Paper, **PM**—Published Months.

Table 1 High efficiency papers

Title	C	PM	E
Current state of research on cross-site scripting (XSS)—a systematic literature review [3]	21	6	1.17
Cloud-enabled prognosis for manufacturing [4]	33	5	1.74
Threat analysis of IoT networks using artificial neural network intrusion detection system [5]	2	22	1.00
A survey of network anomaly detection techniques [6]	13	11	1.00
Design techniques and applications of cyber-physical systems: a survey [7]	62	6	3.44
Health-CPS: healthcare cyber-physical system assisted by cloud and big data [8]	32	7	1.88

Table 2 Average efficiency papers

Title	C	PM	E
Mobile-Sandbox: combining static and dynamic analysis with machine learning techniques	16	6	0.89
A survey of data mining and machine learning methods for cybersecurity intrusion detection	12	9	0.80
MARK-ELM: application of a novel multiple kernel learning framework for improving the robustness of network intrusion detection	10	11	0.77
Malware behavioral detection and vaccine development by using a support vector model classifier	12	11	0.92
Developing a hybrid intrusion detection system using data mining for power systems	13	2	0.59
Exploring threats and vulnerabilities in hacker web: forums, IRC and carding shops	12	6	0.67
Cyber-physical systems for water sustainability: challenges and opportunities	15	4	0.75
Developing an ontology for cybersecurity knowledge graphs	11	3	0.52
Onion bots: subverting privacy infrastructure for cyber-attacks	9	8	0.56
Developing a hybrid intrusion detection system using data mining for power systems	13	2	0.59

Table 3 Low efficiency papers

Title	C	PM	E
Predicting cybersecurity incidents using feature-based characterization of network-level malicious activities	1	2	0.05
Comparative analysis of features based machine learning approaches for phishing detection	1	14	0.10
Use of machine learning in big data analytics for insider threat detection	4	9	0.27
Phishing—the threat that still exists	3	1	0.13
Similarity-based malware classification using hidden Markov model	2	17	0.29
Malware detection via API calls, topic models and machine learning	3	7	0.18
Securing virtual execution environments through machine learning-based intrusion detection	1	10	0.07
Security-aware information classifications using supervised learning-cloud-based cyber risk management in financial big data	2	15	0.22
The security challenges in the IoT enabled cyber-physical systems and opportunities for evolutionary computing and other computational intelligence	1	18	0.17
Security intelligence for industrial control systems	1	18	0.17
An efficient classification model for detecting advanced persistent threat	3	8	0.19
Detecting advanced persistent threats using fractal dimension based machine learning classification	1	14	0.10
Novel approach for detecting network anomalies for substation automation based on IEC 61850	8	2	0.36
A test bed for SCADA cybersecurity and intrusion detection	1	20	0.25
Hive oversight for network intrusion early warning using Diamond: a bee-inspired method for fully distributed cyber defence	1	17	0.14
A kernel machine-based secure data sensing and fusion scheme in wireless sensor networks for the cyber-physical systems	4	11	0.31
TOLA: topic-oriented learning assistance based on cyber-physical system and big data	1	17	0.14
Modeling malicious activities in cyber space	3	11	0.23
Network intrusion detection system using J48 Decision Tree	4	8	0.25
Cyber situational awareness through network anomaly detection: state of the art and new approaches	3	0	0.13
Phase-space detection of cyber events	2	3	0.10
A nifty collaborative intrusion detection and prevention architecture for Smart Grid ecosystems	1	18	0.17
Advanced classification lists (Dirty Word Lists) for automatic security classification	5	9	0.33
A cloud computing based network monitoring and threat detection system for critical infrastructures	1	10	0.07
Mining software component interactions to detect security threats at the architectural level	2	18	0.33
Robust estimation for enhancing the cybersecurity of power state estimation	1	9	0.07
Feature extraction and classification phishing websites based on URL	3	11	0.23

(continued)

Table 3 (continued)

Title	C	PM	E
Multi-granular aggregation of network flows for security analysis	3	6	0.17
A decentralized and proactive architecture based on the cyber-physical system paradigm for smart transmission grids modeling, monitoring and control	2	11	0.15
Guest editors' introduction: special issue on cyber crime	1	14	0.10
Towards a relation extraction framework for cybersecurity concepts	4	3	0.19
A data-driven approach to distinguish cyber-attacks from physical faults in a smart grid	0	9	0.00
Cyber-deception and attribution in capture-the-flag exercises	3	7	0.18
Two-tier network anomaly detection model: a machine learning approach	2	10	0.14
Fighting against phishing attacks: state of the art and future challenges	4	14	0.40
Longitudinal analysis of a large corpus of cyber threat descriptions	4	5	0.21
Systems engineering for industrial cyber–physical systems using aspects	2	14	0.20
XML-AD: detecting anomalous patterns in XML documents	1	6	0.06
Static analysis for web service security—tools and techniques for a secure development life cycle	2	7	0.12
Cybersecurity risk assessment using an interpretable evolutionary fuzzy scoring system	2	12	0.17
A survey of anomaly detection techniques in financial domain	10	0	0.42
Acing the IOC game: toward automatic discovery and analysis of open-source cyber threat intelligence	1	21	0.33
Malware detection using bilayer behavior abstraction and improved one-class support vector machines	5	7	0.29
SNAPS: semantic network traffic analysis through projection and selection	3	10	0.21
Online anomaly detection using dimensionality reduction techniques for HTTP log analysis	1	7	0.06
Advanced temporal-difference learning for intrusion detection	2	8	0.13
Insider attack detection using weak indicators over network flow data	1	11	0.08
A Reflective approach to assessing student performance in cybersecurity exercises	2	14	0.20
Overview of cybersecurity of industrial control system	2	8	0.13
Anomaly detection in IPv4 and IPv6 networks using machine learning	1	14	0.10
Malware traffic detection using tamper resistant features	6	11	0.46
Rise of concerns about AI: reflections and directions	5	9	0.33
Game-theoretic strategies for IDS deployment in peer-to-peer networks	3	6	0.17
Prediction of psychosis using neural oscillations and machine learning in neuroleptic-naïve at-risk patients	6	9	0.40
Predicting vulnerability exploits in the wild	1	12	0.08
An Extreme Learning Machine (ELM) predictor for electric arc furnaces' V–I characteristics	1	12	0.08
A data-driven approach for the science of cybersecurity: challenges and directions	1	11	0.08

(continued)

Table 3 (continued)

Title	C	PM	E
Intrusion detection system using bagging with partial decision tree base classifier	4	5	0.21
A cybersecurity detection framework for supervisory control and data acquisition systems	1	7	0.06
Towards reproducible cybersecurity research through complex node automation	1	8	0.06
An incremental ensemble evolved by using genetic programming to efficiently detect drifts in cybersecurity datasets	1	18	0.17
On the empirical justification of theoretical heuristic transference and learning	1	17	0.14
How to choose from different botnet detection systems?	1	18	0.17
Cyber protection of critical infrastructures using supervised learning	1	7	0.06
Machine learning-based software classification scheme for efficient program similarity analysis	1	9	0.07
A signal processing approach for cyber data classification with deep neural networks	1	9	0.07
Secure and resilient distributed machine learning under adversarial environments	2	8	0.13
A review of the advances in cybersecurity benchmark datasets for evaluating data-driven based intrusion detection systems	1	8	0.06
On textual analysis and machine learning for cyber stalking detection	1	17	0.14
WAMS cyber-physical test bed for power system, cybersecurity study, and data mining	3	14	0.30

For locating efficiency of each paper, the above equation is used for each paper. The citation of the paper is divided by the value obtained by excluding published months from total number of months till the paper published (Till month of December of year 2016). Total 12 months of year 2015 and 12 months of year 2016 that leads us to total of 24 months. According to our dataset we have categorized the results in four categories: (1) High Efficiency Papers, (2) Average Efficiency, (3) Low Efficiency Papers, (4) Zero Efficiency Papers. Table 1 shows the High Efficiency Papers, Table 2 shows the Average Efficiency Papers, Table 3 shows the Low Efficiency Papers and Table 4 shows Zero Efficiency Papers.

Note: The zero efficiency papers have zero citations and so from above equation the value of efficiency is zero.

Table 4 Zero Efficiency Papers

Title	C	PM	E
Adversarial data mining: big data meets cybersecurity	0	21	0.00
Detection of cyber attacks in a water distribution system using machine learning techniques	0	15	0.00
Using bipartite anomaly features for cybersecurity applications	0	11	0.00
Behavior-based attack detection and classification in cyber-physical systems using machine learning	0	16	0.00
Real-time cyber-attack analysis on Hadoop ecosystem using machine learning algorithms	0	16	0.00
Classification of insider threat detection techniques		15	0.00
Behavior analysis of malware using machine learning	0	11	0.00
A big data analytics based approach to anomaly detection	0	23	0.00
Evaluating model drift in machine learning algorithms	0	7	0.00
Machine learning techniques for web intrusion detection—a comparison	0	23	0.00
Flow-based consensus partitions for botnet detection	0	23	0.00
Scalable malware classification with multifaceted content features and threat intelligence	0	18	0.00
Analyzing and assessing the security-related defects	0	19	0.00
A new cybersecurity alert system for Twitter	0	14	0.00
An architecture for semi-automatic collaborative malware analysis for CIs	0	21	0.00
Classification of cyber-attacks based on rough set theory	0	11	0.00
Evolving meta-ensemble of classifiers for handling incomplete and unbalanced datasets in the cybersecurity domain	0	17	0.00
Clustering data with the presence of missing values by ensemble approach	0	14	0.00
Cybersecurity of cyber-physical systems: cyber threats and defence of critical infrastructures	0	14	0.00
Cognitive risk framework for cybersecurity: bounded rationality	0	22	0.00
HAMIDS: hierarchical monitoring intrusion detection system for industrial control systems	0	9	0.00
Risk analysis in cyber situation awareness using Bayesian approach	0	6	0.00
Beyond data: contextual information fusion for cybersecurity analytics	0	15	0.00
Unknown malware detection using network traffic classification	0	11	0.00
Detection of tunnels in PCAP data by random forests	0	15	0.00
Development of cyber situation awareness model	0	6	0.00
Modeling cybersecurity for software-defined networks those grow strong when exposed to threats	0	10	0.00
Understanding taxonomy of cyber risks for cybersecurity insurance of financial industry in cloud computing	0	19	0.00
Ensemble based approach to increase vulnerability assessment and penetration testing accuracy	0	19	0.00
Balancing security and performance for agility in dynamic threat environments	0	20	0.00
Malware task identification: a data-driven approach	0	7	0.00

(continued)

Table 4 (continued)

Title	C	PM	E
A data-driven approach to distinguish cyber-attacks from physical faults in a smart grid	0	9	0.00
Machine-assisted cyber threat analysis using conceptual knowledge discovery	0	23	0.00
Attribution in cyberspace: techniques and legal implications	0	15	0.00
Fuzzy association rule mining using binary particle swarm optimization: Application to cyber fraud analytics	0	14	0.00
Guest editorial on advances in tools and techniques for enabling cyber–physical–social systems—Part I	0	8	0.00
Utilizing network science and honey nets for software induced cyber incident analysis	0	2	0.00
Capture the flag as cybersecurity introduction	0	14	0.00
Data classification and sensitivity estimation for critical asset discovery	0	18	0.00
Real-time situational awareness for critical infrastructure protection	0	14	0.00
Recent advances in learning theory	0	0	0.00
High-speed security analytics powered by in-memory machine learning engine	0	6	0.00
Designing a context-aware cyber-physical system for detecting security threats in motor vehicles	0	8	0.00
"Do You Want to Install an Update of This Application?" a rigorous analysis of updated android applications	0	12	0.00
A classification framework for distinct cyber-attacks based on occurrence patterns	0	8	0.00
Detecting cyber-attacks on wireless mobile networks using multi criterion fuzzy classifier with genetic attribute selection	0	0	0.00
ABDF integratable machine learning algorithms-map reduce implementation	0	7	0.00
Data-driven data center network security	0	14	0.00
Game theory with learning for cybersecurity monitoring	0	14	0.00
Study on implementation of machine learning methods combination for improving attacks detection accuracy on Intrusion Detection System (IDS)	0	12	0.00
Artificial intelligence application for improving cybersecurity acquirement	0	5	0.00
Adapting level of detail in user interfaces for cybersecurity operations	0	18	0.00
Comprehensive literature review on machine learning structures for web spam classification	0	10	0.00
DARPA neurocomputing	0	13	0.00
Fraud prevention framework for electronic business environments: automatic segregation of online phishing attempts	0	21	0.00
Talking about online safety: a qualitative study exploring the cybersecurity learning process of online labor market workers	0	20	0.00
Robust adversarial learning and invariant measures	0	10	0.00
Denial of service lab for experiential cybersecurity learning in primarily undergraduate institutions	0	23	0.00
Future scenarios and challenges for security and privacy	0	22	0.00

(continued)

Table 4 (continued)

Title	C	PM	E
A distributed laboratory architecture for game based learning in cybersecurity and critical infrastructures	0	14	0.00
AD2: Anomaly detection on active directory log data for insider threat monitoring	0	12	0.00
Cybersecurity risks for minors: a taxonomy and a software architecture	0	22	0.00
An autonomous resiliency toolkit—needs, challenges, and concepts for next generation cyber defence platforms	0	23	0.00
Detection of distributed denial of service attacks in software-defined networks	0	22	0.00

3 Conclusion

In this paper, we analyze various bibliometric data of recent papers on machine learning for cybersecurity. It provides justification and suggestion of the most recent findings for researchers, educators and authors. We conclude that most trending objectives from our dataset are detection of malicious activity and defense against adversary from January 2015 to December 2016.

We also concluded that the top countries involved in the research in the field of Machine learning for cybersecurity are *UK and Germany* across the globe. The publisher having maximum research papers published is *IEEE* followed by *ACM* followed by other publishers. Along with the maximum number of research papers published, *IEEE* is having maximum number of citations among any other publisher. Survey is having the maximum citations. Looking forward to refinement of various performance metrics in machine learning for cybersecurity, the field shows a good potential for further research in this area.

References

1. U. Ahmed, I. Raza, S. Asad, H. Amjad, M. Iqbal, and X. Wang, Modelling cyber security for software-defined networks those grow strong when exposed to threats Analysis and propositions," *J. Reliab. Intell. Environ.*, vol. 1, no. 2, pp. 123–146, 2015.
2. Uddin, Shahadat, et al. "Trend and efficiency analysis of co-authorship network," Scientometrics 90.2 (2011): 687–699.
3. A. Dhammi, "Behavior analysis of malware using machine learning," *Contemp. Comput. (IC3), 2015 Eighth Int. Conf. on, IEEE*, 2015.
4. I. B. Analysis, H. Alipour, Y. B. Al-nashif, P. Satam, and S. Hariri, "Wireless Anomaly Detection based on," vol. 6013, no. c, 2015.
5. M. Iannacone, S. Bohn, G. Nakamura, J. Gerth, R. Bridges, E. Ferragut, and J. Goodall, "Developing an Ontology for Cyber Security Knowledge Graphs *," *10th Annu. Cyber Inf. Secur. Res. Conf.*, pp. 4–7, 2015.
6. S. Uddin, L. Hossain, A. Abbasi, and K. Rasmussen, "Trend and efficiency analysis of co-authorship network," no. September 2011, pp. 687–699, 2012.

7. I. Hydara, A. B. Sultan, H. Zulzalil, and N. Admodisastro, "Current state of research on cross-site scripting (XSS) – A systematic literature review," *Inf. Softw. Technol.*, vol. 58, pp. 170–186, 2015.
8. R. Gao, L. Wang, R. Teti, D. Dornfeld, S. Kumara, M. Mori, and M. Helu, "Cloud-enabled prognosis for manufacturing," *CIRP Ann. - Manuf. Technol.*, 2015.

A Comparative Study of Various Routing Technique for Wireless Sensor Network with Sink and Node Mobility

Tejashri Sawant and Sumedha Sirsikar

Abstract Sensor network is made up of a large number of sensor nodes for communication. Some nodes can work as a gateway which transmits data to the sink. These nodes are considered as a bottleneck in multi-hop networks because they send data collected from other nodes, and hence it drains their energy quickly. This problem is known as Hot Spot problem. So, for balancing the energy throughout the network, sink and node mobility can be used. A mobile sink could collect data without the need of gateway, and hence the problem of a bottleneck in the multi-hop network can be minimized. Along with the sink, sensor nodes also can be made mobile. Energy efficiency of WSN is achieved by using mobile sensors and mobile sink with enhanced coverage, and better channel capacity. There are various routing protocols to reduce energy consumption in a mobile WSN. This review paper mainly concentrates on the study of different protocols that are used for efficient energy consumption using the mobile sink and mobile sensor nodes.

Keywords Node mobility · Mobile sink · Energy efficiency

1 Introduction

In recent years, various applications like health monitoring [1], environment monitoring, object tracking, military area, alarm system [2], pollution detection [3] and smart-city [4] uses WSN for communication and computation. WSN contains a huge number of sensor nodes and sink or base station for collection of data [5]. For most of the applications sensor nodes and sink remain static for performing monitoring function in the region of interest. Sensor nodes sense data in the network and

T. Sawant (✉) · S. Sirsikar
Department of Information Technology,
Maharashtra Institute of Technology, Pune, India
e-mail: tejashrisawant1@gmail.com

S. Sirsikar
e-mail: sumedha.sirsikar@mitpune.edu.in

© Springer Nature Singapore Pte. Ltd. 2018   227
Y.-C. Hu et al. (eds.), *Intelligent Communication and Computational Technologies*, Lecture Notes in Networks and Systems 19,
https://doi.org/10.1007/978-981-10-5523-2_21

forward it to sink over single-hop or a multi-hop manner [6]. The sink collects data from the entire network and sends it for further processing. If the position of a sink is static, a sensor node forwards the data in a multi-hop manner. In this case the nodes near to the sink forward their data as well as the data from other nodes in the network and require more energy as compared to other nodes in the sensor network. So, nodes near to sink die early and cause network partitioning problem known as 'Hot Spot problem' [7, 8] or 'energy hole problem.' The energy hole problem can be resolved by making the sink node mobile [9].

WSN with mobile sink decreases the possibility of network partitioning. In mobile sink network [10], by moving around the network, the sink gathers data from sensor nodes. It not only minimizes the Hot Spot problem but also does load balancing, and better handling to the disconnected network. According to the requirements of application the sensor nodes and the sink can be static or mobile. Capability and flexibility of WSN are enhanced by using the node mobility, which supports multiple tasks and handles the existing problems.

There are various advantages [11] of mobile WSN such as better energy efficiency, better coverage, enhances target tracking and superior channel capacity. By using the node mobility, the total energy required for data transmission is reduced which results in reducing the delay [12] and prolong to network lifetime. Also, it reduces the hop count required for transmission of data. The operation and the performance of WSN can get affected significantly by the nature of the movement of mobile nodes, its direction, speed and rate of change. Shorter data distribution path gives longer network lifetime by increasing throughput and decreasing energy consumption.

The energy consumption can be minimized by changing the structure of sensor network or the mobility pattern [13] or manipulating the data transmission approach. Key challenges in WSN are the lack of energy efficiency in the network and small network lifetime due to the failure of the sensor as they have limited energy. This paper mainly consists of a review of various approaches like LEACH-M [14], LEACH-ME [15], CBR [16], ECBR-MWSN [17], MBC [18], E2R2 [19], MBEENISH, iMBEENISH [20], etc., to provide energy efficiency, network lifetime and throughput. The mobile sink is used along with the static or mobile sensors to enhance the performance of the network lifetime.

Figure 1 shows the MWSN (Mobile Wireless Sensor Network) structure which gives the position of a mobile node at time different timeslot such as t, (t + 1), and (t + 2) are shown as A, B, and C, respectively.

The remaining paper is categorized as follows: Sect. 2 contains the literature review. Section 3 explains the proposed system model, Sect. 4 Network model and 5 gives the methodology used followed by Sect. 6 that gives an example for WSN with mobile nodes and Sect. 7 gives the summary of paper.

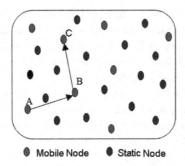

Fig. 1 Mobile wireless sensor network

2 Literature Review

There is a large number of current work and work that are on the go, for the development of routing protocol. These protocols are developed based on application needs. Also some factors are taken into consideration for development of routing protocols such as energy efficiency that affects the extension of the network lifetime. This section reviews different approaches that move sink or sensor nodes around or within the network area to collect the data from active sensor nodes based on time-driven [21] and event-driven style [21]. There are different surveys in the lifetime on routing protocol in WSN.

In CBR-M [16] technique CH allocates a time slot to each cluster member (CM) and receives the data packet from all the CMs in the network in given time slot. CH assigns a time slot to CM by TDMA. If a node enters any cluster during free time slots, then every CH acts like free cluster head one after the other [22]. TDMA scheduling is changed accordingly, by the traffic and mobility of the sensor network by the CBR Mobile WSN. In this protocol data transmission is done by the received signal strength.

In ECBR-MWSN [17] CH selection is made on the basis of parameters like lowest mobility, highest residual energy, and least distance from the base station. This protocol is an improvement in CBR-M protocol [16] regarding energy efficient CH selection. For a selection of new CHs, the BS periodically runs this. It is designed to prolong the network lifetime by balancing the energy depletion of the nodes. This protocol gives improved performance regarding throughput, higher packet delivery ratio, routing overhead, energy consumption, and delay.

The improved LEACH-M [14] protocol, i.e., LEACH-ME is suitable for mobile WSN. In this protocol, CH selection is on the basis of mobility of node. The idea of this protocol is to make sure that the CHs are from the group of mobile nodes with minimum node mobility. Or the CHs are in a group with other mobile CMs. The extra time slot is required by this protocol. This time slot is in TDMA schedule, required for mobility calculation based on a how many numbers of time a node move from one cluster to another. The LEACH-ME protocol supports mobility and reduces the packet loss compare to LEACH and LEACH-M protocol.

Energy-efficient and reliable routing protocol is a novel energy efficient routing protocol. E2R2 [19, 23] protocol structure is hierarchical and cluster based [22]. It is suitable for data delivery at Base Station (BS). It considers mobility at sensor node and BS. The primary objective of the E2R2 protocol is to achieve energy-efficient network and provide connectivity to the nodes. The routing decision is carried out for mobility of the nodes. The objective is to move data packet through suitable route in spite of node mobility. The objective of this protocol is to enhance the throughput at the time of high data rate. It gives a better output in the case of throughput and lifetime of the network.

In mobility-based clustering (MBC) [18] protocol for WSN with mobile nodes, by residual energy and mobility, a sensor node selects itself as a CH. CHs allocates timeslots to each non-CH node based on TDMA schedule for data transmission. Timeslot for each node is in increasing order of TDMA schedule based on the evaluated connection time. In the respective timeslot, a sensor node transfers its sensed data and broadcasts a joint request message. By using this request message, a node can join a new cluster. There is the possibility of packet loss when the connection with CHs is lost or is going to lose. This packet loss can be avoided by using the join request message. This protocol outperforms both the CBR [16] protocol and the LEACH-M [14] protocol in respect of average energy consumption and control overhead. This protocol can perform better with high node mobility environment.

In this paper, protocols for heterogeneous WSNs like BEENISH, improved BEENISH iBEENISH, MBEENISH, and iMBEENISH are given. BEENISH protocol considers four energy levels of nodes for CHs selection. Also, it considers levels of residual energy for nodes and average energy level of the network. In the iBEENISH protocol, the probability of CHs selection varies dynamically in an efficient manner. Through this CHs selection method network lifetime is increased compared to previous protocol. By using a mobile sink in BEENISH (MBEENISH)

Table 1 Comparison of the clustering protocols for heterogeneous wireless sensor networks

Routing technique	Mobile element	Network stability	Throughput	Protocol objective	Energy efficiency
E2R2	Sensor nodes and sink	Moderate	Moderate	Energy efficiency and connectivity of nodes	Good
LEACH-M	Sensor nodes	Good	Low	Maximizing lifetime support mobility	Low
LEACH-ME	Sensor nodes	Good	Good	Improve packet delivery rate	Good
CBR-M	Sensor nodes	Good	Good	Improve packet delivery rate	Moderate
ECBR-MWSN	Sensor nodes	Moderate	Moderate	Extend network lifetime	Good
MBC	Sensor nodes	Good	Moderate	Improve packet delivery rate	Moderate

and iBEENISH (iMBEENISH), the resulting protocols outperform previous protocols in terms of throughput, stability period, and a lifetime of the network (Table 1).

3 Proposed System Model

Some of the existing routing protocols such as LEACH-M, E2R2, CBR, etc., use static nodes and mobile sink with static sensor nodes. These protocols are energy-efficient, but there are some problems like energy hole problem in static WSN when only sink considered as mobile it increases the load on the sink as well as increases its energy consumption. So, to improve these factors in proposed work some mobile nodes are used with a mobile sink. Use of these mobile nodes decreases the load on the sink and increases energy efficiency. Also proposed work increases network lifetime.

4 Network Model

In this paper, a WSN is considered that contains static and mobile sensor nodes with the mobile sink. Static sensors are enormously located in the network to sense data from the environment and forward it to CH. Mobile nodes and mobile sink are used to collect the data from these CH nodes. Each sensor has adequate memory to store data. On the other hand, mobile sink and mobile nodes are comparatively high-priced and can periodically be recharged (Fig. 2).

5 Mathematical Model

A mathematical model for proposed work is described as follows:

$$S = \{\text{set of all static nodes in the network}\}$$
$$S = \{s1, s2, s3, \ldots\}$$

All nodes in this set are static which are used for cluster formation in the network.

$$CH = Cluster\ Head$$

CH monitors the cluster's functionality.

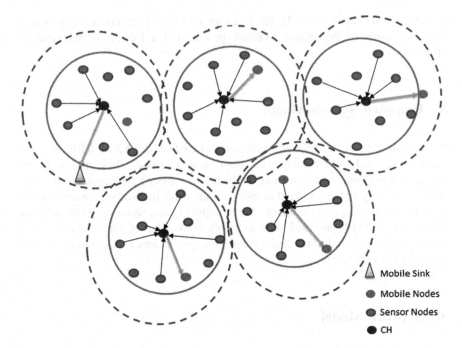

Fig. 2 Network model

1. Energy Consumption

$$Ea = (I - R)/S$$

Ea   Average energy consumption
I   Initial energy of each sensor node
R   Residual energy of each sensor node
(I − R)   Total energy consumption

2. Network Lifetime

The time interval gives network lifetime from the start of operation of the sensor network to the death of the last alive sensor node in the network.

6 Methodology

As in existing protocol, the time slot assigned to the cluster members by CH and sink assign time slot to the CH in the network. In the particular time slot, all nodes are kept to be active and send data to CH. And mobile sink travels to collect data from respective CHs. There are different approaches for sink traveling pattern such as PLMA [24], ELMA [24], MESS [25], MSRP, NADC, etc. Sink have to travel more locations for data collection. Due to which energy consumption by sink increases as well as load on sink increases. This energy consumption will affect the network lifetime. So the proposed work can help to increase the network lifetime and energy efficiency.

By using this work, the load on the mobile sink can be decreased so the energy consumption by sink can get reduced, which will decrease the total energy consumption of the network (Fig. 3).

Initially, all the sensor nodes are with the equal amount of energy, the energy level of a node may be left with unequal after some time of operation. In the system, the sink and some sensor nodes are taken as mobile. The nodes are randomly placed in the network. CH selection is made by node degree, node distance and residual energy of nodes [26]. Then cluster formation is done. The mobile sink and mobile nodes assign a time slot to CHs and CHs assign a time slot to cluster nodes [27]. In the data collection phase, the CH collects data from cluster nodes, aggregate that data to remove the redundancies from data. Select location for the mobile sink and mobile nodes, for the collection of data. Mobile sink and mobile nodes go to respective CH location and collect data from CHs.

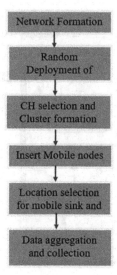

Fig. 3 System framework

7 Example for WSN with Node Mobility

Figure 4 describes a remote healthcare and smart home network using Internet of Things (IoT). In this applications, different sensors are used which can be static or mobile. By using the mobility of nodes, we can reduce the required number of nodes. For remote healthcare [28, 29], application network must be compatible with mobile nodes. Some biosensor nodes are taken as mobile, due to node mobility doctors can remotely access the particular patient. Also, it can provide many advantages regarding financial benefits supervised, many long-term chronic diseases and improved quality of life for patients.

Divide the home network into some clusters. Each cluster contains Cluster Head (CH), which performs data aggregation after collecting all the raw data from its members. Different sensors are used for the smart home [30, 31] network like cameras, biosensors, RFID, etc. Cameras are installed at various locations such as in the main room and dining room; as the camera does not suit in the living room. Biosensors are attached to the human body to collect physical information like blood pressure, senses the blood oxygen saturation, body temperature, the heart rate and breathing, etc. Some nodes also can install on the wall, chair, dining table which are static or some nodes are attached to the body which is mobile. These nodes do a collection of raw data from various sensor nodes. There are some devices like a cell phone that are used as bridge or gateway server for data transformation between both inside and outside devices through a wired or wireless communication. Such that it will receive commands from users and deliver requests to the certain sensor or sink nodes.

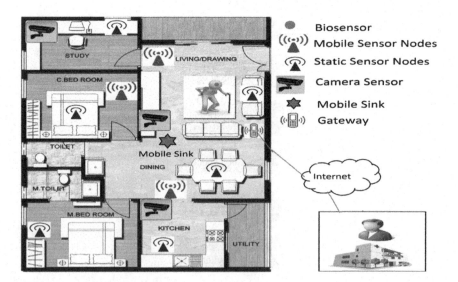

Fig. 4 Smart home network with IoT

8 Conclusion

In this chapter, a wide literature review of various algorithms that achieve node mobility and thus improve the lifetime of the network is studied. The proposed system uses mobile nodes with a mobile sink. By using which the energy consumption by the node is minimized. Also use of mobile nodes is beneficial for reducing data collection load on the mobile sink. The advantage of using node mobility in the network is for reducing the hop count for data transmission. Minimization of the hop count directly affects the network lifetime. This makes the protocol energy-efficient regarding network lifetime than the existing protocols. According to the requirements of applications nodes are mobile or static in the network. Some applications of mobile network are military, smart city, pollution control, remote healthcare, etc. Here the examples of remote healthcare and smart home using IoT are discussed.

References

1. Wan Aida Nadia binti Wan Abdullah, N. Yaakob, R Badlishah, A Amir, Siti Asilah binti Yah, "On the effectiveness of Congestion Control Mechanisms for Remote Healthcare Monitoring System in IoT Environment – A Review", 2016 3rd International Conference on Electronic Design (ICED), August 11–12, 2016, Phuket, Thailand
2. Suwarna Latambale, Sumedha Sirsikar, "A Survey of various Sink Mobility based Techniques in Wireless Sensor Network", Indore, India © 2016 ACM. ISBN 978-1-4503-4278-0/16/03
3. Suganya E, Vijayashaarathi S, "Smart Vehicle Monitoring System for Air Pollution Detection using Wsn", International Conference on Communication and Signal Processing, April 6–8, 2016, India
4. Jin Wang, Yue Yin, Jianwei Zhang, Sungyoung Lee, and R. Simon Sherratt, "Mobility based Energy Efficient and Multi-Sink Algorithms for Consumer Home Networks", 0098 3063/13/ $20.00 © 2013 IEEE
5. Sulata Mitra, Anup Roy, "Communication Void Free Routing Protocol in Wireless Sensor Network", Wireless Personal Communications, February 2015
6. Junhu Zhang, Zhenhua Sun, "Assessing Multi-Hop Performance of Reactive Routing Protocols in Wireless Sensor Networks", 2016 8th IEEE International Conference on Communication Software and Networks
7. Majid I. Khan, Wilfried N. Gansterer, Guenter Haring, 'Static vs. mobile sink: The influence of basic parameters on energy efficiency in wireless sensor networks', IEEE, 2013
8. Chenglin Zhao, Yongxing Wang, Xuebin Sun and Ting Jiang, "An Energy Saving Cluster Routing for Wireless Sensor Network with Mobile Sink," International Conference on Advanced Intelligence and Awareness Internet, pp 113–117, 2010
9. Majid I. Khan, Wilfried N. Gansterer and Guenter Haring, "Static vs. mobile sink: The influence of basic parameters on energy efficiency in wireless sensor networks," Elsevier Computer Communications, November 2012
10. Yu Gu, Member, Fuji Ren Senior Member, Yusheng Ji, and Jie Li, "The Evolution of Sink Mobility Management in Wireless Sensor Networks: A Survey", IEEE COMMUNICATIONS SURVEYS AND TUTORIALS, 2015

11. Divya Bharti, Manjeet Behniwal, Ajay Kumar Sharma, "Performance Analysis and Mobility Management in Wireless Sensor Network," Volume 3, Issue 7, July 2013 ISSN: 2277 128X © 2013, IJARCSSE
12. Md. Sharif Hossen, Muhammad Sajjadur Rahim, "Impact of Mobile Nodes for Few Mobility Models on Delay-Tolerant Network Routing Protocols" 978-1-5090-0203-0/16/$31.00 ©2016 IEEE
13. Ioannis Chatzigiannakis, Athanasios Kinalis, Sotiris Nikoletseas, 'Efficient data propagation strategies in wireless sensor networks using a single mobile sink,' Computer Communication ELSEVIER, 2007
14. Renugadevi G., Sumithra M.G., "An Analysis on LEACH-Mobile Protocol for Mobile Wireless Sensor Networks," International Journal of Computer Applications (0975 – 8887) Volume 65– No. 21, March 2013
15. G. Santhosh Kumar, Vinu Paul M V, G. Athithan3 and K Poulose Jacob, "Routing Protocol Enhancement for handling Node Mobility in Wireless Sensor Networks," TENCON 2008 - 2008 IEEE Region 10 Conference
16. S.A. B. Awwad, Chee K. Ng, Nor K. Noordin, and Mohd. Fadlee A. Rasid, "Cluster Based Routing Protocol for Mobile Nodes in Wireless Sensor Network," Int. Symp. On Collaborative Technologies and Systems, pp. 233–241, ©2009 IEEE
17. R.U. Anitha, Dr. P. Kamalakkannan, "Enhanced Cluster Based Routing Protocol for Mobile Nodes in Wireless Sensor Network," 978-1-4673-5845-3/13/$31.00©2013 IEEEI.S.
18. S. Deng, J. Li, L. Shen, "Mobility-based clustering protocol for wireless sensor networks with mobile nodes," IET Wirel. Sens. Syst., 2011, Vol. 1, Iss. 1, pp. 39–47 doi:10.1049/iet-wss. 2010.0084.
19. Hiren Kumar Deva Sarma, Rajib Mall, Avijit Kar, "E2R2: Energy-Efficient and Reliable Routing for Mobile Wireless Sensor Networks", 1932-8184 © 2015 IEEE.
20. Mariam Akbar, Nadeem Javaid, Muhammad Imran, Naeem Amjad, Majid Iqbal Khan and Mohsen Guizani, "Sink mobility aware energy-efficient network integrated super heterogeneous protocol for WSNs", 2016 Springer
21. Tariq A. A. Alsbouí, Mohammad Hammoudeh, Zuhair Bandar, Andy Nisbet, "An Overview and Classification of Approaches to Information Extraction in Wireless Sensor Networks", IARIA, 2011. ISBN: 978-1-61208-144-1, SENSORCOMM 2011
22. Jaideep Lakhotia, Rajeev Kumar, "cluster based routing protocols for mobile wireless sensor network: a review," (IJARCET) Volume 3 Issue 7, July 2014
23. Hiran Kumar Deva Sarma, Avijit Kar, Rajib Mall, "Energy Efficient and Reliable Routing for Mobile Wireless Sensor Networks." IEEE 2010, pp. 1–6
24. Metin Koc, Ibrahim Korpeoglu, "Traffic and Energy Load Based Sink Mobility Algorithms for Wireless Sensor Networks", International Journal Sensor Networks, 2015
25. Ying C, Lau V.K.N., Wang R, Huang H, Zhang S., "A Survey on Delay-Aware Resource Control for Wireless System Large Deviation Theory, Stochastic Lyapunov Drift, and Distributed Stochastic Learning", IEEE Transaction on Information Theory, 2012
26. Abhishek Chunawale, Sumedha Sirsikar, "Formation of Clusters Using Residual Energy and Node Distance in Wireless Sensor Networks", Cyber Times International Journal of Technology and Management, Vol. 7 Issue 1, October 2013–March 2014
27. Adelcio Biazi, César Marcon, Fauzi Shubeita, Letícia Poehls, Thais Webber, Fabian Vargas, "A Dynamic TDMA-Based Sleep Scheduling to Minimize WSN Energy Consumption", 2016 IEEE
28. Lin Yang, Yanhong Ge, Wenfeng Li, Wenbi Rao, Weiming Shen, "A Home Mobile Healthcare System for Wheelchair Users", 2014 IEEE
29. Sendra S, Granell E, Lloret J, "Smart Collaborative Mobile System for Taking Care of Disabled and Elderly People," Mobile Networks and Applications, 2013, pp. 1–16
30. Freddy K Santoso, and Nicholas C H Vun, "Securing IoT for Smart Home System", 2015 IEEE International Symposium on Consumer Electronics (ISCE)
31. Dominik Kovac, Jiri Hosek, Pavel Masek, and Martin Stusek, "Keeping Eyes on your Home: Open-source Network Monitoring Center for Mobile Devices", 2015 IEEE

Review of Hierarchical Routing Protocols for Wireless Sensor Networks

Misbahul Haque, Tauseef Ahmad and Mohd. Imran

Abstract Recent few years has seen a tremendous advancement in wireless communication technology. This advancement has opened a door for researchers to work in the area of wireless sensor networks (WSNs) for use in a broad array of real-life applications. An enormous number of remotely deployed autonomous sensors gather data from their vicinity and communicate it to the base station after processing. The sensors communicate through some wireless strategies governed by routing protocols, which has a great impact on the performance of sensor networks. With this insight, we extensively surveyed routing protocols for WSNs. The network structure leads to the broad classification of WSNs' protocols in three foremost classes: flat, hierarchical, and location based routing. Cluster-based routing provides certain advantages over others like scalability, increased network lifetime and efficient data aggregation. In this work, we study and provide a detailed survey of famous hierarchical routing protocols, a taxonomy of hierarchical routing protocols along with the design challenges and also present a comparative analysis based on their traits and limitations.

Keywords Wireless sensor networks · Sensor · Hierarchical routing protocols

M. Haque (✉) · T. Ahmad · Mohd. Imran
Department of Computer Engineering, ZHCET,
Aligarh Muslim University, Aligarh, UP, India
e-mail: misbahul.haque@zhcet.ac.in

T. Ahmad
e-mail: tauseefahmad@zhcet.ac.in

Mohd. Imran
e-mail: mimran.ce@amu.ac.in

© Springer Nature Singapore Pte. Ltd. 2018
Y.-C. Hu et al. (eds.), *Intelligent Communication and Computational
Technologies*, Lecture Notes in Networks and Systems 19,
https://doi.org/10.1007/978-981-10-5523-2_22

1 Introduction

Wireless sensor networks (WSNs) is an infrastructure-less network that consists of at least one base station (BS) and a plenteous amount of sovereign sensors having capabilities like data processing and communication, disseminated to cover a large geographical region. These BS receives the data from these disseminated sensor nodes. Because of wireless nature, WSNs are easily deployable. The evolution of WSNs was initially stimulated by armed forces for supervision in conflict zones, tracking militancy, tracking opponents movement [1], but now the application is extended to medical and health, industrial infrastructure, calamity management, habitat monitoring etc. [2], thereby connecting the three distinct world, i.e., the physical realm, the computing world, and the human society. A typical wireless sensor network's sensor nodes are composed of a radio transceiver having an antenna, a microcontroller, a sensor interfacing electronic circuit and a power supply, usually a battery. The capability of a single sensor node is limited which makes them inadequate in congregating valuable information from a particular domain [3]. To accomplish the data congregation process, hundreds or even thousands of sensors are deployed to work collectively. WSNs do not have need of a central organization and are self-configurable. Since WSNs mostly have a dense deployment of sensor nodes, this provides the ability to withstand harsh environmental conditions without network failure [4].

WSNs catch their application in very diverse areas. Having a look at the projects involving WSNs around the globe, we can broadly identify different types of WSNs [5], like structure sensor network (SSN), transport and logistics sensor network (TSN), body sensor network (BSN), environmental sensor network (ESN) and participatory sensor network (PSN).

The sensor nodes in wireless sensor networks perform the task of collecting raw data from deployed region along with data storage, some local data processing and routing [6]. The processed information is then passed to the intended base station. In doing so, the sensor nodes consume energy. Mostly, sensors are powered by small batteries which cannot be replaced or even recharged because of the deployment of sensors in unattended environments. So, energy constraints must be taken into account in WSNs design goal. A large number of sensors and the energy constraints provoke for some energy-aware routing algorithms and data gathering protocol which can offer an extended lifetime of sensors and scalability [7]. To achieve high energy efficiency, extended lifetime and scalability objective, the research community has widely adopted the idea of grouping sensor nodes into clusters in large-scale WSN environments.

The hierarchical network structure has a two-level hierarchy. Each cluster in the network elects a particular sensor node as the cluster head (CH) for coordinating the data gathering and aggregation process in the cluster. The CHs nodes form the upper level and all other nodes in the cluster constitute the subsequent level. The CH node accumulates data from sensor nodes within its neighborhood and passes them after processing and aggregation to the base station through other

Fig. 1 Hierarchical network
architecture of a WSN

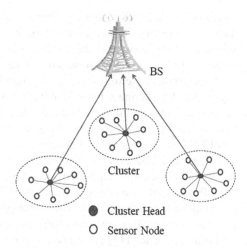

intermediate CH nodes or directly. The energy drain rate for CH nodes is higher than the ordinary sensor nodes because CH nodes transmit data over a long range (CH to base station) while other nodes communicate only with CH nodes within their cluster. Figure 1 shows typical hierarchical network architecture for a wireless sensor network. The figure shows a number of sensor nodes organized in different clusters and having a particular node as their respective cluster head (CH) which gather data from other plane nodes within the cluster and transmit it to the sink or base station after necessary processing. In order to optimize the energy consumption, one can switch the CH responsibility among other sensor nodes in the cluster by periodically re-electing new CHs in the cluster [7]. Clustering provides certain advantages such as reduction in communication overhead, eradication of data redundancy and increased efficiency of data transmission with drawbacks like overheads in cluster formation and election of CHs [6].

In this chapter, we have provided a broad survey of some existing hierarchical routing protocols proposed recently. Based on certain metrics, we also plan to compare the performances of these protocols.

The rest of the paper is organized as follows: Sect. 2 covers a brief description of research that has been carried out in this area. The overview and classification of WSN hierarchical routing protocols along with the design challenges that must be taken care are discussed in Sect. 3. Section 4 presents the comparison of performances based on certain attributes. The last section concludes the paper.

2 Related Work

Several research studies have been done in the context of classifying and comparing the routing protocols for wireless sensor networks. These comparisons give an idea about their behavior and effectiveness. Deosarkar et al. [8] presented a detailed discussion of different clustering schemes emphasizing mainly on the taxonomy of adaptive, deterministic and combine metric scheme based cluster head selection strategies. They compared the CH selection cost with that of cluster formation, creation of clusters, and distribution of CHs.

A survey on clustering algorithms by Jiang et al. [9] was presented by giving the taxonomy of clustering schemes for WSN based on certain clustering attributes and discussing some prominent advantages like less overhead, easy maintenance and more scalability for WSNs. Clustering algorithms like LEACH, HEED, PEGASIS, and EEUC were analyzed and compared.

Abbasi and Younis [10] surveyed the present clustering algorithms and give a taxonomy of clustering algorithms. They presented a summary of WSN clustering algorithms based on convergence time, highlighting their features and complexity. Based on certain metrics like cluster overlapping, stability, rate of convergence and mobility support, they compared these clustering approaches.

Yadav and Rana [11] presented a survey on cluster based routing strategies in WSNs suggesting a taxonomy of the clustering protocols. They have discussed in detail the merits and limitations of various cluster based protocols like GAF, SLGC, HGMR, TSC, PEGASIS, HCTE, BCDCP, MWBLA, and LEACH-VF and also compared these protocols for their performances based on certain performance attributes like load balancing, algorithm complexity, delivery delay, etc., thereby concluding that cluster based routing strategies are much more efficient than other schemes in performance enhancement of WSNs.

A general classification of various cluster-based protocols for WSNs based on CH selection and cluster formation parameters is given by Kumarawadu et al. [12]. They have discussed some design challenges and performance issues of proba-bilistic cluster based, neighborhood information based, biologically inspired clus-tering and identity-based clustering algorithms.

Maimour et al. [13] discussed clustering routing protocols from the perspective of achieving energy efficiency and presented a review from data routing perspec-tive, proposing a simple categorization of routing protocols for clustering in WSNs. Pre-established and on-demand clustering routing protocols are discussed along with nine other clustering protocols.

Wei et al. [14] presented a review of state-of-the-art routing methods for wireless sensor networks outlining the clustering architecture. Based on attributes like the hop count between CH and nodes, parameters for CH selection and the existence of centralized control during cluster formation, they have given a simple classification of clustering routing protocols. Some design challenges were also discussed.

A comparative analysis to improve the network lifetime for certain WSN clus-tering routing algorithms is presented by Haneef and Zhongliang [15] along with the design challenges that comes into the way and affect the design of WSN's routing protocols. The authors presented a taxonomy of routing protocols and a comparative analysis of many efficient clustering based routing protocols is given.

Lotf et al. [16] surveyed some clustering protocols and discuss their operations along with their advantages and limitations. The authors compared clustering algorithms like EECS, TEEN, APTEEN, and LEACH on the basis of network lifetime and energy consumptions.

A brief introduction of design goals of clustering and overview of operations of proposed clustering algorithms are given by Dechene et al. [17]. The authors have examined the performances of heuristic, weighted, hierarchical, and grid-based clustering algorithms from two aspects: power, energy and network lifetime; and quality and reliability of links.

A simple survey by Xu and Gao [18] of clustering routing protocols is presented. The authors have described only six clustering algorithms. Comparison of these routing algorithms has been done based on certain performance factors like data aggregation, robustness, network lifespan, energy conservation, scalability, and security.

A detailed survey of chain-based routing protocols has been presented by Marhoon et al. [19] highlighting the characteristics of the chain based routing and discussing the advantages and disadvantages of these protocols over other classes of protocols by explaining the functioning of many of the chain-based routing protocols. They have discussed the protocols like PEGASIS, CRBCC, REC+, BCBRP, RPB, and PCCR.

3   Overview of Hierarchical Routing Protocols

In WSN, the responsibility of routing implementation for the data coming to the network is done by the network layer. In single-hop networks, the source node can directly communicate with the sink but this is not the case in multi-hop networks. In multi-hop networks, the data packets transmitted by source node are relayed by the intermediate hops in the network so that the data packets can reach the sink. In all these scenarios, routing tables has to be maintained for smooth operation and are governed by some routing protocols. Network structure, communication initiator, routing path establishment, protocol operation and selection of next hop are some criteria to classify the routing strategies in WSNs. Further, the routing protocols based on wireless network architecture can be categorized into three subcategories namely flat, hierarchical and location based routing protocols. The role of sensor nodes is same in flat routing in comparison to that in hierarchical routing. As routing decisions are inherently localized, the location-based routing allows the network size to be scalable without a significant increase in signaling overhead. Our main focus in this work is on network structure based hierarchical routing protocols.

The selection of nodes makes hierarchical routing energy-efficient in a way that sensing information is assigned to the nodes with low energy while data processing and transmission task are assigned to nodes with high energy. Thus, increased lifetime, scalability and energy minimization can be achieved. The hierarchical routing can also be called as cluster based routing. Block, grid, and chain cluster based routing protocols are the typical classification of hierarchical routing

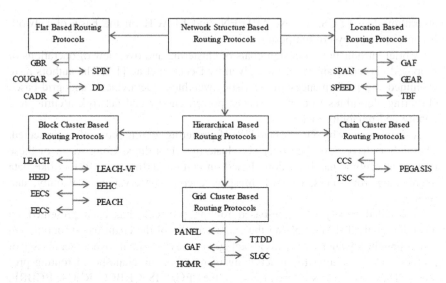

Fig. 2 Taxonomy of network structure based routing protocols

protocols [20, 21]. Figure 2 shows a taxonomy of network structure based routing protocols.

3.1 Challenges for Hierarchical/Clustering Protocols

In WSNs, clustering seems to play an important role. Clustering in WSN improves bandwidth utilization, thereby reducing the useful energy consumption and it also reduces the wasteful energy consumption as a result of reduced overhead [22]. However, besides several advantages, clustering scheme must take into account certain key limitations which are of particular importance in WSNs [23].

Network Lifetime: The energy limitations on sensor nodes greatly affect the network lifespan for sensor nodes in a wireless network. Effective clustering aids in reducing energy usage in intra-cluster and inter-cluster communication, thereby increasing network lifetime.

Limited Energy: Sensor nodes in WSNs are operated by small size battery, so their energy storage has a limit. This limited energy must be used efficiently, and must be taken into consideration as overall energy consumed in the network can greatly be reduced by applying proper clustering scheme.

Limited Capabilities: Numerous abilities of sensor nodes like processing, communication range, storage, and memory get limited by the small amount of stored energy and small physical size of sensor nodes. It is possible to make efficient use of shared resources within an organizational structure by applying good clustering algorithm, simultaneously taking the limitations of sensor nodes into account.

Cost of Clustering: Clustering plays a key role, but at some cost. Certain resources like processing tasks and communication are always required in creating and maintaining the clustering topology. Costs involved in these tasks are overhead as these resources will not be used for sensing or transmitting data.

Cluster Formation and Selection of CHs: The physical dimension of a cluster or the number of sensor nodes within a cluster may play a vital role in the functioning of a cluster for a particular application. Therefore, the designers have to examine cautiously the cluster formation in a network while designing for a particular application. These criteria also have an impact on election and re-election of cluster heads (CHs) within the cluster.

Scalability: In WSN, the coverage range of the nodes is limited. This leads to the deployment of thousands of sensor nodes where a relatively larger area has to be covered. Therefore, the routing protocols in such scenarios must be capable of handling a vast amount of sensor nodes. In a network with a massive number of nodes, it is not possible to preserve the global information of network topology for every node in the network.

Data Aggregation: The larger wireless sensor networks are often densely populated. In such scenario, there is always a possibility that multiple nodes sense similar information. Therefore, there must be some mechanism which can eradicate data duplication. Data aggregation is a technique which differentiates useful data from sensed data. Data aggregation capabilities are being provided by many clustering algorithms. So, while selecting a clustering approach, the requirement for data aggregation must be carefully considered.

Synchronization: Limited energy capacity of sensor nodes has an adverse impact on the performance of wireless sensor networks. Energy usage can be minimized by allowing sensor nodes to repeatedly schedule sleep interludes through particular slotted transmission scheme such a TDMA. To have a proper setup and maintainable transmission schedule, such schemes require certain synchronization mechanisms. Thus, synchronization and scheduling will have a great impact on overall performance of WSNs while considering a clustering scheme.

Secure Communication: In hierarchical routing protocols, communication takes place within the cluster as well as outside world. An energy efficient and secure inter-cluster as well as intra-cluster communication is one of the most important challenges for clustering protocol design.

Repair Mechanisms: Because of the absence of static structure, the wireless sensor networks are frequently prone to node movement, delay, interference and node demise. A link failure can occur as a result of these situations. Therefore, while looking for clustering schemes, link recovery and reliable data communication mechanism must be considered.

Quality of Service: Quality of service (QoS) requirements in WSNs is very important aspect from an overall network standpoint. These services are prompted by the functionalities and applications of the network. Some application-dependent QoS requirements are packet loss tolerance, acceptable delay, and precision. The main focus of most existing clustering routing algorithms is to provide energy efficient network utilization rather than QoS support. The design process must consider the QoS metrics for better network performance.

4 Comparison of Hierarchical Routing Protocols

In this section, we present the comparison between some popular hierarchical routing protocols like LEACH, HEED, EECS, EEHC, LEACH-VF, PEACH, CCM, PANEL, TTDD, GAF, SLGC, HGMR, CCS, PEGASIS, and TSC in WSN based on important metrics like cluster stability, scalability, mobility, energy efficiency, data aggregation, and delivery delay in Table 1.
From this comparative study, we observe that:

- Block cluster based routing protocols provide better cluster stability than others.
- Very few (e.g., HGMR) provides very high scalability and other provides moderate to low scalability.
- Most of the protocols provide no mobility (e.g., EECS, EEHC) while few protocols provide limited mobility (e.g., HEED, CCM).
- Block cluster based protocols provide better energy efficiency than others.
- Most of the grid based protocols does not support data aggregation.
- The delivery delay varies from very low (e.g., SLGC) to very high (e.g., TTDD, PEGASIS).

Table 1 Comparison between different hierarchical routing protocols in WSNs

Protocol name	Cluster stability	Scalability	Mobility	Energy efficiency	Data aggregation	Delivery delay
LEACH	Moderate	Very low	Limited	Very low	Yes	Low
HEED	High	Moderate	Limited	Moderate	Yes	Moderate
EECS	High	Low	No	Moderate	Yes	Low
EEHC	High	Moderate	No	High	Yes	Low
LEACH-VF	High	Very low	Limited	Moderate	Yes	Low
PEACH	High	Moderate	Yes	Very high	Yes	Moderate
CCM	High	Very low	Limited	Low	Yes	Low
PANEL	Low	Low	No	Moderate	No	Moderate
TTDD	Very high	Low	Yes	Very low	No	Very high
GAF	Moderate	High	Limited	Moderate	No	Low
SLGC	Moderate	Very low	No	Moderate	No	Very low
HGMR	High	Very high	No	Low	No	Moderate
CCS	Low	Low	No	Low	No	High
PEGASIS	Low	Very low	No	Low	Yes	Very high
TSC	Moderate	Moderate	No	Moderate	Yes	Moderate

5 Conclusion

Wireless sensor networks have received much attraction in recent years and find their application in an extensively broad spectrum like environmental monitoring, security surveillance, and military applications. In WSNs what is more challenging is the design of routing protocols which can support robustness, effectiveness, and scalability. The hierarchical based routing protocols can well match the challenges and constraints of WSNs.

In this paper, we have provided an in-depth analysis of protocols for hierarchical routing used in WSNs and also established a taxonomy of network structure based routing protocols. In our work, we have focused on certain merits and limitations of some popular network structure based hierarchical routing protocols based on certain attributes and presented the result in a tabular form. The comparison analysis reflects that application of hierarchical routing to wireless sensor networks improves their performances up to a great extent. In future, the information provided in this paper can be used by researchers willing to devise their own hierarchical routing protocol.

References

1. Akyildiz, I.F., Su, W., Sankarasubramaniam, Y., Cyirci, E.: Wireless Sensor Networks: A Survey. Computer Networks, vol. 38, no. 4, pp. 393–422. (2002)
2. Liu, X.: A Survey on Clustering Routing Protocols in Wireless Sensor Networks. Sensors, vol. 12, pp. 11113–11153. (2012)
3. Patil, P., Kulkarni, U., Ayachit, N.H.: Some Issues in Clustering Algorithms for Wireless Sensor Networks. 2nd National Conference - Computing, Communication and Sensor Networks (IJCA Special Issue). (2011)
4. Ibriq, J., Mahgoub, I.: Cluster based Routing in Wireless Sensor Networks: Issues and Challenges. In: 2004 Symposium on Performance Evaluation of Computer Telecommunication Systems (SPECTS). (2004)
5. Meratnia, N., Zwaag, B.J., Dijk, H.W., Bijwaard, D.J.A., Havinga, P.J.M.: Sensor Networks in the Low Lands. Sensors, vol. 10, pp. 8504–8525. (2010)
6. Boyinbode, O., Le, H., Mbogho, A., Takizawa, M., Poliah, R.: A Survey on Clustering Algorithms for Wireless Sensor Networks. In: 13th IEEE International Conference on Network-Based Information System, pp. 358–364. (2010)
7. Mamalis, B., Gavalas, D., Konstantopoulos, C., Pantziou, G.: Clustering in Wireless Sensor Networks. RFID and Sensor Networks, pp. 323–354. (2009)
8. Deosarkar, B. P., Yada, N.S., Yadav, R.P.: Cluster Head Selection in Clustering Algorithms for Wireless Sensor Networks: A Survey. In: 2008 International Conference on Computing, Communication and Networking, pp. 1–8. Virgin Island, USA. (2008)
9. Jiang, C., Yuan, D., Zhao, Y.: Towards Clustering Algorithms in Wireless Sensor Networks: A Survey. In: IEEE Wireless Communications and Networking Conference, pp. 1–6, Budapest, Hungary. (2009)
10. Abbasi, A.A., Younis, M.: A Survey on Clustering Algorithms for Wireless Sensor Networks. Computer Communications, vol. 30, pp. 2826–2841. (2007)

11. Yadav, A. K., Rana, P.: Cluster Based Routing Schemes in Wireless Sensor Networks: A Comparative Study. International Journal of Computer Applications, vol. 125, no. 13, pp. 31–36. (2015)
12. Kumarawadu, P., Dechene, D., Luccini, M., Sauer, A.: Algorithms for Node Clustering in Wireless Sensor Networks: A Survey. In: 4th International Conference on Information and Automation for Sustainability, pp. 295–300. Colombo, Sri Lanka. (2008)
13. Maimour, M., Zeghilet, H., Lepage, F.: Cluster-Based Routing Protocols for Energy-Efficiency in Wireless Sensor Networks. Sustainable Wireless Sensor Networks, pp. 167–188. (2010)
14. Wei, C., Yang, J., Gao, J., Zhang, Z.: Cluster-Based Routing Protocols in Wireless Sensor Networks: A Survey. In: 2011 International Conference on Computer Science and Network Technology, pp. 1659–1663. Harbin, China. (2011)
15. Haneef, M., Zhongliang, D.: Design Challenges and Comparative Analysis of Cluster-Based Routing Protocols used in wireless Sensor Networks for Improving Network Lifetime. Adv. Inf. Sci. Serv. Sci., vol. 4, pp. 450–459. (2012)
16. Lotf, J., Hosseinzadeh, M., Alguliev, R.M.: Hierarchical Routing in Wireless Sensor Networks: A Survey. In: 2010 2nd International Conference on Computer Engineering and Technology, pp. 650–654. China. (2010)
17. Dechene, D.J., Jardali, A., Luccini, M., Sauer, A.: A Survey of Clustering Algorithms for Wireless Sensor Networks. Computer Communications, Butterworth-Heinemann Newton, MA USA. (2007)
18. Xu, D., Gao, J.: Comparison Study of Hierarchical Routing Protocols in Wireless Sensor Networks. Procedia Environ. Sci., vol. 10, pp. 595–600. (2011)
19. Marhoon, H. A., Mahmuddin, M., Nor, S. A.: Chain-Based Routing Protocols in Wireless Sensor Networks: A Survey. ARPN Journal of Engineering and Applied Sciences, vol. 10, no. 3, pp. 1389–1398. (2015)
20. Naeimi, S., Ghafghazi, H., Chow, C.O., Ishi, H.: A Survey on the Taxonomy of Cluster-Based Routing Protocols for Homogeneous Wireless Sensor Networks. Sensors, vol. 12, no. 6, pp. 7350–7409. (2012)
21. Tnag, F., You, I., Gou, S., Gou, M., Ma, Y.: A Chain-Cluster based Routing Algorithm for Wireless Sensor Networks. J. Int. Man. vol. 23, no. 4, pp. 1305–1313. (2010)
22. Ding, P., Holliday, J., Celik, A.: Distributed Energy-Efficient Hierarchical Clustering for Wireless Sensor Networks. In: 8th IEEE International Conference on Distributed Computing in Sensor Systems, pp. 322–339, USA. (2005)
23. Xiaoyan, M.: Study and Design on Cluster Routing Protocols of Wireless Sensor Networks. Ph.D. Thesis. Zhejiang University, China. (2006)

A 3.432 GHz Low-Power High-Gain Down-Conversion Gilbert Cell Mixer in 0.18 μm CMOS Technology for UWB Application

Gaurav Bansal and Abhay Chaturvedi

Abstract This work presents low-power- and high-gain CMOS down-conversion Gilbert cell mixer for ultrawide band application, designed in 0.18 μm CMOS technology. Inductive source degeneration is used at the RF trans-conductance stage of Gilbert cell mixer to increase its linearity. Differential LC matching is used at RF and LO stage to increase the conversion gain of mixer. Output buffer is used at the load to achieve 50 ohm impedance match to improve the return-loss of the mixer. The proposed mixer shows low reflection coefficient up to −19 dB for entire band ranges from 3.168 to 3.696 GHz frequency. The proposed mixer works at DC supply of 1.5 V with low power consumption. Simulation results show that the mixer achieves the conversion gain of 10.60 dB, 1 dB compression point of −10.596, IIP_3 is +0.056 dBm and matched RF (input) port impedance of 50.7 ohm.

Keywords Differential LC-matching network · Gilbert mixer · Down-conversion mixer · CMOS

1 Introduction

Federal communication commission introduced the ultrawide band (UWB) technology in the frequency range of 3.1–10.6 GHz. The UWB plan is shown in Fig. 1.

G. Bansal (✉) · A. Chaturvedi
Department of Electronics and Communication Engineering,
GLA University, Mathura, India
e-mail: gauravbansal076@gmail.com

A. Chaturvedi
e-mail: abhay.chaturvedi@gla.ac.in

© Springer Nature Singapore Pte. Ltd. 2018
Y.-C. Hu et al. (eds.), *Intelligent Communication and Computational Technologies*, Lecture Notes in Networks and Systems 19,
https://doi.org/10.1007/978-981-10-5523-2_23

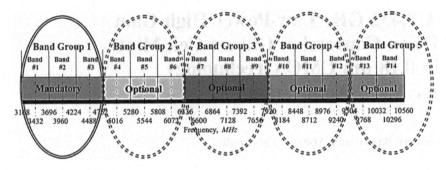

Fig. 1 Band plan for UWB technology [12]

Fig. 2 Frequency conversion
using mixer [13]

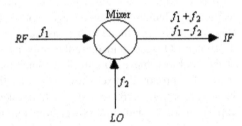

UWB technology capable of transmitting information over high bandwidth (>500 MHz). UWB technology includes high data rate, low cost, and consumes less power. This paper presents the designing of down-conversion Gilbert cell mixer for the UWB applications. RF mixer performs the frequency translation to the lower frequency signal (IF signal) after multiplying two input signal called RF and LO signals as shown in Fig. 2.

In the receiver, down-conversion mixer is used, in which IF frequency is lower than the RF frequency. Otherwise in transmitter the IF frequency is higher than the RF frequency as shown in Fig. 2. Double-balanced active mixer is preferred among all the mixers which is also called Gilbert mixer [1]. The main advantages of Gilbert cell mixer are: good port–port isolation, high linearity, and low noise figure. Gilbert mixer provides reduced second-order harmonic distortion and improved immunity to LO leakage [2]. Achieving all the parameters simultaneously at acceptable levels is not an easy task. [3]. LC differential matching is used both at LO and RF stage. Inductive source degeneration [4] is used at the RF stage transistors to increase the conversion gain and linearity of the mixer. Generally trans-conductance stage decides the linearity of the mixer [5]. Common source transistors at the load stage perform as an output buffer to attain the impedance match of 50 ohm [6].

Designing of proposed mixer is discussed in Sect. 2, simulation results are discussed in Sects. 3 and 4 discussed the conclusions.

2 Circuit Design

The basic Gilbert cell mixer [7] has three stages, differential trans-conductance stage (M5 and M6) and LO stage (M1–M4) and output signal is taken from the IF load stage. A double-balanced Gilbert cell mixer is shown in Fig. 3.

The presented mixer is shown in Fig. 4, which uses conventional Gilbert cell mixer for mixing operation. The input RF signal is provided to the transistors M5 and M6 (bottom two transistors) which convert voltage signal into the current signal. This current signal is multiplied with LO signal across M1–M4 which provide the switching function. Output IF voltages is produced by load current [8]. By adding inductive source degeneration at the RF stage transistors M5 and M6, the conversion gain and the linearity of the mixer are improved.

The multiplication of two input signal is expressed as

$$a = A \sin(w_1 t + \phi_1) \tag{1}$$

$$b = B \sin(w_2 t + \phi_2) \tag{2}$$

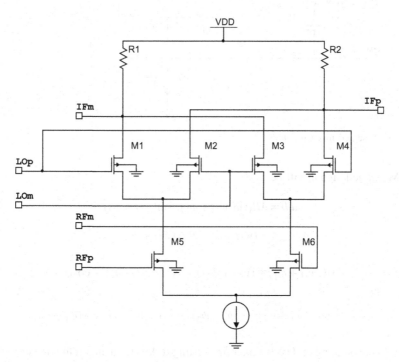

Fig. 3 Double-balanced Gilbert cell architecture [7]

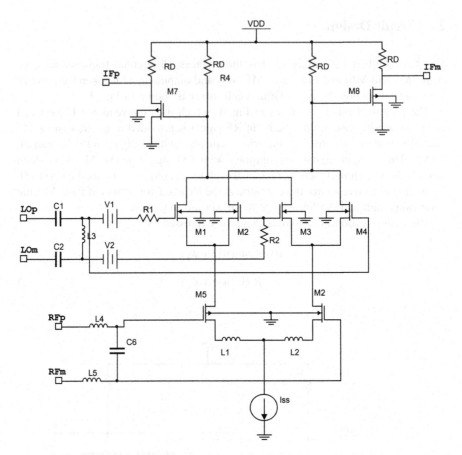

Fig. 4 Presented mixer topology

Multiplied signal will be

$$a.b = AB(\sin w_1 t + \phi_1). \sin(w_2 t + \phi_2) \tag{3}$$

$$A = (w_1 t + \phi_1),\ B = (w_2 t + \phi_2) \tag{4}$$

$$ab = \frac{-AB}{2}[\cos((w_1 t + \phi_1) + (w_2 t + \phi_2)) - \cos((w_1 t + \phi_1) - \cos(w_2 t + \phi_2))] \tag{5}$$

$$ab = \frac{-AB}{2}[\cos((w_1 + w_2)t + (\phi_1 + \phi_2)) - \cos((w_1 - \phi_1)t - (\phi_1 - \phi_2))] \tag{6}$$

While designing an UWB mixer, interesting task is to maintain the linearity, i.e., third-order intercept point, conversion gain, and noise figure over the channel bandwidth >500 MHz [9]. The conversion gain varies with the load resistor, therefore to achieve maximum conversion gain the value of resistors should be optimized [10].

$$CG = \frac{2}{\pi} \left(\frac{R_L}{j\omega L_s + \frac{1}{g_{m\_RF}}} \right), \tag{7}$$

where, L_S is the degeneration inductor used at the source of the transistor at trans-conductance stage and R_L is the load resistor.

$$g_{m\_RF} = \sqrt{2k_n \frac{W}{L}} \cdot \sqrt{I_D} \ and \ k_n = \frac{1}{2} \mu_n C_{ox} \left(\frac{W}{L} \right), \tag{8}$$

where, W and L implies the effective channel width and channel length of the transistor respectively, I_D is the drain current, μ_n represents surface mobility channel of the transistor, Cox represents gate oxide capacitance per unit area. [10]. To match the load resistance with 50 ohm source resistance LC impedance matching network [11] is used.

For $R_L < R_S$, L section matching network is shown in Fig. 5
Where

$$Z_L = R_L + jX_L \tag{9}$$

Solution for LC-Matching Network [11] is given below.

$$X = \pm \sqrt{(Z_0 - R_L)R_L} - X_L \tag{10}$$

$$B = \pm \frac{1}{Z_0} \sqrt{\frac{Z_0 - R_L}{R_L}} \tag{11}$$

$$L = \frac{X}{2\pi f} \tag{12}$$

$$C = \frac{B}{2\pi f} \tag{13}$$

Fig. 5 LC-matching network [11]

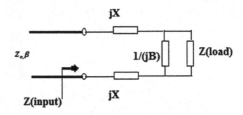

For differential pair,

$$I_{D1} - I_{D2} = \frac{1}{2}\mu_n C_{ox}\left(\frac{W}{L}\right)(V_{in})\sqrt{((4I_{ss}/\mu_n C_{ox}\left(\frac{W}{L}\right)) - (V_{in})^2)} \qquad (14)$$

3 Simulation Results

The simulation of presented mixer is prepared in 0.18 µm CMOS technology using Advanced Design System (ADS). The presented mixer operates with a 1.5 V supply, LO signal of 3.696 GHz, RF signal of 3.432 GHz and IF of 264 MHz. The conversion gain verses RF power curve is shown in Fig. 6 with conversion gain of 10.600 dB. The relationship between the input (RF) power and output (IF) power is shown in Fig. 7. This shows that conversion gain of the presented mixer does not depend on the RF power, i.e., constant (linear). Conversion gain decreases when the input RF power becomes too large. Figure 8 shows the curve between input RF frequency and the SSB-NF. Figure 9 indicates reflection coefficient up to −19 dB for the entire ultrawide band ranges from (3.168–3.696) GHz frequency with 528 MHz bandwidth **(Fig. 10).

Fig. 6 Conversion gain with input RF power

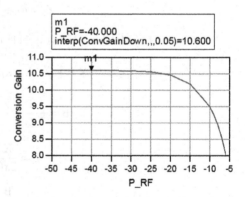

Fig. 7 Output power versus RF input power

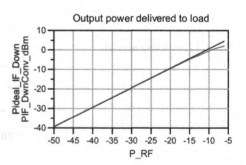

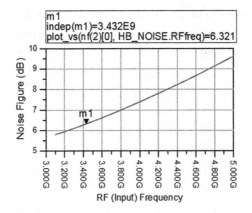

Fig. 8 SSB-NF with the input RF frequency

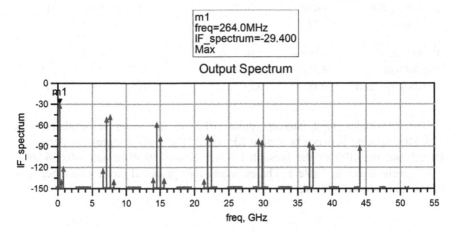

Fig. 9 IF-spectrum versus input RF frequency

Fig. 10 Plot between S11 and input RF frequency

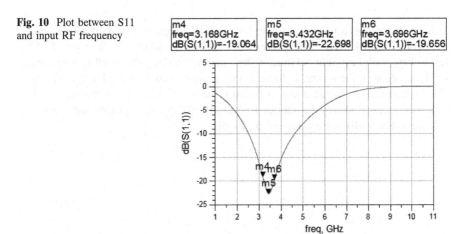

Table 1 Simulation results for IIP3, 1 dB gain compression point, SSB-NF and conversion gain

Third-order input intercept point (dBm)	+0.056
1 dB compression point (dBm)	−10.596
SSB-NF (dB)	6.321
Conversion gain	10.600

Table 2 Comparison of presented mixer with other work

Parameters	This work	[14]	[15]	[16]	[17]
RF Freq. (GHz)	3.432	1–10	2.4	0.7	3.1–10.6
IF Freq. (MHz)	264	100–1000	30	0	264
Supply voltage (V)	1.5	1.2	1.8	3.3	1.2
Conversion gain (dB)	10.600	3–8	9	7.3	9.8
Input IP3 (dBm)	+0.056	−7~−4	−14.5	8.69	−11
S11 (dB)	−22	>10 dB (Return-loss)	–	–	–
P1-dB (dBm)	−10.596	−16	–	−1.28	−24
SSB-NF (dB)	6.321	11.3–15	11	8.2	17.5
CMOS Tech. (μm)	0.18	0.13	0.18	0.18	0.13

The third-order input intercept point (IIP3) of the mixer is simulated with 528 MHz spacing between two input RF signal. The 1 dB compression point, i.e., P1-dB and the IIP3 of the presented mixer is −10.59 dBm and +0.056 dBm respectively shown in Table 1. The comparison of proposed mixer with other state of the art is shown in Table 2.

4 Conclusion

This work presents a proposed mixer for input RF frequency of 3.432 GHz and IF frequency of 264 MHz. The presented mixer is simulated in 0.18 μm CMOS technology with key sight Advanced Designed Software. Gilbert cell architecture is used as a core of the mixer. Inductive source degeneration is used at the RF input stage transistors to increase the conversion gain and linearity of the mixer. The proposed mixer achieves the gain of 10.600 dB, +0.056 dBm of IIP3 and the single sideband (SSB-NF) noise figure of 6.321 dB.

References

1. B. Razavi, Design of analog CMOS Integrated Circuits, McGraw Companies, 2001.
2. Xiongliang Lai and F ei Yuan, "A Comparative Study of Low-Power CMOS Gilbert Mixers in Weak and Strong Inversion" 978-1-61284-857-0111/$26.00 ©2011 IEEE.
3. T. Anh, K. Chang-Wan, K. Min-Suk and L. Sang-Gug, "A High Performance CMOS Direct Down Conversion Mixer for UWB system" Proceedings of the 14th ACM Great Lakes Symposium on VLSI, 2004.
4. T. H. Lee, The Design of CMOS Radio-Frequency Integrated Circuits, Second edition, Cambridge University Press, New York, NY, 1998.
5. D. Selvathi, M. Pown and S. Manjula, "Design and Analysis of UWB Down-Conversion Mixer with Linearization Techniques," wseas transactions on circuits and systems Volume 13, 2014.
6. K.-H. Liang, H.-Y. Chang, and Y.-J. Chan, "A 0.5–7.5 GHz ultra low-voltage low-power mixer using bulk-injection method by 0.18-_m CMOS technology.
7. B. Gilbert, "The MICROMIXER: A Highly Linear Variant of the Gilbert Mixer using a Bisymmetric Class-AB Input Stage," IEEE Journal on Solid-State Circuits, vol. 32, no. 9, pp. 1412–1423, Sep. 1997.
8. M. Jouri, A. Golmakani, M. Yahyabadi and H. Khosrowjerdi, "Design and simulation of a down-conversion CMOS mixer for UWB applications," International Conference on Electrical Engineering/Electronics Computer Telecommunications and Information Technology (ECTI-CON), pp. 937–940, 2010.
9. P. Paliwoda and M. Hella, "An Optimized CMOS Gilbert Mixer Using Inter-Stage Inductance for Ultra Wideband Receivers," IEEE 49th International Midwest Symposium on Circuits and Systems (MWSCAS), vol. 1, pp. 362–365, 2006.
10. B. IjiAyobami, F. Zhu and M. Heimlich, "A Down Converter Active Mixer, in 0.25 μm CMOS Process for Ultra Wide-Band Applications," International Symposium on Communications and Information Technologies (ISCIT), 2012, pp. 28–31, 2012.
11. R. E. Collin, Foundations for Microwave Engineering, Second Edition, McGraw-Hill, N.Y., 1992.
12. T. Anh, K. Chang-Wan, K. Min-Suk and L. Sang-Gug, "A High Performance CMOS Direct Down Conversion Mixer for UWB system" Proceedings of the 14th ACM Great Lakes Symposium on VLSI, 2004.
13. Fong-Cheng Chang, Ping-Cheng Huang, Shih-Fong Chao, and Huei Wang, "A Low Power Folded Mixer for UWB System Applications in 0.18 μm CMOS Technology".
14. Hu Zijie, *Student Member,* IEEE, and Koen Mouthaan, Member, IEEE, IEEE TRANSACTIONS on circuits and systems-II: express briefs, VOL. 60, NO. 11, NOVEMBER 2013.
15. A. Do, C. Boon, M. Do, K. Yeo, and A. Cabuk, "A weak-inversion low-power active mixer for 2.4 GHz ISM band applications," IEEE Microwave and Wireless Components Letters, Vol. 19, No. 11, pp. 719–721, November 2009.
16. G. O. Barraza Wolf § F. H. Gregorio † J. E. Cousseau ‡, "Low Noise Design of a CMOS Gilbert Mixer Cell in 700 MHz".
17. J. Seo, J. Kim, H. Sun, and T. Yun, "A low-power and high-gain mixer for UWB systems," IEEE Microwave and Wireless Components Letters, Vol. 18, No. 2, pp. 803–805, December 2008.

Part IV
Big Data and Cloud Computing

GCC-Git Change Classifier for Extraction and Classification of Changes in Software Systems

Arvinder Kaur and Deepti Chopra

Abstract Software repositories are used for many purposes like version control, source code management, bug and issue tracking and change log management. GitHub is one of the popular software repositories. GitHub contains commit history of software that lists all changes recorded in the software system, but it does not classify the changes according to the reason for change. In this study a mechanism for extraction and classification of changes is proposed and Git Change Classifier (GCC) tool is developed. The tool uses regular expression to extract changes and employs Text Mining to determine the type of change. GCC Tool reports the year-wise number of changes for a file and classifies the changes into three types: (a) Bug Repairing Changes (BRC), (b) Feature Introducing Changes (FIC) and (c) General Changes (GC). This classification is useful for predicting the effort required for new changes, tracking the resolution of bugs in software and understanding the evolution of the software as it may depend on the type of change.

Keywords Software change classification · Software repositories
Text mining · GitHub

1 Introduction

GitHub [1] is a popular software repository with web-based graphical user interface. It was launched in 2009 and since then is majorly used for management of source code and version control. But also provides other features like commit history, issue tracking, pull requests, feature requests and documentation.

A. Kaur · D. Chopra (✉)
University School of Information and Communication Technology,
Guru Gobind Singh Indraprastha University, New Delhi, India
e-mail: dchopra27@gmail.com

A. Kaur
e-mail: arvinder70@gmail.com

© Springer Nature Singapore Pte. Ltd. 2018
Y.-C. Hu et al. (eds.), *Intelligent Communication and Computational
Technologies*, Lecture Notes in Networks and Systems 19,
https://doi.org/10.1007/978-981-10-5523-2_24

Although it provides a list of commits for each file in the software system. It does not have a reporting system that reports to the user the type of changes. Most researchers require not only the number of changes but need to analyze the changes. Changes in software systems can be due to different reasons such as for adding features to the software, correcting bugs in the software, improving the performance of software and refactoring among others.

The information about changes is analyzed to guide the software development and maintenance process. The various application areas where an analysis of change history may be required include change prediction, bug prediction, software evolution, code reuse, and code cloning.

In this paper we describe Git Change Classifier (GCC) that extracts and classifies the changes in the software system. Section 2 describes related work. Section 3 describes our tool. The results of the GCC Tool are described in Sect. 4. Section 5 describes the threats to validity and the results of the study are concluded in Sect. 6.

2 Related Work

A lot of work has been done in the field of software engineering that analyzes changes in software systems. Software change history is often analyzed for suggesting improvements in the software development process and guiding software maintenance activities. In [2] Xie et al. analyze the software change history of three software systems to study code genealogy, i.e., evolution of code clones along the revisions in software system. Mani et al. [3] analyzed software commits to identify essential non-committers who act as bug resolution catalysts but are not directly responsible for fixing the bugs. Vandecruys et al. [4] analyze both bug and change repository of software to predict erroneous software modules using Ant Colony Optimization (ACO). Spinellis et al. [5] evaluate the quality of open-source software using both product and process metrics derived from software change and bug repositories.

There are many applications of software repositories and for a long time researchers in this field have extracted this data either manually or through tools. For instance, D'Ambros and Lanza [6] developed a tool Churrasco for collaborative software evolution analysis, which also provided modules for data extraction from Bugzilla and Subversion (SVN). In [7] Malhotra and Agrawal describe the framework for a simple Configuration Management System (CMS) tool for extracting data using log files from Concurrent Versioning System (CVS). Bajracharya et al. [8] developed an infrastructure for collection and analysis of source code of open source software systems. It also extracts structural information from the source code and its repository. It provides a basic framework on which other applications and services can be built.

Most tools extract only the number of changes and defects but sometimes it may be necessary to analyze the types of changes and defects. For instance in [9] Hassan predicted faults using complexity of code changes. Hassan describes that there are different types of changes and that how each change is not equally complex. In this study a mechanism is proposed for extracting and classifying different types of software changes. GCC tool is described that reports the year-wise number of changes of each type from GitHub [1] repository.

3 GCC Tool

GCC tool is used to extract and classify the commits from GitHub [1] repository. We describe the features of this tool, mechanism of the tool, structure of GitHub change log and classification scheme in the following subsections.

3.1 Utility of GCC

The main purpose of Git Change Classifier (GCC) is extraction and classification of changes from GitHub repository.

The major features of the tool include:

- Automated Extraction of Changes
- Calculating Year-wise number of changes
- Classification of Changes
- Exporting results to MS-EXCEL

3.2 Working Mechanism

Figure 1 shows a flowchart depicting the working of the GCC Tool. The software file whose changes are to be analyzed is identified. The URL containing its commit history is provided to the tool. The commit history at the URL is extracted. From this content the date and description of each commit is extracted using a regular expression. Regular expression is derived by analyzing the structure of change log in GitHub. The change description is classified using a text mining based classification scheme. The structure of GitHub repository and the classification scheme used by our tool is described in the following sections. This data is also exported in MS-Excel.

Fig. 1 Working of GCC tool

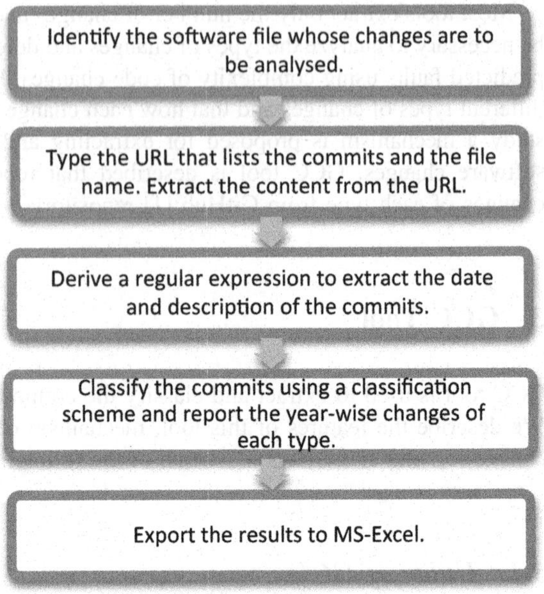

Identify the software file whose changes are to be analysed.

Type the URL that lists the commits and the file name. Extract the content from the URL.

Derive a regular expression to extract the date and description of the commits.

Classify the commits using a classification scheme and report the year-wise changes of each type.

Export the results to MS-Excel.

Fig. 2 Structure of Git change log

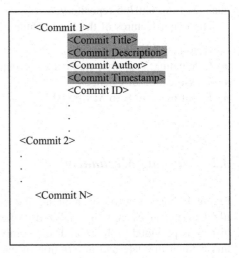

```
<Commit 1>
        <Commit Title>
        <Commit Description>
        <Commit Author>
        <Commit Timestamp>
        <Commit ID>
            .
            .
            .
<Commit 2>
    .
    .
    .
    <Commit N>
```

3.3 Structure of GitHub Change Log

GitHub stores the change history in the form of commits. Figure 2 shows the structure of GitHub change log.

Each commit consists of the following information:

- **Commit Title**: This is a brief title of the change.
- **Commit Description**: provides the detailed description regarding the change including the reason of change.
- **Commit Author**: It provides information about the contributors of the commit.
- **Commit Timestamp**: includes the date and time when the commit was submitted by the contributor.

GCC Tool extracts the Commit Title and description along with the timestamp of the commit which have been highlighted in Fig. 2.

3.4 Classification Scheme

GCC Tool classifies the changes into three different types depending on the reason for the change [9]:

- **Bug Repairing Changes (BRC)** are the changes that are done to resolve an issue or a bug. They generally correspond to Bug reports in a bug repository.
- **Feature Introducing Changes (FIC)** are the changes that are done to fulfill feature requests of the users or to enhance the functions of the software.
- **General Changes (GC)** include all other changes that are carried out to improve the performance and reliability of the software.

The reason for a commit not only helps in understanding the evolution of the software, but different changes have different impact on the quality of the software. For instance, GC are less likely to produce bugs in the software than BRC or FIC. Also there is a greater probability of making faults while introducing a new feature than when rectifying an existing fault in the software.

The classification scheme takes the change description as the input and uses text mining to classify the change into one of the above three types. The classification scheme applies a keyword based searching mechanism on the change description. The classification approach used in GCC is depicted in Fig. 3. In the next section we describe the results of GCC tool.

3.5 Output of GCC Tool

The GCC tool is provided with the URL to the webpage containing the change history of the file. The user mentions the years for which the analysis is required and the output consisting of year-wise classification of changes is obtained. The same is exported to an excel sheet.

Figure 4 depicts a sample output of GCC tool. GCC takes the URL of commit history of "mod_dir.c" file of "apache/httpd/modules/mappers" subsystem from the user, and gives year-wise classification of changes between the year 2004–2015.

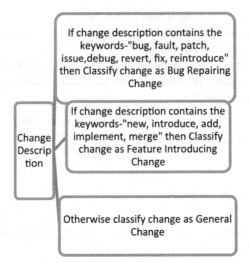

Fig. 3　Classification approach

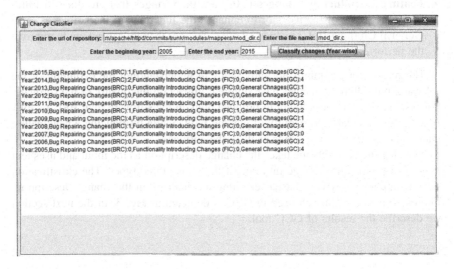

Fig. 4　Output of GCC tool

4　Performance Analysis

The performance of any classification scheme is measured in terms of the following performance measures [10]:

Accuracy: Accuracy of a classification scheme is defined as the ratio of correctly classified instances to the total number of instances classified as given in Eq. (1).

$$Accuracy = \frac{Number\ of\ correct\ classifications}{Total\ number\ of\ classifications} \qquad (1)$$

Precision: Precision for a particular classification class say X is defined as the ratio of instances correctly classified as X to the total number of instances classified as X. It is given in Eq. (2).

$$Precision = \frac{Number\ of\ instances\ correctly\ classified\ as\ X}{Total\ number\ of\ instances\ classified\ as\ X} \qquad (2)$$

Recall: Recall for a particular classification class X is defined as the ratio of instances correctly classified as X to the number of instances actually belonging to class X as specified in Eq. (3).

$$Recall = \frac{Number\ of\ instances\ correctly\ classified\ as\ X}{Total\ number\ of\ instances\ of\ class\ X} \qquad (3)$$

To calculate these performance measures, we classified the commits of "apache/httpd/modules/mappers" available in GitHub [1] repository over the time period 2010–2015 using our GCC Tool and manually. The confusion matrix obtained between manual classification and GCC classification is derived. Using Eq. (1) the accuracy of GCC tool is calculated to be 0.931 (93.1%). The Precision and Recall values for types of changes were calculated using Eqs. (2) and (3), respectively. The confusion matrix along with precision and recall values is given in Table 1. GCC Tool has accuracy, precision and recall greater than 90%, thus indicating very little chances of misclassification.

Table 1 Confusion matrix, precision and recall for GCC

GCC classification	Manually classified				Precision	Recall
	BRC	FIC	GC			
BRC	80	5	3	Classified BRC: 88	0.91 (91%)	0.964 (96.4%)
FIC	1	48	2	Classified FIC: 51	0.941 (94.1%)	0.906 (90.6%)
GC	2	0	48	Classified GC: 50	0.96 (96%)	0.906 (90.6%)
	Total BRC: 83	Total FIC: 53	Total GC: 53	Total: 189		

5   Threats to Validity

The correctness of results reported by GCC tool are constrained by the internal and external threats to validity as described in the following subsections.

5.1   External Threats to Validity

The threats to external validity of the results reported by GCC include:

- The GCC Tool works only for GitHub repository. It may be extended for other repositories by studying the structure of commits in those repositories. In its present form the tool only classifies changes for software projects that use GitHub repository.
- The tool needs to be updated if the layout or structure of GitHub is changed. The regular expression needs to be revised according to the structure of commits and the layout of GitHub webpage.

5.2   Internal Threats to Validity

Internal validity of the result may be affected by the fact that the classification is based on the change description provided by the committer of the change. The results depend on how correctly and completely the change is described by the committer. The classification results are dependent on the quality of commit description.

6   Conclusion and Future Work

The developed GCC Tool is useful for extraction of year-wise changes. It not only extracts the year-wise number of changes from the commit history but also classifies the changes depending on the reason of change with an accuracy of 93.1% and very high (greater than 90%) precision and recall values. The results are also exported to MS-Excel. The classification of changes into: (a) Bug Reporting Changes (BRC), (b) Feature Introducing Changes (FIC) and (c) General Changes (GC) is beneficial, since each type of changes are not equally complex. For instance, FIC are more complex and time consuming than GC. It is also possible to conduct studies to track how bugs are resolved by analyzing change history especially BRC.

The tool extracts this information from GitHub which a popular repository for many open source as well as proprietary software systems. Since GitHub is the most popular repository used, this tool can be used to analyze the change history of most software systems. This study also details the working mechanism of the tool which can be adopted for other software repositories as well.

References

1. https://github.com/.
2. Xie, S., Khomh, F., & Zou, Y. (2013, May). An empirical study of the fault-proneness of clone mutation and clone migration. In Proceedings of the 10th Working Conference on Mining Software Repositories (pp. 149–158). IEEE Press.
3. Mani, S., Nagar, S., Mukherjee, D., Narayanam, R., Sinha, V. S., & Nanavati, A. A. (2013, May). Bug resolution catalysts: Identifying essential non-committers from bug repositories. In Proceedings of the 10th Working Conference on Mining Software Repositories (pp. 193–202). IEEE Press.
4. Vandecruys, O., Martens, D., Baesens, B., Mues, C., De Backer, M., & Haesen, R. (2008). Mining software repositories for comprehensible software fault prediction models. Journal of Systems and software, 81(5), 823–839.
5. Spinellis, D., Gousios, G., Karakoidas, V., Louridas, P., Adams, P. J., Samoladas, I., & Stamelos, I. (2009). Evaluating the quality of open source software. Electronic Notes in Theoretical Computer Science, 233, 5–28.
6. D'Ambros, M., & Lanza, M. (2010). Distributed and collaborative software evolution analysis with churrasco. Science of Computer Programming, 75(4), 276–287.
7. Malhotra, R., & Agrawal, A. (2014). CMS tool: calculating defect and change data from software project repositories. ACM SIGSOFT Software Engineering Notes, 39(1), 1–5.
8. Bajracharya, Sushil, Joel Ossher, and Cristina Lopes. "Sourcerer: An infrastructure for large-scale collection and analysis of open-source code." *Science of Computer Programming* 79 (2014): 241–259.
9. Hassan, A. E. (2009, May). Predicting faults using the complexity of code changes. In Proceedings of the 31st International Conference on Software Engineering (pp. 78–88). IEEE Computer Society.
10. Junker, Markus, Rainer Hoch, and Andreas Dengel. "On the evaluation of document analysis components by recall, precision, and accuracy." *Document Analysis and Recognition, 1999. ICDAR'99. Proceedings of the Fifth International Conference on.* IEEE, 1999.

ECDMPSO: A Modified MPSO Technique Using Exponential Cumulative Distribution

Narinder Singh and S.B. Singh

Abstract Various efficient computational optimization techniques have been developed during last few decades but quest for a new and better one still continues. In this article, a modified version of Mean Particle Swarm Optimization (MPSO) called Exponential Cumulative Distribution mean PSO (ECDMPSO) has been proposed for finding the global or near to global optimal solution of the optimization problem. The objective of this article is to make an effort to remove the premature stagnation of the SPSO and to extend the approach of Mean PSO. The approach has been verified through several benchmark functions as well as mathematical models of some constrained, fractional, multimodal problems.

Keywords Standard particle swarm optimization · Mean PSO · Exponential cumulative distribution

1 Introduction

Many researchers have modified the standard variant of PSO during the last decades and proposed the new hybrid and probabilistic meta-heuristic variants of PSO like PSO-E [1], Mean PSO [2], THPSO [3]. In these modified variant, researchers have improved the efficiency of standard version of PSO in terms of convergence speed and global optima. Kim, B.I. and Son, S.J. [4] proposed a simple particle swarm optimization technique for solving the CVRP. This variant uses a probability matrix as the main device for particle decoding and encoding. On the basis of experimental results it was shown that this version is effective compared to other meta-heuristics. Kanoh, H. and Chen, S. [5] developed a new algorithm to solve university course timetabling problems using PSO and introduced transition probability into PSO to

N. Singh (✉) · S.B. Singh
Department of Mathematics, Punjabi University, Patiala 147002, Punjab, India
e-mail: narindersinghgoria@ymail.com

S.B. Singh
e-mail: sbsingh69@yahoo.com

© Springer Nature Singapore Pte. Ltd. 2018
Y.-C. Hu et al. (eds.), *Intelligent Communication and Computational Technologies*, Lecture Notes in Networks and Systems 19,
https://doi.org/10.1007/978-981-10-5523-2_25

settle this problem. The authors concluded that proposed algorithm got more efficient solution than an ES (Evolution Strategy) for time table problem of the University of Tsukuba.

2   Mean PSO

Deep and Bansal [2] proposed a modified approach of PSO known as Mean PSO. This variant was constructed by replacing two terms of basic update velocity equation of SPSO by two new terms based on the linear combination of personal best and global best. The updated mathematical equations of velocity and position as proposed by authors

$$v_{ij}^{k+1} = w \times v_{ij}^k + c_1 r_{1j}^k \times \left(\frac{p_{best,\,i}^k + g_{best}}{2} - x_{ij}^k \right) + c_2 r_{2j}^k \times \left(\frac{p_{best,\,i}^k - g_{best}}{2} - x_{ij}^k \right) \quad (1)$$

$$x_{ij}^{k+1} = x_{ij}^k + v_{ij}^{k+1} \quad\quad\quad\quad\quad (2)$$

3   PSO-E

Krohling, R.A. et al. [1] proposed exponential distribution based particle swarm optimization algorithm. The authors used an exponential distribution to generate random numbers and claimed that it may provide a more effective way to avoid local optima. The method was tested on a set of test problems and the results obtained were compared with results obtained through standard PSO. The performance of the proposed algorithm was observed to be satisfactory as compared. In this paper, it is proposed to use the exponential approach to generate x_{ij}^{k+1} in (Eq. 2) instead of using it in generating random number in v_{ij}^{k+1}.

4   Modified PSO-E

In our proposed approach, we replace Eq. (2) with the following Eq. (3) keeping other equation unchanged. The idea is to replace the conventional generation of x_{ij}^{k+1} by exponential cumulative distribution ($f(x) = 1 - e^{-\lambda x}$ $x \geq 0$, 0 for $x < 0$). The rest of the operations are same as in Mean PSO. The main objective of this modification is to make a balance between the local and global exploration abilities using the exponential cumulative distribution.

In the proposed exponential cumulative distribution mean particle swarm optimization algorithm (ECDMPSO) following equations are used:

$$x_{ij}^{k+1} = 1 - \exp^{-\left(x_{ij}^k + v_{ij}^{k+1}\right)} \tag{3}$$

where $x_{ij}^{k+1} \in x \geq 0$, $x_{ij}^{k+1} \notin x < 0$ and $\lambda = 1$.

$$v_{ij}^{k+1} = w \times v_{ij}^k + c_1 r_{1j}^k \times \left(\frac{p_{best,i}^k + g_{best}}{2} - x_{ij}^k\right) + c_2 r_{2j}^k \times \left(\frac{p_{best,i}^k - g_{best}}{2} - x_{ij}^k\right) \tag{4}$$

In ECDMPSO technique, the movement of each particle in the global search space is updated by exponential cumulative distribution $(1 - \exp^{-x})$ when $x_{ij}^{k+1} \in x \geq 0$. This condition helps in assigning the best position of each member of the crowd in the search area.

5 Computational Steps of ECDMPSO

Step I: Initialize the population of particles uniformly distributed random vector: $x_i - U(M, m)$, where m and M are upper and lower boundaries of the search area.

Step II: Evaluate fitness of individual particles

Step III: If a particle's present position is superior than update its old best position.

Step IV: Determine the best particle (according to the particle's previous best positions)

Step V: Update particle's velocities by using Eq. (4)

Step VI: Move particles to their new positions by Eq. (3)

Step VII: Go to step 2 until stopping criterion is satisfied

Computer code

Initialize the population // initialize all particles
For every candidate (or particle) in the swarm do
If $f(x_i) < f(p_{best,i})$
    Then

$$x_i = p_{best,i}$$

End if
If $f(p_{best,i}) < f(g_{best})$ then $g_{best} = p_{best,i}$
End if
   End for
   // update particle's velocity and position using Eq. (4) and (3)
   For every candidate (or particle) in the swarm do
   For every j-dimension in D do
   End for
   End for
   Iteration = Iteration + 1
   Until > Max_Iterations

6 Testing Functions

The performance of the proposed version has been tested on 16 standard benchmark functions with different ranges of the solution space. All these standard functions are described in the following Table 1 [6].

Table 1 Benchmark functions

Benchmark function	Function name	Range	Optimal solution
1	Ackley	[−30, 30]	0
2	Function 6	[−30, 30]	0
3	Zakharov's	[−4.12, 4.12]	0
4	Parabola (Sphere)	[−4.12, 4.12]	0
5	Step function	[−4.12, 4.12]	0
6	Clerc's f1, Alpine function	[−1, 1]	0
7	Axis parallel hyper ellipsoid	[−4.12, 4.12]	0
8	Schwefel 3	[−10, 10]	0
9	Dejong	[−10, 10]	0
10	Becker and lago	[−10, 10]	0
11	Bohachevsky 1	[−50, 50]	0
12	Bohachevsky 2	[−50, 50]	0
13	Periodic	[−10, 10]	0.9
14	Camel back-3	[−5, 5]	0
15	Camel back-6	[−5, 5]	−1.0316
16	Aluffi-Pentini's	[−10, 10]	0.3523

Fig. 1 Comparison of performances of ECDMPSO algorithm with Mean PSO algorithm by minimum objective value

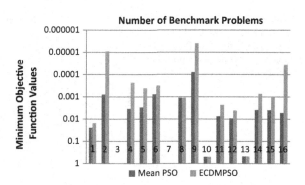

7 Experimental Setup

The ECDMPSO and Mean PSO pseudo code are coded in Microsoft Visual C++ 6.0 and implemented on Intel HD Graphics, 15.6″ 3 GB Memory, i5 Processor 430 M, 16.9 HD LCD, Pentium-Intel Core (TM) and 320 GB HDD. Number of search agents (30), maximum number of iterations (100), maximum number of evaluation for each run (10,000) $c_1 = c_2 = c_3 = 1.4$, $w = 0.6$ and $l \in [2, 0]$ all these parameter settings are applied to test the quality of meta-heuristics (Figs. 1, 2, 3 and 4).

8 Results and Discussion

The performance of proposed variant has been tested on 16 well-known benchmark functions and some practical functions. The results obtained using the proposed approach has been compared with the results obtained through Mean PSO. A set of 16 classical problems including fractional, constrained, unconstrained, multi-objective and mix integer programming problems are given in Tables 1 and 5. To check the performance of various these problems have been solved using ECDMPSO, PSO-E and Mean PSO. The results obtained through these algorithms have been compared in the Tables 2, 6, 7 and 8. In addition, the performance of proposed variant has also been compared with SPSO and PSO-E in Table 8. The performance has been compared on the basis of minimum function value, rate of success rate (%), mean and standard deviation values. On the basis of accuracy, convergence rate and reliability in results obtained, it is observed that the modified variant is more efficient and superior to Mean PSO and PSO-E. Table 2 shows the simulated results obtained using modified variant and Mean PSO in the terms of minimum function value, accuracy, convergence rate, rate of success and errors. Clearly proposed variant presents the superior quality of results for all the benchmark functions as compared to Mean PSO. Tables 3 and 4 illustrate the performance of each particle of the swarm in the global search space. This stepwise movement of each particle from one position to other has also been plotted in Figs. 5, 6, 7 and 8.

Fig. 2 Comparison of
performance of ECDMPSO
algorithm with Mean PSO
algorithm by mean of optimal
solutions

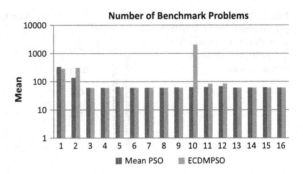

Fig. 3 Comparison of
performance of ECDMPSO
algorithm with Mean PSO
algorithm by standard
deviation of optimal solutions

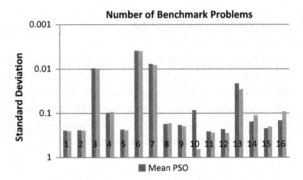

Fig. 4 Comparison of
performance of ECDMPSO
algorithm with Mean PSO
algorithm by amount of error

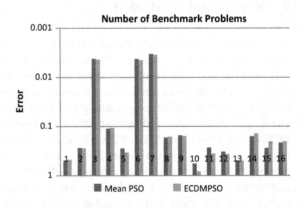

The performance of the modified variant has also been tested on some practical function. The simulated results using modified variant and Mean PSO have been shown in Tables 6 and 7. The results assert that the modified variant helped in improving the performance of Mean PSO in terms of convergence rate and global optima. Furthermore the capability of proposed variant has also been tested on some problem having more than one optimum in Table 8. The experimental results and analysis shows that the ECDMPSO outperformed SPSO and PSO-E in the

Table 2 Comparison of performances of ECDMPSO algorithm and Mean PSO algorithm tested on benchmark functions

Benchmark function	Minimum of optimal solution		Mean of optimal solution		Standard deviation of optimal solution		Error		Success rate (%)	
	Mean PSO	ECDMPSO	Mean PSO	ECDMPSO	Mean PSO	ECDMPSO	Mean PSO	ECDMPSO	Mean PSO	ECDMPSO
1	0.024623	0.015358	327.90000	286.20000	0.245343	0.251057	0.485073	0.470973	100.00	100.00
2	0.000791	0.000009	136.80000	304.80000	0.240821	0.245913	0.272594	0.275246	100.00	100.00
3	0.000000	0.000000	60.000000	60.000000	0.009630	0.010214	0.004183	0.004336	100.00	100.00
4	0.003470	0.000230	60.000000	60.000000	0.097447	0.093586	0.110531	0.106110	100.00	100.00
5	0.002974	0.000409	64.800000	63.300000	0.234480	0.243649	0.278543	0.337322	100.00	100.00
6	0.000776	0.000317	60.000000	60.000000	0.003889	0.003940	0.004235	0.004471	100.00	100.00
7	0.000000	0.000000	60.000000	60.000000	0.007704	0.008171	0.003347	0.003468	100.00	100.00
8	0.001106	0.001106	60.300000	60.000000	0.175687	0.169383	0.167440	0.161525	100.00	100.00
9	0.000079	0.000004	62.100000	60.000000	0.185341	0.199435	0.151650	0.157508	100.00	100.00
10	0.500034	0.500012	62.400000	2052.0000	0.086527	0.651071	0.570506	0.827946	100.00	100.00
11	0.007578	0.002272	63.600000	82.800000	0.258545	0.274359	0.268784	0.355339	100.00	100.00
12	0.009279	0.004074	67.800000	84.300000	0.229583	0.280910	0.325561	0.365841	100.00	100.00
13	0.480468	0.480465	60.000000	60.000000	0.021465	0.028887	0.493992	0.502431	100.00	100.00
14	0.003868	0.000738	60.000000	60.000000	0.155977	0.111656	0.158498	0.138854	100.00	100.00
15	0.003951	0.001073	61.500000	60.300000	0.220210	0.201875	0.273662	0.202319	100.00	100.00
16	0.005224	0.000038	60.000000	60.000000	0.146680	0.092072	0.214411	0.200202	100.00	100.00

Table 3 Comparison of each particle's position in the search space on benchmark functions Ackley and Function 6

Iteration no. ↓	Mean PSO		ECDMPSO		Mean PSO		ECDMPSO	
Ben. funct. →	Ackley				Function 6			
Particles →	P = 1	P = 2	P = 1	P = 2	P = 1	P = 2	P = 1	P = 2
90	0.044706	-0.021300	0.080692	0.009811	-0.631887	-4.694966	-0.585235	0.417889
91	0.073631	0.064561	-0.006933	0.092412	1.638371	3.285418	-0.651675	0.997546
92	-0.037270	0.028777	0.065368	0.030586	1.653842	3.148173	-0.637657	-30.000000
93	-0.088640	0.067865	0.065267	-0.071184	-0.620885	-2.411096	-0.624761	0.406703
94	-0.066064	-0.066726	0.032276	0.023797	-0.612679	-20.482975	-0.630000	0.974806
95	-0.093481	0.033441	-0.114080	0.072773	-0.589599	16.962541	-0.618166	-0.202088
96	0.070775	0.061718	0.016371	0.142510	1.616190	-11.886456	1.650746	-27.868587
97	0.028312	0.078560	-0.052043	0.129793	-0.604566	0.785241	-0.605611	-30.000000
98	-0.054725	-0.039990	-0.125084	-0.049227	-0.631663	2.735353	-0.615375	0.939542
99	-0.037705	0.086615	0.077911	0.097908	-0.629761	8.088413	-0.610598	-0.344339
100	-0.050448	0.019726	0.003945	0.036722	1.641218	-8.118771	-0.597493	-30.000000

Table 4 Comparison of each particle's position in the search space on benchmark functions Zakharov's and Parabola (Sphere)

Iteration no. ↓ Bench. funct. → Particles →	Mean PSO Zakharov's		ECDMPSO		Mean PSO Parabola (Sphere)		ECDMPSO	
	P = 1	P = 2	P = 1	P = 2	P = 1	P = 2	P = 1	P = 2
90	2.238234	0.033469	-3.433157	0.008061	-0.067271	0.441201	-0.069585	0.356737
91	0.393186	0.013043	0.325097	0.012958	0.058636	-0.081634	0.056950	-0.085058
92	1.362075	-0.012073	0.743871	-0.012146	0.097829	0.336973	0.093196	0.286072
93	0.175934	-0.047948	0.161327	-0.049116	0.175934	-0.047948	0.161327	-0.049116
94	-2.252412	-0.007031	-2.252412	-0.007031	0.392974	-0.489355	0.502073	0.035772
95	-0.655804	-0.068640	-0.926692	-0.071050	0.034584	-0.198299	0.089069	0.367757
96	-2.832743	0.052345	-2.832743	0.052345	0.423782	-0.077576	0.345434	-0.080664
97	0.949024	-0.011482	0.583305	-0.011346	-0.342380	-0.073757	-0.542111	0.251900
98	0.246912	0.011138	0.218790	0.011077	0.246912	0.011138	0.218790	0.011077
99	4.420604	-0.038595	4.420604	-0.038595	-0.365801	0.350642	0.244488	0.396116
100	2.235120	0.129104	-0.243075	0.013756	-0.240769	0.092164	-0.243075	0.013756

Fig. 5 Comparison the updated path of ECDMPSO and Mean PSO for each particle per iteration in the search space by testing on benchmark function (Ackley)

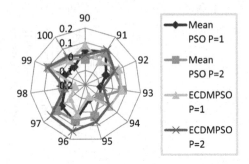

Fig. 6 Comparison the updated path of ECDMPSO and Mean PSO for each particle per iteration in the search space by testing on benchmark function (Function 6)

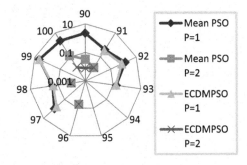

Fig. 7 Comparison the updated path of ECDMPSO and Mean PSO for each particle per iteration in the search space by testing on benchmark function (Zakharov's)

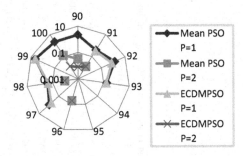

Fig. 8 Comparison the updated path of ECDMPSO and Mean PSO for each particle per iteration in the search space by testing on benchmark function (Parabola (Sphere))

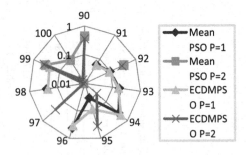

terms of global optima and computational efforts. The ability of ECDMPSO and Mean PSO variants has also been tested on 07 scalable and 09 non-scalable problems and the results are compared in Table 9. It is clear that the proposed variant works well on all scalable and non-scalable problems and outperforms Mean PSO. Table 10 demonstrates the performances of Mean PSO and proposed variant tested on seven practical problems. As seen, both the variants give a equally good quality of solutions on 03 problems but the proposed variant provide a better quality of global optima solutions on other 04 problems. On the other hand, Mean PSO does not show better quality of solution on any of the practical problems as compared to ECDMPSO variant. The comparison of performances of SPSO, PSO-E and ECDMPSO algorithms in solving three multimodal functions has been shown in the Table 11. It is observed that all these variants provide equivalent results for two problems but the proposed variant is better on the maximum number of problems compared to other meta-heuristics. The numerical experiments revealed that the proposed variant is capable of finding the better quality of solutions on maximum number of standard benchmark and real-life optimization problems as compared to SPSP, PSO-E and Mean PSO.

9 Tested Practical Problems

In this study, six minimization problems including fractional, constrained, unconstrained, multi-objective and mix integer programming problems have been solved using all the three approaches. These problems are listed in the following Table 5 and the solutions are listed in Table 6. Conley problem (Mohan, C. 2011) has been solved in Table 7 using all these techniques. The performance of the proposed algorithm has also been tested by solving three multimodal problems (Table 8), 16 scalable and non-scalable problems (Table 9). The performance of the all the three algorithms has been checked in Table 7, 9, 10 and 11 for Conley problem, multimodal problems, scalable, non-scalable problems and practical functions respectively.

PROBLEM-VII: Conley (1984): This problem taken from the book of (Mohan, C. 2011).

Objective Function is

$$
\begin{aligned}
Maxz = {}& 50y_1 + 150y_2 + 100y_3 + 92y_4 + 55y_5 + 12y_6 + 11y_7 + 10y_8 + 8y_9 + 3y_{10} + 11y_{11} + 90y_{12} + 87y_{13} \\
& + 91y_{14} + 58y_{15} + 16y_{16} + 19y_{17} + 22y_{18} + 21y_{19} + 32y_{20} + 53y_{21} + 56y_{22} + 118y_{23} + 192y_{24} + 52y_{25} \\
& + 204y_{26} + 250y_{27} + 295y_{28} + 82y_{29} + 30y_{30} + 29y_{31} + 2y_{32}y_{32} + 9y_{33}y_{33} + 94y_{34} + 15y_{35}^3 + 17y_{36}^2 \\
& - 15y_{37} - 2y_{38} + y_{39} + 3y_{40}^4 + 52y_{41} + 57y_{42}^2 - y_{43}^2 + 12y_{44} + 21y_{45} + 6y_{46} + 7y_{47} - y_{48} + y_{49} + y_{50} \\
& + 119y_{51} + 82y_{52} + 75y_{53} + 18y_{54} + 16y_{55} + 12y_{56} + 6y_{57} + 7y_{58} + 3y_{59} + 6y_{60} + 12y_{61} + 13y_{62} + 18y_{63} \\
& + 7y_{64} + 3y_{65} + 19y_{66} + 22y_{67} + 3y_{68} + 12y_{69} + 9y_{70} + 18y_{71} + 19y_{72} + 12y_{73} + 8y_{74} + 5y_{75} + 2y_{76} + 16y_{77} \\
& + 17y_{78} + 11y_{79} + 12y_{80} + 9y_{81} + 12y_{82} + 11y_{83} + 14y_{84} + 16y_{85} + 3y_{86} + 9y_{87} + 10y_{88} + 3y_{89} + y_{90} + 12y_{91} \\
& + 3y_{92} + 12y_{93} + 2y_{94}^2 - y_{95} + 6y_{96} + 7y_{97} + 4y_{98} + y_{99} + 2y_{100}
\end{aligned}
$$

Table 5 Practical problems

Problem	Functions	Optimal value $(x_1, x_2; f)$
I (Jiao, H. 2006)	Objective function: $Min f(x_1,x_2) = \dfrac{(x_1+x_2+1)^{1.1}}{(x_1+x_2+2)^{1.1}} \times \dfrac{(x_1+x_2+3)^{1.2}}{(x_1+x_2+4)^{1.2}} + \dfrac{(x_1+x_2+5)^{1.1}}{(x_1+x_2+6)^{1.1}} \times \dfrac{(x_1+x_2+7)^{1.2}}{(x_1+x_2+8)^{1.2}}$ s.t., $x_1 x_2^2 + x_1^2 x_2 \leq 10$ range: $1 \leq (x_1,x_2) \leq 2$	$(1, 1; 1.346382)$
II (Jiao, H. 2006)	Objective function: $Min f(x_1,x_2) = \dfrac{(x_1+x_2+1)^{1.1}}{(x_1+x_2+2)^{1.1}} \times \dfrac{(x_1+x_2+3)^{1.2}}{(x_1+x_2+4)^{1.2}} + \dfrac{(x_1+x_2+6)^{1.1}}{(x_1+x_2+5)^{1.1}} \times \dfrac{(x_1+x_2+8)^{1.2}}{(x_1+x_2+7)^{1.2}}$ s.t. $x_1 x_2^{0.5} + x_1 x_2 \leq 4$ Range: $1 \leq (x_1,x_2) \leq 2$	$(1, 1; -0.72882929647);$
III (Sohrab, E. et al. 2012)	Objective function: $Min f(x_1,x_2) = \dfrac{(7x_1+x_2, 7x_1+x_2+3)}{(3x_1+4x_2+12, 3x_1+4x_2+36)}$ s.t. $x_1 + x_2 \leq 7$, $4x_1 - 9x_2 \leq 3$, $x_1 + 2x_2 \geq 1.5$ Range: $x_1, x_2 \geq 0$	$(1.000000, 1.000000; -0.72882929929647)$
IV (Swarup, K. 2011)	Objective function: $Max z = 2x_1 + 3x_2$ s.t. $5x_1 + 7x_2 \leq 35$, $4x_1 + 9x_2 \leq 36$ Range: $x_1, x_2 \geq 0$	–
V (Li 1994)	Objective function: $Min z = x_1 x_2 x_3 (x_4^2 - x_4 x_5 + x_5^2) - x_1 x_5^{0.5} + x_2$ s.t. $3x_1 + x_2 x_5 = 5$, $5x_2 + x_4 \geq 6$, $x_3 + x_1 \geq 1$; Range $0 \leq x_4, x_5 \leq 2$	x_1, x_2, x_3 Values 0 or 1 only but x_4, x_5 have real values. $(X, Y) = (1, 1, 0, 1, 2)$ $f_{min} = 1 - \sqrt{2}$
VI (Mohan, C. 2011)	Objective function: $Min \ (f_1(X), f_2(X))$ *where* $f_1(X) = x_1^{-1} x_2^{-1} x_3^{-1} + 25x_1 x_2^{-1}$, $f_2(X) = 0.1 x_1 x_2$ s.t., $5x_1 x_2^{-1} + x_2 x_3^{-1} \leq 1$ Range: $x_1, x_2, x_3 > 0$	$(0.7219, 28.8737, 33.4851)$ with $f_1(X) = 2.1567$ $f_2(X) = 2.1567$

Table 6 Comparison of performances of ECDMPSO and Mean PSO tested on practical problems

Problem	Mean PSO Optimal point					ECDMPSO Optimal point					Mean PSO Fitness value	ECDMPSO
	x_1	x_2	x_3	x_4	x_5	x_1	x_2	x_3	x_4	x_5	f	f
I	-0.350116	-0.417266	–	–	–	-0.500000	-0.500000	–	–	–	0.664624	0.650224
II	0.552477	0.846004	–	–	–	0.552477	0.846004	–	–	–	0.721109	0.721109
III	-0.290384	0.940626	–	–	–	-0.290384	0.940626	–	–	–	0.000173	0.000173
IV	–	–	–	–	–	–	–	–	–	–	0.004569	0.000013
V	-0.014153	0.210173	1.443369	-0.143601	-0.629217	0.001586	-2.000000	0.662838	-1.969537	-0.590458	0.000000	0.000000
VI	-0.050212	0.533822	-0.079383	–	–	0.253442	-2.000000	-1.704826	–	–	$f_1 = 0.011978$, $f_2 = -0.00268042$	$f_1 = 0.001633$, $f_2 = -0.0506884$

Table 7 Comparison of Mean PSO and ECDMPSO variants tested on Conley problem

Optimal solution		Standard deviation		Mean value		Success of rate		Total clocks	
Mean PSO	ECDMPSO	Mean PSO	ECDMPSO	Mean PSO	ECDMPSO	Mean PSO	ECDMPSO	Mean PSO	ECDMPSO
0.000147	0.000057	0.022193	0.037754	60.000	60.000	100%	100%	3055	1836

Table 8 Comparison of performances of SPSO, PSO-E and ECDMPSO variants tested on multi modal problems

Problem	N	Range	Algorithm	Best value	Mean	S.D.
$\sum_{i=1}^{N}\{x_i^2 - 10\cos(2\pi x_i) + 10\}$	30	(−4.12, 4.12)	SPSO	27.8538	88.9534	28.3976
			PSO-E	11.9345	30.8792	16.5274
			ECDMPSO	0.000087	82.0000	0.266211
$\frac{1}{4000}\sum_{i=1}^{N}x_i^2 - \prod_{i=1}^{M}\cos(\frac{x_i}{\sqrt{i}}) + 1$	30	(−600, 600)	SPSO	0.0	0.0123	11.0715
			PSO-E	0.0	0.0148	0.0349
			ECDMPSO	0.0	0.04979	0.026912
$0.1\left\{\sin^2(3\pi x_i) + \cdot\frac{1}{N}\sum_{i=1}^{N-1}(x_i-1)^2\{1+\sin^2(3\pi x_{i+1})\} + (x_N-1)^2\{1+\sin^2(3\pi x_{i+1})\}\right\} + \sum_{i=1}^{N}u(x,5,100,4)$	30	(−50, 50)	SPSO	0.0	0.292	0.4001
			PSO-E	0.0	0.1195	0.1982
			ECDMPSO	0.0	0.1066	0.1546

Table 9 Comparison of Mean PSO and ECDMPSO variants tested on sixteen classical functions

Problems	Total	Mean PSO			ECDMPSO		
		Best	Bad	Fail	Best	Bad	Fail
Scalable	07	00	07	00	07	00	00
Non-scalable	09	00	09	00	09	00	00

Table 10 Comparison of Mean PSO and ECDMPSO variants tested on seven practical problems

Problems	Total	Mean PSO			ECDMPSO		
		Best	Bad	Equal	Best	Bad	Equal
Practical problems	07	00	04	03	04	00	03

Table 11 Comparison of SPSO, PSO-E and ECDMPSO variants tested on three multi modal problems

Problems	Total	SPSO			PSO-E			ECDMPSO		
		Best	Bad	Equal	Best	Bad	Equal	Best	Bad	Equal
Multi modal problems	03	00	01	02	00	01	02	01	00	02

Subject to Constraints

$$\sum_{i=1}^{100} y_i \leq 7500; \quad \sum_{i=1}^{50} 10y_i + \sum_{i=1}^{100} y_i \leq 42000; \quad 0 \leq y_i \leq 99, \quad i = 1, 2, \ldots, 100$$

10　Conclusions

Standard PSO (SPSO) is a heuristic optimization technique, which is being frequently applied in engineering applications, computer science and different fields of applied mathematics. However SPSO suffers from problem of premature termination at a local optimal point. To overcome this difficulty an alternative version of SPSO called exponential cumulative distribution mean PSO (ECDMPSO) has been proposed in this article for an optimization problem having more than one optimal solution. The objective of using exponential cumulative distribution in the proposed algorithm is to improve the performance of mean particle swarm optimization variant. The main aim of the proposed variant is to improve the performance of each member of the entire swarm in the global search space. The quality and performance of ECDMPSO algorithm have been tested on several benchmark and practical problems. On the basis of results obtained it is concluded that the proposed variant performs better than SPSO, Mean PSO and PSO-E.

References

1. Krohling, R.A., Leandro dos Santos Coelho, L.D.S.: PSO-E: Particle Swarm with Exponential Distribution. IEEE congress on Evolutionary Computation, Sheraton Vancouver Wall Centre Hotel, Vancouver, BC, Canada (2006).
2. Deep, K., Bansal, J.C.: Mean particle swarm optimization for function optimization. International Journal of Computational Intelligence Studies. 1(1), 72–92 (2009).
3. Singh, S.B., Singh, N.: Triple Hybrid of Particle Swarm Optimization Algorithm. In proceeding of ERCICA: Emerging Research in Computing, Information, Communication and Applications, Elsevier, ISBN: 9789351071020.508–515 (2013).
4. Kim, B.I., Son, S.J.: A probability matrix based particle swarm optimization for the capacitated vehicle routing problem. Journal of Intelligent Manufacturing. 23(4), 1119–1126 (2012).
5. Kanoh, H., Chen, S.: Particle Swarm Optimization with Transition Probability for Timetabling Problems. Adaptive and Natural Computing Algorithms, Lecture Notes in Computer Science, LNCS 7824. 256–264 (2013).
6. Liang J., Suganthan P., Deb K.: Novel composition test functions for numerical global optimization. In proceeding of Swarm intelligence symposium, SIS 2005, IEEE 2005. 68–75 (2005).

Allocation of Resource Using Penny Auction in Cloud Computing

Aditya Kumar Naik and Gaurav Baranwal

Abstract Allocation of resource using dynamic pricing is an important challenge in Cloud computing. Auction can be used for this problem. Auctioneer is considered as an entity that performs auction. In most of the variants of auction, auctioneer does not get any benefits to perform auction, i.e., there is no motivation to encourage auctioneer. In this work, a model is proposed using penny auction for allocation of resources and their pricing in such a way that auctioneer also gets some benefit. In proposed work, customers have to pay a non-refundable fee to participate in auction. Proposed work is elaborated by an example and future direction is also provided.

Keywords Cloud computing · Resource allocation · Penny auction
Dynamic pricing

1 Introduction

Cloud computing is a large scale distributed computing paradigm in which a pool of computing resources is available to users (called Cloud customers) via internet. Cloud customers need computing resources as a service from Cloud service providers to run jobs. Cloud service providers fulfill the need of customers by provisioning their computing resources using certain policies.

Since dynamism is an important characteristic of provisioning of computing resources, providers have to select various policies of provisioning very carefully.

A.K. Naik (✉) · G. Baranwal
Department of Computer Science and Engineering, Madan Mohan Malaviya
University of Technology, Gorakhpur, Uttar Pradesh, India
e-mail: sunnynaik96@gmail.com

G. Baranwal
e-mail: gaurav.vag@gmail.com

© Springer Nature Singapore Pte. Ltd. 2018
Y.-C. Hu et al. (eds.), *Intelligent Communication and Computational
Technologies*, Lecture Notes in Networks and Systems 19,
https://doi.org/10.1007/978-981-10-5523-2_26

Since Cloud is business model, price can be used as tool to design various policies, according to behavior of customer and to maximize the expectation of customers and providers. Generally, two types of pricing methods are found in literature of Cloud, i.e., Static pricing and Dynamic Pricing. Amazon offers resources in three ways: On Demand, Long Term Reservation, and Spot [1]. On demand and long term reservation follow static pricing while spot follows dynamic pricing.

Currently, only few Cloud service providers such as Amazon and Google are using dynamic pricing for allocation of resources. Auction and negotiation, both are widely accepted techniques for implementation of dynamic pricing. Recently auction got a great attention by researchers in the Cloud for allocation of computing resources. Auctioneer is an entity that performs auction. In most of the existing works in Cloud, auctioneer gets no benefit to perform auction. To resolve this issue, a model is proposed using penny auction for allocation of computing resources and their pricing in this work.

2 Related Work

Among various variants of auction such as forward auction [2], double auction [3], and reverse auction [4], forward auction is widely accepted not only in the literature of Cloud but practically as well. Amazon offers computing resources, i.e., virtual machines (VMs) and calls them instances [1]. As discussed before Amazon provides instances using spot pricing. Spot pricing is basically a forward auction mechanism. Amazon claims that one can save money by using spot instances. Only the main issue is unreliability of Spot instances. Forward auction can be defined as an auction where multiple customers place their bids. Few recent research works related to forward auction are as follows.

In [2], W. Xing-Wei and W. Xue-Yi (2012) introduced an improved periodical auction model. This model was introduced to solve the limitation of English combinatorial auction. Based on this auction the trading price of resource is determined between resource buyer and resource provider. After conducting the experiment the result showed that this model is both feasible and effective which can increase the revenue of seller and reduce the execution time for the winner.

In [5], Sharukh Zaman and D. Grosu (2013) introduced two new methods CA-LP (Combinatorial Auction-Linear Programming) and CA-GREEDY (Combinatorial Auction-Greedy) for VM allocation with a combinatorial auction mechanism and compared them with static pricing. The proposed method helped in capturing the market demand effectively and concluded that combinatorial forward auction is a better choice for allocation of resources in Cloud. CA-LP performs better for less number of users and when auction intervals are large. While CA-GREEDY can be a good choice for any number of users with short auction intervals.

In [6], Sharukh Zaman and D. Grosu (2013) proposed a modified method of his previous proposed method that is CA-PROVISION. After conducting the experiment, it is found that this method helped in capturing the market demand, helped in matching the computing resource with the demanded resource and last but not the least it generated higher revenue than CA-GREEDY method.

In [7], Lena Mashayekh and Daniel Grosu (2015) introduced a new method Polynomial-Time Approximation Scheme (PTAS). The reason for proposing this method was for easy allocation of resources from the provider's to the needy user who has requested for that particular resource. Although the main goal was to increase the revenue of social welfare in order to lead the system in an equilibrium state.

In [8], Mahyar Movahed Nejad and Lena Mashayekhy (2015) introduced a new method G-VMPAC that is truthful optimal and greedy mechanism for solving the VMPAC (VM provisioning and allocation in clouds) problem. After conducting the experiment the result showed that the proposed technique helped in easily acquiring the market demand and the execution time is very less. G-VMPAC is the optimal choice for the Cloud provider and earned the highest revenue.

In [9], Y. Choi and Y. Lin (2015) introduced a new method that is winner determination mechanism along with considering the penalty cost. The reason for establishing this method was to maximize the performance of the provider and helps in the easy utilization of the resource. The result showed that it generated more revenue when considering dynamic pricing.

Although in [2, 5–9], all the authors have discussed about increasing the revenue for the provider, but auctioneer does not get any profit for conducting the auction. Auctioneer is an important entity in auction. Auctioneer is a market maker that not only conducts the auction but also helps in improving customer–provider relationship. Without getting some benefit, auctioneer cannot be considered as independent entity. Forward auction does not give any guarantee to all bidders whether they will win or lose and does not provide any mechanism to attract new customers. In this work, we adopted penny auction [10] for allocation and pricing of computing resources in Cloud that not only helps in improvement of revenue of provider but also provides certain profit to both provider and auctioneer, and winning customer has not to overbid.

3 Proposed Model

In the proposed model in this work, computing resources are allocated using penny auction. Penny auction is also known as bidding fee auction. It is a type of auction similar to all-pay-auction. In this type of auction, all the participants are required to pay a small non-refundable fee in order to place a bid in each round. Each round generally last after a certain time and generally the last person who has placed the bid before the clock turns zero wins the item as the last bid is generally the highest bid. In this Cloud auction, there are three main entities: Cloud Customer, Cloud Provider, and Cloud Auctioneer. Cloud customers are Cloud bidders, i.e., users who are interested in assimilating the Cloud services. Cloud service sellers are Cloud providers who are interested in providing the computing resources to the customers. Cloud Auctioneer is the most important entity for the auction to take place. It simply acts as the mediator between the Cloud customer and the cloud provider. It simply accepts the offering of resource by Cloud provider and accepts bids from Cloud customers. After acceptance, it decides the winner; it is basically a market maker.

Start of Auction: After getting information about computing resource offered by provider, the Cloud auctioneer starts the auction and conveys the message to Cloud customers to bid. The auctioneer gives the information about the computing resource, which is being placed for bidding, to the customers.

Submission of Bids: Let n be the number of bidders, i.e., $(b_1, b_2, b_3, \ldots, b_n)$ which are ready to bid for single resource offered by provider in the Cloud market. Each bidder who wants to place a bid has to pay a non-refundable fee, i.e., registration fee. Let $RegFee_r$ be the registration fee for rth auction round and bv_i be the value of bid of b_i in current auction round.

Determination of the Winner: The winner of the auction is determined at the end of the auction and the Cloud resource for which bid was being done is handed over to the winner. There are two cases of determining the winner. First is one when the auction ends after the clock runs out and the person who has placed the last bid wins the item and he has to pay the final price of the item along with the non-refundable fee. Second is one when no bid is placed and the last person who has bided is declared as winner and collects the Cloud resource by paying the final bid to the auctioneer.

Closing of Participation in Auction: Here the job of the auctioneer is to make aware about the time allotted for bidding for the bidders so that they can place non-refundable participation fee. The auctioneer defines the time and all the interested player or bidder can register by paying a non-refundable fee for a current round of auction. Once the time is over the auctioneer closes the registration and no participant is allowed to register.

Let b_i be the ith bidder, r be the rth auction round, w be the index of winner, i.e., if b_i is bidder then $w = i$, hbv be the highest bid value in current auction round, $RegFee_r$ be the registration fee for rth auction round, $TotalRegFee_r$ be the total registration fee collected by auctioneer in rth round of auction, $TotalRegFee$ be the variable to total registration fee collected in the auction, $FinalCost$ be the total cost obtained by auctioneer including the cost that to be paid to the provider, i.e., $hbv + TotalRegFee$. The algorithm for proposed model is as follows.

```
TotalRegFee = 0
w = −1
r = 1
Flag a = true
hbv = 0
while (a == true){
        flag x = false
        TotalRegFee_r = 0
        Start the Clock
        for all bidder b_i, do {
                if (i ≠ w)and (bv_i > hbv){
                        w = i
                        b_i pays RegFee_r to Auctioneer
                        hbv = bv_i
                        TotalRegFee_r = TotalRegFee_r + RegFee_r
                        x = true
                } end if
                if (Clock ends)
                        Exit from the for loop i.e.Termination of current rth round
                end if
        } end for
        TotalRegFee = TotalRegFee + TotalRegFee_r
        if (x == false ){
                a = false
        } end if
        r = r + 1
} end while
FinalCost = hbv + TotalRegFee
Bidder b_w pays bv_w or hbv to provider
```

4 Numerical Example

Let us assume that Amazon is auctioning a virtual machine resource that is M4.10XLarge whose on-demand price is $2.394 per hour [1] and there are 5 cloud customers who want to place a bid in order to get this particular resource. Currently, there is no winner. The auction starts and all the 5 cloud customers need to pay a non-refundable fee, i.e., registration fee in each round to bid. We have fixed registration fee for first round which is $0.01 and in subsequent round, we are increasing this registration fee by $0.01, i.e., $RegFee_1 = 0.01$, $RegFee_2 = (0.01 + 0.01) = 0.02$ and so on. In each round of auction, cloud customer who bids first becomes the leader of that round. Table 1 gives the exact summary after completion of first round.

After the completion of first round, auctioneer checks that whether there is a winner or not. If there is a winner, again a request is made to every customer that whoever wants to bid again has to pay registration fee for second round one and the

Table 1 First round of auction

Bidder	Bid value	Registration fee ($RegFee_1$)	Leader (w)
b_1	$bv_1 = \$0.06$	$0.01	–
b_2	$bv_2 = \$0.08$	$0.01	–
b_3	$bv_3 = \$0.13$	$0.01	–
b_4	$bv_4 = \$0.15$	$0.01	–
b_5	$bv_5 = \$0.20$	$0.01	5
Winner of first round = b_5			
Total registration fee for this round ($TotalRegFee_1$) = \$0.05 $TotalRegFee$ = \$0.05	Total cost = \$0.20		
Total collection = \$0.05 + \$0.20 = \$0.25			

Table 2 Second round of auction

Bidder	Bid value	Registration fee ($RegFee_2$)	Leader (w)
b_1	$bv_1 = \$0.21$	$0.02	–
b_2	$bv_2 = \$0.23$	$0.02	–
b_3	$bv_3 = \$0.31$	$0.02	3
Winner of second round = b_3			
Total registration fee for this round ($TotalRegFee_2$) = \$0.06 $TotalRegFee$ = \$0.05 + \$0.06 = \$0.11	Total cost = \$0.31		
Total collection = \$0.11 + \$0.31 = \$0.42			

leader of first round is not allowed to bid at first so that all cloud customers may get the chance to become the leader. Table 2 gives the exact summary after the second round is completed.

After round two if customers want to bid then again the same procedure will be followed else if nobody wants to bid than the last person who placed the bid is declared as winner. If say they want to bid again exact summary after the completion of third round of auction is given in Table 3.

As we assumed that this was the last round and the last person who has placed the bid is declared as the winner. After the auction is completed, the final cost is \$0.76, i.e., summing of the total registration fee and bid value collected from the winner. The cloud auctioneer gets the total registration fees (\$0.17) which has been collected during auction. At the end the *bidder b_2*, i.e., cloud customer pays the final cost (\$0.59) to cloud provider and gets the resource.

Table 3 Third round of auction

Bidder	Bid value	Registration fee ($RegFee_3$)	Leader (w)
b_1	$bv_1 = \$0.45$	\$0.03	–
b_2	$bv_2 = \$0.59$	\$0.03	2
Winner of third round = b_2			
Total registration fee for this round ($TotalRegFee_3$) = \$0.06 $TotalRegFee$ = \$0.11 + \$0.06 = \$0.17	Total cost = \$0.59		
Total collection = \$0.17 + \$0.59 = \$0.76			

5 Discussion and Conclusion

Recently a concept of Cloud market has been introduced where computing resources are provided on pay per use basis. Benefits of Cloud computing have attracted not only customers but providers also for their contribution in recent past. A great increment in number of providers and customers has been observed which in turns increased a healthy competition in the Cloud market. In nascent age, service providers offered resources using static pricing. But because of increasing competition in Cloud market, Cloud service providers have been started offering computing resources using dynamic pricing too to tackle the shortcomings of static pricing and to increase their revenue and satisfaction of their customers. Negotiation and auction are widely accepted methods to implement dynamic pricing. In this work, auction is considered to implement dynamic pricing.

There is availability of various variants of auction. In literature of Cloud, Forward auction has got attention not only by researchers but Amazon also implemented it as spot pricing. In this work, a framework for allocation of computing resources and their pricing using penny auction is provided.

The contributions of proposed work are discussed below. Related works of forward auction are explained and various shortcomings are identified and discussed. It is observed that auctioneer is a main entity that performs auction but does not get any benefit from the auction in most of the existing works in the Cloud. Proposed work provides a model to allocate computing resource to customers in Cloud using penny auction in such a way that auctioneer also gets some benefit in terms of money. In this auction, all the participating customers are required to pay a small non-refundable fee in order to place a bid in each round.

In the literature of economics, various advantages of Penny Auction [10] for allocation and pricing of resources are observed which are as follows:

1. In this auction, generally, the winner pays less money than the original value of the item and revenue is generated from the bid fee rather than the winning price.

2. Penny Auction provides a winning guarantee that is by default at-least one auction to be won by its bidder. Generally, penny auction offers set of auction for new bidders only so that new bidder can win few auctions.
3. Penny Auction is keen on acquiring new bidders by providing multiple offers to the new bidders rather than the existing bidders.
4. In Penny auction, the auctioneer makes money in two ways. First is through the total collected bid that is the non-refundable fee. Second is the final bid that the customer has paid in order to gain that particular item

As discussed before, there are various advantages of penny auction in economics. One can say that limitation of penny auction is that bidders have to pay the participation fee. But if auctioneers have to get some profit, participation fee may be a better approach. Quibids [11] is a retailer website which is running penny auction successfully. Since Cloud computing is also a business model, future work will concentrate on the implementation part to evaluate penny auction for allocation of resource in Cloud in order to study the performance of the proposed framework.

References

1. Amazon EC2, https://aws.amazon.com/ec2/, 2017.
2. X. W. Wang, X. Y. Wang and M. Huang, "A resource allocation method based on the limited English combinatorial auction under cloud computing environment", In Fuzzy Systems and Knowledge Discovery (FSKD), 2012 9th International Conference on, pp. 905–909, May 2012, IEEE.
3. G. Baranwal and D. P. Vidyarthi. "A Truthful and Fair Multi-Attribute Combinatorial Reverse Auction for Resource Procurement in Cloud Computing." IEEE Transactions on Services Computing, 2016.
4. D. Kumar, G. Baranwal, Z. Raza and D. P. Vidyarthi "A Systematic Study of Double Auction Mechanisms in Cloud Computing." Journal of Systems and Software, vol. 125, pp. 234–255, 2017.
5. S. Zaman and D. Grosu, Combinatorial auction-based dynamic vm provisioning and allocation in clouds. In Cloud Computing Technology and Science (CloudCom), 2011 IEEE Third International Conference on, pp. 107–114, November 2011, IEEE.
6. S. Zaman and D. Grosu, "A Combinatorial Auction-Based Mechanism for Dynamic VM Provisioning and Allocation in Clouds," IEEE Transactions on Cloud Computing, vol. 1, no. 2, pp. 129–141, 2013.
7. L. Mashayekhy, M. M. Nejad and D. Grosu, "A PTAS Mechanism for Provisioning and Allocation of Heterogeneous Cloud Resources," IEEE Transactions on Parallel and Distributed Systems, vol. 26, no. 9, pp. 2386–2399, 2015.
8. M. M. Nejad, L. Mashayekhy and D. Grosu, "Truthful Greedy Mechanisms for Dynamic Virtual Machine Provisioning and Allocation in Clouds," IEEE Transactions on Parallel and Distributed Systems, vol. 26, no. 2, pp. 594–603, 2015.
9. Y. Choi and Y. Lim, "A Framework for Optimizing Resource Allocation in Clouds", In 2015 3rd International Conference on Computer, Information and Application, pp. 10–14, May 2015, IEEE.
10. Hinnosaar, "Penny auctions are unpredictable. Unpublished manuscript", 2010, Available at http://toomas.hinnosaar.net/pennyauctions.pdf.
11. QuiBids, http://quibids.com/en/, 2017.

Optimized Location-Specific Movie-Reviewing System Using Tweets

Vijay Singh, Bhasker Pant and Devesh Pratap Singh

Abstract Day-by-day, the density of the mobile phone is increasing throughout the world, and consequently, the social media become more powerful than before. People share almost everything on the social media and this data can be further used to make various kinds of recommendation, prediction, and data analysis task. One of the main domains in this scenario is reviewing system. In this article, we review currently running movies based on dynamically collected tweets. Tweets are collected based on specific location so that the system can review the movie status on that location. The experimental results show that the system performs well and the same system can be applicable to other similar topics like song-reviewing system and product-reviewing system.

Keywords Data analysis · Movie-reviewing system · Recommendation system

1 Introduction

Social network becomes the new friend of everyone's life. People share almost every aspect of their life on the social networks. This generates the huge amount of data per second containing lots of information. Various research papers have been published in this domain; how to extract useful information from this data. When the data size becomes big, the conventional algorithm does not perform well and optimization required. Many applications and decisions systems have been developed using these social networks. Various websites IMDB (http://www.imdb.com) and reviews of the movies based on users post about the movies. When this kind of reviewing systems does not exist in the past, people used to refer another people go and watch this and that movie. The first part of the movie is not good. Now people want more precise information in local as well as global scenario and increased the dependencies on these sites. The purpose behind the reviewing system is how to

V. Singh (✉) · B. Pant · D.P. Singh
Graphic Era University, Dehradun, India
e-mail: vijaysingh_agra@hotmail.com

© Springer Nature Singapore Pte. Ltd. 2018
Y.-C. Hu et al. (eds.), *Intelligent Communication and Computational Technologies*, Lecture Notes in Networks and Systems 19,
https://doi.org/10.1007/978-981-10-5523-2_27

produce sentiment summary from the text. Text can be web-post, document, or tweets. Most of the similar work is focused on the product review. In this paper, we focused on reviews collected from twitter based on specific location. Twitter is one of the easiest ways to share information limiting to 140 characters. The paper is organized as follows: In the next section, background details and related work of reviewing system are shown. In Sect. 3, the methodology proposed for location-based reviewing system is shown. Section 4 described how much the system is optimized with the experimental results. Section 5 concludes the work and discusses the future improvements.

2 Background Details and Related Work

Word of mouth has been considering as an effective way to recommending and suggesting products or movies, and this technique is still working in a limited scenario. Nowadays, people exchange their contents on different websites more frequently and freely. Those who share contents (reviews, article, or blog) generally eager to know what other think about the product (movie, song, product) or collectively. Increased number of current studies focused on the economic importance of reviews and established the relationship between reviews of the product (movies, songs, restaurants, etc.) and their sales [1]. Xiaohui Yu et al. proposed a quality aware methodology for anticipating the future sales of the product. Marco Bonzanini et al. used Luhn's [2] approach for detecting the significance of the sentence. Further, they did a sentiment classification on the collected reviews from IMDB [3]. With such huge number of tweets, there is huge demand as well for technologies that enable users to get summarized from the tweets sentiments [4]. Jonghyup Lee et al. proposed a method in which they investigate the study of the effect of review sentiment by using entropy level, and established the relation between entropy level and movie sales [5]. In recent times, location services play an important role in social networking. People can share location-specific contents on these websites. Murale Narayanan and Aswani Kumar Cherukui, works on recommender system for location-based social networks (LBSN) [6]. Choochart Haruechaiyasak et al. proposed a method of text classification by predefined classes using fuzzy association technique. Further, they compare fuzzy association with vector space model to find the effect of keyword selection [7]. Megha Dawar and Aruna Tiwari proposed a technique of text classification using fast fuzzy feature clustering. By using this method, they reduced the number of iterations to find the desired number of the cluster [8]. Anil kumar reddy tetali, B P N Madhu kumar, and K. Chandr Kumar proposed an idea of text classification using fuzzy-based incremental feature selection algorithm. They did similarity test for grouping the documents [9]. Abinash Tripathy, Ankit Agarwal, and Santanu Kumar Rath proposed a comparison study of machine learning algorithm, support vector machine, and Naive Bayes for sentiment classification. Their findings show that support vector machine performs very well in the context of review classification [10]. Deepa Anand and Deepan

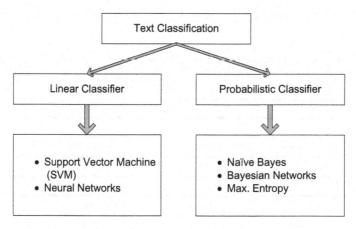

Fig. 1 Text classification techniques

Naorem work on movie-reviewing system, using aspect-based sentiment analysis. They aggregate opinions across multiple review statements and then apply clustering algorithm for automatic classes for aspect category mapping [11]. Mainly there are two types of techniques used for text classification: linear classification and probabilistic classification. Further, linear classification is classified as support vector machines and neural networks, and probabilistic classifiers are classified as Naive Bayes, Bayesian networks, and maximum entropy as shown in Fig. 1. In this article, we use Naive Bayes classification to convert token matrix to polarity matrix.

3   Proposed Approach

In this research, we have divided the whole task into following steps:

- Python-based tweepy library is used to download the location-specific tweets.
- Preprocessing techniques are used to remove noisy data from the collected tweets so that the efficiency of the system can be improved.
- After removal of noisy data, rest of the tokens are arranged into a token matrix.
- By using Naive Bayes, token matrix is converted into polarity matrix.
- Now polarity matrix is arranged as enough groups and groups are converted into single value.
- At last, the single value is left, that is the output of the system either positive or negative.

Python-based tweepy library is used to download the tweets from the twitter. Tweets were downloaded in the Indian region only. The Indian geographically location is divided into four regions, North, South, East, and West. Respectively, GUI developed where the user can choose the geographical region, where he or she wants to know the status of the movie using drop-down shown in Fig. 2.

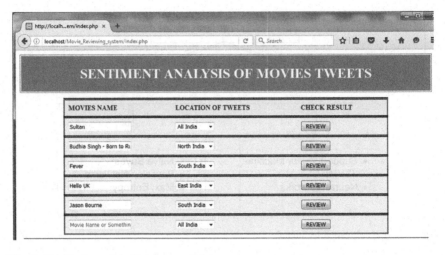

Fig. 2 Graphical user interface for user input

Architectural overview of movie reviewing system is shown in Fig. 3. After collecting location-based reviews from the twitter, by preprocessing, noisy information are filtered out. The steps involved in the proposed methodology is shown in Fig. 3.

- Data cleaning.
- Data transformation.
- Data reduction.

Whatever data are downloaded from the social network, there is the high probability that the collected data is a noisy data and various special characters, which does not contain any meaningful information. The attribute of choice may not always be there. So data cleaning is required, like filling missing values, smoothing or removing unnecessary information, and removing outliers. In data transformation, normalization, aggregation, and generalization of data are performed, whereas in data reduction technique, data size is reduced without losing critical information. After preprocessing of data, tokens are left. In the next step, convert these tokens into token matrix. Then apply Naïve Bayes classifiers to convert these tokens into positive and negative values, 1 and 0, respectively. Now we have a huge matrix containing 1 and 0 s. Algorithm 1 shows step-by-step procedure to generate final decision using polarity matrix. In Fig. 4, polarity matrix reduction technique is shown. In Fig. 4, the input matrix size is 8×8 containing 64 elements. In the next step, four groups are selected, and count the number of the positive element. If the positive elements are more than 50% of extracted group size, then the group is stored as a single value that is 1 otherwise set it to 0. When the threshold value achieved, then there is no need to further search. This saves ample amount of computing time. The same procedure is repeated until the final decision gets. The proposed algorithm is implemented using PHP, python, and R. The output screens of case 1 (when the final output is positive) and case 2 (when the final output is Negative) are shown in Figs. 5 and 6.

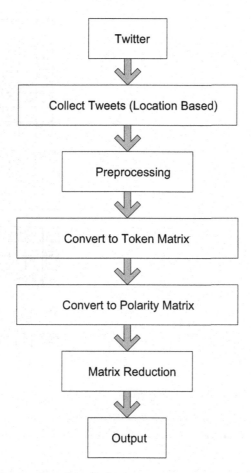

Fig. 3 Architectural steps of movie-reviewing system

Fig. 4 Polarity matrix
reduction technique

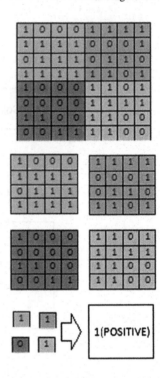

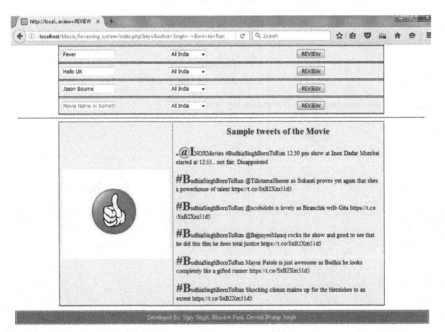

Fig. 5 Output screen of case_1

Algorithm 1: Generate decision using polarity Matrix
INPUT: Matrix (a) of n x n
// z: Dimensions of the sub Matrix
1. x = z
2.   for i = 1 to n
3.   for j = 1 to n
4.   if (j <= k) then
5.   if (i<= k) then
6.   M [i, j] = a [i, j]
7.   if (i = k and j = k) then
8.   val = cal_polarity(M)
9.   val_x = val_x + val
10.   k = k + z
11.   if (val_x >= (k x k)/ 2)
12.   for i = 1 to n/k
13.   for j = 1 to n/k
14.   mat [i, j] = 1
15.   else
16.   mat [i, j] = 0
12.cal_polarity(M)
13. n = rows(M)
14.   x = 0
15.   for i = 1 to n do
16.   for j = 1 to n do
17.   if (M [I, j] = 1) then
18.   x = x + 1
19.   return x

4 Experimental Setup and Results

In this section, we compare two approaches, without matrix reduction (WMR) and with matrix reduction (MR). For experimental setup, we take three samples, sample 1, sample 2, and sample containing 1024(32 × 32) elements with different polarity values. 1024 values represent token polarity value randomly. If matrix reduction technique is not applied, then we must process all the values; this issue is overcome and it proposed the matrix reduction technique, as shown in Fig. 7.

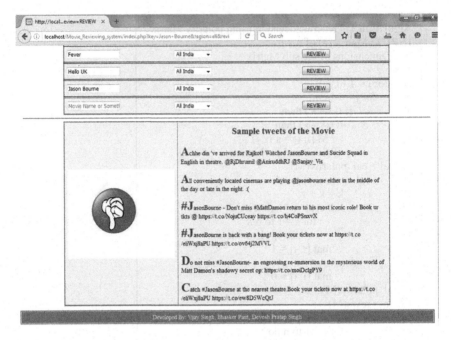

Fig. 6 Output screen of case_2

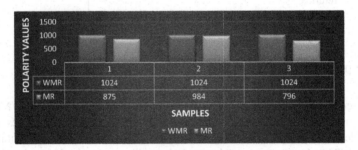

Fig. 7 Comparison analysis of matrix reduction (MR) technique with and without matrix reduction technique (WMR)

5 Conclusions

In this article, we proposed a novel approach to developing optimized movie-reviewing system using location-based sentiment classification and matrix reduction technique. Results are pleasing if compared with, without using matrix reduction technique. It is observed that matrix reduction technique outperforms in this kind of scenario. The same application can be applicable to other similar domain like product-reviewing system and hotel-reviewing system. In this study, we proposed a methodology for single processor machine. In future work, we shall develop a system for the parallel computing environment.

References

1. Xiaohui Yu, Yang Liu, Xiangji Huang and Aijun An, "A Quality-Aware Model for sales prediction Using Reviews", WWW2010, ACM 978-1- 60558-799-8/10/04.
2. H.P. Luhn, "The automatic creation of literature abstracts", IBM journal of Research and Development, 2(2) (1958):159–165.
3. Marco Bonzanini, Miguel Matinez-Alvarez and Thomas Roellaka, "Opinion Summarisation through Sentence Extraction: an invertigation with Movie Reviews", SIGIR:12, ACM 978-1-4503-1472- 5/12/08.
4. Anan Liu, Yongdong Zhang, Jitao Li, "Personalized Movie Recommendation", MM:09, ACM 978-1-60558-608-3/09/10.
5. Jonghyup Lee, Jaehong Park and Sunho Jung, "The impact of the Entropy of Review text sentiments on movie box office sales", ICES'15 ACM ISBN 978-1-4503-3461-7/15/08.
6. Marule Narayanan and Aswani Kumar Cherukuri, "A Study and analysis of recommendation systems for location-based Social Networks (LBSN) with Big data", IIIM Management Review (2016) 28, 25–30.
7. Choochart Haruechaiyasak, Mei-Ling Shyu and Shu-Ching Chen, "Web Document Classification Based on Fuzzy Association", Computer Software and Application Conference, 2002, pages 487–492.
8. Megha Dawar and Arjun Tiwari, "Fast Fuzzt Feature Clustering for Text Classification" Academy and Industry Research Collaboration Centre, Computer Science and Information technology, Vol.2, pages 167–172, 2012.
9. Anil Kumar Reddy Tetali, BPN Madhu Kumar and K.Chandrakumar, "Classification of text Using Fuzzy Based Incremental feature Classification algorithms" International Journal of Advanced Research In computer Engineering and technology, Vol.1, Issue.5, 2012.
10. Abinash Anand, Ankit Agarwal and Santanu Kumar rath, "Classification of Sentiment Reviews using machine Learning Techniques", 3rd International Conference on Recent trends in Computing, Procedia Computer Science, Vol.57, pages 821–829, 2015.
11. Deepa Anand and Deepam Naorem, "Semi-Supervised Aspect based Sentiment Analysis for Movies Using Review Filtering", 7th International Conference on Intelligent Computer Interaction, IHCL, Procedia Computer Science, Vol. 84, 2016.

Adaptive System for Handling Variety in Big Text

Shantanu Pathak and D. Rajeshwar Rao

Abstract Today in every corporate, banking, judicial, or medical ecosystem varieties of text are generated like customer reviews, product manuals, white papers, system logs, and usage data. They vary in language, size, context, and formats. Handling such text using a single system is still a challenge. Traditionally, systems exist to handle each specific part of generated text, separately. So this work proposes a concrete step toward integrated solution to the challenge. The proposed system handles text with different formats, sizes, languages, and context seamlessly, encompassing text generated across the ecosystem. Implementation over heterogeneous dataset of text shows promising results. This integrated approach empowers analytics with an extra edge to learn hidden relational and contextual patterns over complete system.

Keywords Big text · Variety · Text model · Text analytics · Big data

1 Introduction

Big Data field has seen a lot of experiments on volume [1]. Variety has been a neglected part of Big Data. Even a small quantity of data with variety possesses challenges [2]. Variety is not only in source of data and format but also in size, languages, etc. Variety in text data brings about different sets of challenges and can be termed as Big Text.

Variety in text can be based on many parameters like source of text, format, size, and languages. Texts from various sources like databases, tables, sheets, web pages, and social media have variety in purpose and formats. Such variety in structured text in terms of sources and formats has been well handled in literature as separate cases. On the other hand, variety in unstructured text is handled by various systems. Such systems handle these differences as separate cases [3–5].

S. Pathak (✉) · D. Rajeshwar Rao
CSE Department, K L University (K L Education Foundation), Vijayawada, India
e-mail: shantanuspathak@gmail.com

D. Rajeshwar Rao
e-mail: rajeshduvvada@kluniversity.in

© Springer Nature Singapore Pte. Ltd. 2018
Y.-C. Hu et al. (eds.), *Intelligent Communication and Computational Technologies*, Lecture Notes in Networks and Systems 19,
https://doi.org/10.1007/978-981-10-5523-2_28

305

This paper is organized into five sections. Literature survey describes existing models and research in the concerned area. Next, mathematical model is described followed by the proposed system. Then experimentation, results, and conclusion follow one after the other.

2   Literature Survey

Kaisler in [1] described all Vs of Big Data along with challenges and issues in this field. It clearly points out that Volume is the most addressed V in Big Data. At the same time other V's, variety and velocity, also need focus for future applications of Big data. On contrary, [6] discusses different controversies faced by Big data. Important issues like Big data as a tool for Hadoop promotion and ethical access of data are mentioned here.

There are few works which address variety aspect in Big Data. For a mix of both numerical and categorical data, [7] proposes an efficient method based on Adjusted Self-Organizing Incremental Neural Network (ASOINN). Work [2] discusses the strategies for handling Big Data volume and variety. It showcases a variety of data as a serious challenge. Case of medical data containing more than 90k medical codes and 31k medicine names is elaborated. With such variety in the text, determination of patterns in relations, anomalies, and errors is difficult. In Big Text, [5] handles text with multiple languages as variety. This work introduces a method for training the classifier with multilingual text. On the other hand, authors in [8] discuss challenges in handling social media text. Although the focus is on validity and representativeness, a variety of data as per platforms is also described here. It handles methodological issues in social media data. Work [9] handles social media data for mood sensing. Additionally, [10] describes the intrusion detection system over heterogeneous Big Data and [11] describes it for twitter data.

Heterogeneous text from the web is handled in [12]. Here text from the web is segmented and segments are processed for classification. The text handled contains judicial data in multiple languages including Slovak language. Most frequent terms from the segments are used here for improving classification accuracy. Its application in transcription and dictation system has shown significant improvements. Similar is the case with [13], where trend analysis of a technical website stack overflow is done. Another work [4] handles Chinese web pages by using size and place of the phrase for feature selection.

Further, [14] describes the large-scale heterogeneous text networks. These networks are based on the word-to-word or word-to-document or word-to-class mappings. This technique is applied efficiently on the predictive text embedding. Results are comparable with the state-of-the-art supervised systems. On the other hand, [15] works on streaming log data for abnormality analysis.

In a matter of handling different sizes of text, splitting and extension methods are popular. For handling documents with large size [16] and with multiple topics [17], splitting document into micro-documents is given in [18]. On the other hand,

[3, 19] use extension method with the relevant model on short texts to improve text categorization. Also, [20, 21] discuss the model to add relevant features to enrich the short text with context. Next, TermCut method in [22] discusses core term-based clustering for short texts with claims of better performance than relevant methods of clustering. Short text classification is also discussed in [23, 24].

To improve text categorization, [25] proposed Term Ranking Identifier (TRI) measure. This measure helps in identifying the semantic value of each term in the document. Along with term frequency it also takes into consideration the number of documents containing the term and belonging to the same class. Additionally, [26] uses space density with class index. Author badawi in [27] proposes novel framework for term selection for binary text classification. Next, [28] gives improved Term Frequency-Inverted Document Frequency (tf-idf) for micro-blog handling. Another approach is Term Occurrence Ratio Weighing proposed by chirawichitchai in [29]. A novel approach of prototyping for classification of text is presented in [30]. Survey of such methods for text representation is presented in [31].

To sum up, in studied literature there is scope for improvement in terms of single model for representing varieties of text like short text, paragraphed text, document, monolingual, and multilingual text. The proposed system is an attempt to address this gap. This model can act as an integrating factor for handling various text sources by a single system.

3 Mathematical Model

3.1 Measure of Variety

In this system three parameters, size of the document s, number of languages in the document l, and topics in the document top, are used to measure variety. As all the three parameters are directly proportional to the variety, a combined feature v is formed. Standard deviation σ_v and mean μ_v of variety are calculated based on v:

$$v = s \times l \times top. \tag{1}$$

3.2 Coefficient of Variation

In statistics, CV is measure for Variation in numeric data. It is defined as division of standard deviation σ and mean μ. Here, CV is applied to text data to measure variations in provided dataset with m documents:

$$CV = \frac{\sqrt{\sum_{i=0}^{m}(v_i - \mu_v)}}{\sum_{i=0}^{m} v_i}. \tag{2}$$

3.3 Thresholds

System needs to define minimum and maximum limits on size of the input text:

$$t_{smin} = \mu_v - \sigma_v \tag{3}$$

$$t_{smax} = \mu_v + \sigma_v. \tag{4}$$

Additionally, threshold on number of languages t_l and number of topics t_t are set to default value of 1. All these thresholds help select appropriate processing steps in the system.

4 Proposed System

Prominent purpose of proposed system is to process varieties of texts. It can handle text from various sources like review comments, product documentation, or emails. Input text can be of either short or paragraph or document size. Input text expressed using multiple languages is well handled by the model. System is independent of the format of texts like xml, pdf, txt, html, or any other text format.

Major steps in the system are preprocessing and text model building. Preprocessing has four sequences of substeps. First sequence of substeps is splitting, dimensionality reduction, and pre-clustering. It is useful for processing large size text or multilingual text. Another sequence is extension, dimensionality reduction, and pre-clustering. This is useful for short-sized text. In case of multilingual short text, splitting, extension, dimensionality reduction, and pre-clustering are used. For single language paragraphed text, within limits of threshold of size, dimensionality reduction and pre-clustering are included. Standard cleaning substep, including stemming and stop word removal, is common to all the sequences. For any input text, selector module selects an appropriate sequence.

Next step, text model building works on the preprocessed text. Proposed parameters of the model are term frequency, document frequency, class frequency, and space density. First two parameters are taken from popular tf-idf model [32]. They are considered as most important parameters to represent text. Next, class frequency

Fig. 1 Proposed system

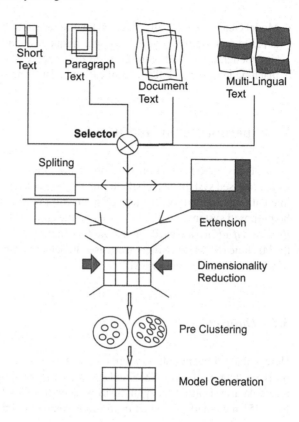

and space density [26] are recently proved parameters for efficient text representation. Implementation of space density parameter will be done in future.

Explanation for each part of the system as shown in Fig. 1 is as follows:

- Selector: This component is responsible for directing the cleaned text to appropriate sequence of preprocessing. Selection of sequence is based on size thresholds t_{smin} and t_{smax}, number of language threshold t_l, and number of topics threshold t_t.
- Splitting: Text containing multiple fragments is splitted here. Fragments can be identified based on the size, context, and language [18].
- Extension: Enriching short text is the responsibility of this part. Text with tokens less than t_{smin} is extended with relevant tokens. Related tokens are added using tools like Wordnet as implemented by Man [21] and Sarker [3].
- Dimensionality Reduction: This is a very common step in preprocessing of text. Here, most important and contextual features or dimensions are selected from input text.
- Pre-clustering: This step has been added to preprocessing by Zhou [33] and Jun [34]. In this work pre-clustering is used for two purposes, namely to select the relevant features and to reduce sparseness.

- Model Generation: Finally, selected features are used to generate the model. Model is based on term frequency, document frequency, class frequency, and space density. Generated model is in the form of a matrix. It can be directly used for any machine learning task like classification or clustering.

5 Experimentation and Results

Goal of this experimentation is to prove the effectiveness of the model generated from proposed system on classification of variegated text. Scope of the current experimentation is limited to detection of spam from pure English text of different sizes. So proposed system is evaluated based on the composite dataset of SMS and emails. *tf-idf* and *tf-idf-icf* are used for text model generation. Naive Bayes and Support Vector Machine (SVM) classifiers were used to detect the spam or ham text, irrespective of whether it is an email or SMS.

5.1 Dataset

Here, a dataset containing mixture of varieties of text is needed. So, for experiment purpose, dataset of spam SMS and spam emails are mixed together to form a composite dataset. Dataset for spam SMS is taken from UCI Machine Learning Repository [35] and emails are taken from Lings spam email dataset [36].

5.2 Results

Classification results for model generated from the proposed system are shown in Fig. 2. Here results are evaluated using two benchmark models, tf-idf and

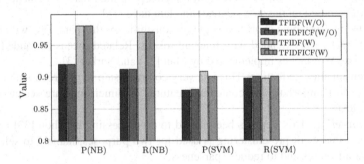

Fig. 2 Comparison of classification results

tf-idf-icf. Precision and recall on Naive Bayes (NB) and SVM are compared here. TFIDF(W/O) shows precision and recall for plain tf-idf model, and similar entry is TFIDFICF(W/O) for plain tf-idf-icf model. On the other hand, TFIDFICF(W) and TFIDFICF(W) show precision and recall using the proposed system. In both cases precision is improved using this system on NB and SVM. Additionally, recall in NB is also improved for TFIDFICF(W) and TFIDFICF(W). Only exception here is in case of recall of SVM. It is observed to be maintained by the system if not improved.

6 Conclusion

Proposed system handles a variety of text in a single model. Text from various sources, formats, sizes, and languages are successfully handled by the system. Proposed system is tested on a composite dataset of spam emails and spam SMS. Text model, generated by the system, improved results of classification on both Naive Bayes and SVM classifier.

It is prominent that application will be to integrate product reviews (from websources like twitter), customer email queries, and product enquiries (from help desk logs) of a single organization in a single text model. Such model will reveal new patterns, contexts, and learnings for better decision-making in a single window. Another application is to provide a single SPAM filter for any variety of text. Here system has been proved to combinedly handle email and SMS spams. Taking it further, a single window for any text analytics over variety of text is possible using this system to build the text model.

In future, same system can be tested over multilingual text with multiple languages. Handling short-hand text in SMS is another area for exploration.

References

1. S. Kaisler, F. Armour, J. A. Espinosa, and W. Money, "Big data: Issues and challenges moving forward," in *System Sciences (HICSS), 2013 46th Hawaii International Conference on*. IEEE, 2013, pp. 995–1004.
2. B. Shneiderman and C. Plaisant, "Sharpening analytic focus to cope with big data volume and variety," *Computer Graphics and Applications, IEEE*, vol. 35, no. 3, pp. 10–14, 2015.
3. A. Sarker and G. Gonzalez, "Portable automatic text classification for adverse drug reaction detection via multi-corpus training," *Journal of biomedical informatics*, vol. 53, pp. 196–207, 2015.
4. Y. Zheng, W. Han, and C. Zhu, "A novel feature selection method based on category distribution and phrase attributes," in *Trustworthy Computing and Services*. Springer, 2014, pp. 25–32.
5. C.-P. Wei, C.-S. Yang, C.-H. Lee, H. Shi, and C. C. Yang, "Exploiting poly-lingual documents for improving text categorization effectiveness," *Decision Support Systems*, vol. 57, pp. 64–76, 2014.
6. W. Fan and A. Bifet, "Mining big data: current status, and forecast to the future," *ACM sIGKDD Explorations Newsletter*, vol. 14, no. 2, pp. 1–5, 2013.

7. F. Noorbehbahani, S. R. Mousavi, and A. Mirzaei, "An incremental mixed data clustering method using a new distance measure," *Soft Computing*, vol. 19, no. 3, pp. 731–743, 2015.

8. Z. Tufekci, "Big questions for social media big data: Representativeness, validity and other methodological pitfalls," *arXiv preprint* arXiv:1403.7400, 2014.

9. T. Nguyen, D. Phung, B. Adams, and S. Venkatesh, "Mood sensing from social media texts and its applications," *Knowledge and information systems*, vol. 39, no. 3, pp. 667–702, 2014.

10. R. Zuech, T. M. Khoshgoftaar, and R. Wald, "Intrusion detection and big heterogeneous data: A survey," *Journal of Big Data*, vol. 2, no. 1, pp. 1–41, 2015.

11. Z. Miller, B. Dickinson, W. Deitrick, W. Hu, and A. H. Wang, "Twitter spammer detection using data stream clustering," *Information Sciences*, vol. 260, pp. 64–73, 2014.

12. J. Staš, J. Juhár, and D. Hládek, "Classification of heterogeneous text data for robust domain-specific language modeling," *EURASIP Journal on Audio, Speech, and Music Processing*, vol. 2014, no. 1, pp. 1–12, 2014.

13. A. Barua, S. W. Thomas, and A. E. Hassan, "What are developers talking about? an analysis of topics and trends in stack overflow," *Empirical Software Engineering*, vol. 19, no. 3, pp. 619–654, 2014.

14. J. Tang, M. Qu, and Q. Mei, "Pte: Predictive text embedding through large-scale heterogeneous text networks," in *Proceedings of the 21th ACM SIGKDD International Conference on Knowledge Discovery and Data Mining*. ACM, 2015, pp. 1165–1174.

15. A. N. Harutyunyan, A. V. Poghosyan, N. M. Grigoryan, and M. A. Marvasti, "Abnormality analysis of streamed log data," in *Network Operations and Management Symposium (NOMS), 2014 IEEE*. IEEE, 2014, pp. 1–7.

16. S. Baccianella, A. Esuli, and F. Sebastiani, "Using micro-documents for feature selection: The case of ordinal text classification," *Expert Systems with Applications*, vol. 40, no. 11, pp. 4687–4696, 2013.

17. Q. Wang, Y. Qian, R. Song, Z. Dou, F. Zhang, T. Sakai, and Q. Zheng, "Mining subtopics from text fragments for a web query," *Information retrieval*, vol. 16, no. 4, pp. 484–503, 2013.

18. A. Tagarelli and G. Karypis, "A segment-based approach to clustering multi-topic documents," *Knowledge and information systems*, vol. 34, no. 3, pp. 563–595, 2013.

19. A. Awajan, "Semantic similarity based approach for reducing arabic texts dimensionality," *International Journal of Speech Technology*, pp. 1–11, 2015.

20. J. Tang, X. Wang, H. Gao, X. Hu, and H. Liu, "Enriching short text representation in microblog for clustering," *Frontiers of Computer Science*, vol. 6, no. 1, pp. 88–101, 2012.

21. Y. Man, "Feature extension for short text categorization using frequent term sets," *Procedia Computer Science*, vol. 31, pp. 663–670, 2014.

22. X. Ni, X. Quan, Z. Lu, L. Wenyin, and B. Hua, "Short text clustering by finding core terms," *Knowledge and information systems*, vol. 27, no. 3, pp. 345–365, 2011.

23. B.-k. Wang, Y.-f. Huang, W.-x. Yang, and X. Li, "Short text classification based on strong feature thesaurus," *Journal of Zhejiang University SCIENCE C*, vol. 13, no. 9, pp. 649–659, 2012.

24. D. D. R. R. S Pathak, "Message manager (mm): A novel sms classification system," *International Journal of Advanced Computer Communications and Control*, vol. 02, no. 02, p. 2, april 2014.

25. K. P. Chand and G. Narsimha, "An integrated approach to improve the text categorization using semantic measures," in *Computational Intelligence in Data Mining-Volume 2*. Springer, 2015, pp. 39–47.

26. F. Ren and M. G. Sohrab, "Class-indexing-based term weighting for automatic text classification," *Information Sciences*, vol. 236, pp. 109–125, 2013.

27. D. Badawi and H. Altınçay, "A novel framework for termset selection and weighting in binary text classification," *Engineering Applications of Artificial Intelligence*, vol. 35, pp. 38–53, 2014.

28. X. Huang and Q. Wu, "Micro-blog commercial word extraction based on improved tf-idf algorithm," in *TENCON 2013-2013 IEEE Region 10 Conference (31194)*. IEEE, 2013, pp. 1–5.

29. N. Chirawichitchai, "Developing term weighting scheme based on term occurrence ratio for sentiment analysis," in *Information Science and Applications*. Springer, 2015, pp. 737–744.

30. J. Zhang, L. Chen, and G. Guo, "Projected-prototype based classifier for text categorization," *Knowledge-Based Systems*, vol. 49, pp. 179 189, 2013.

31. D. D. R. R. S Pathak, "Extensive study on text representation models in text mining," *IJAER*, vol. 10, no. 13, pp. 32 967–32 973, Oct 2015.

32. G. Salton and C. Buckley, "Term-weighting approaches in automatic text retrieval," *Information processing & management*, vol. 24, no. 5, pp. 513–523, 1988.

33. X. Zhou, Y. Hu, and L. Guo, "Text categorization based on clustering feature selection," *Procedia Computer Science*, vol. 31, pp. 398–405, 2014.

34. S. Jun, S.-S. Park, and D.-S. Jang, "Document clustering method using dimension reduction and support vector clustering to overcome sparseness," *Expert Systems with Applications*, vol. 41, no. 7, pp. 3204–3212, 2014.

35. T. A. Almeida, J. M. G. Hidalgo, and A. Yamakami, "Contributions to the study of sms spam filtering: New collection and results," in *Proceedings of the 11th ACM Symposium on Document Engineering*, ser. DocEng '11. New York, NY, USA: ACM, 2011, pp. 259–262. [Online]. Available: doi:10.1145/2034691.2034742

36. I. Androutsopoulos, J. Koutsias, K. V. Chandrinos, G. Paliouras, and C. D. Spyropoulos, "An evaluation of naive bayesian anti-spam filtering," arXiv preprint arXiv:cs/0006013, 2000.

Storage Size Estimation for Schemaless Big Data Applications: A JSON-Based Overview

Devang Swami and Bibhudatta Sahoo

Abstract Numerous technologies have been proposed for storing big data on the Cloud platform. However, choice of these technologies is always application specific. Determining a strong model is a perplexing task which makes it necessary for the architects and designers to review the requirements and choose a solution. This paper presents 14 data models available in the market to choose from. Above all, there are more than 45 database solutions available in the market, which can be categorized into one of the data models each of which is applicable to its own set of use cases (However, there are few products which could not be categorized into any of these 14 data models). Contributors have figured out that while storing schemaless information, the size of data stored in the database is higher than the original size. Metadata information and physical schema are the two responsible factors for such a high amount of storage requirement. Mathematical models and experimental evaluations conducted show that MongoDB requires storage space many times more than the original size of data. A storage space estimation equation for JSON-based solutions has been suggested, which can compare the storage requirement size using space required by CSV as a base. This may be used to decide an approximate amount of storage space required by the application, before buying a storage space in the Cloud environment.

Keywords Big data · Schemaless data · Cloud · Storage

1 Introduction

Big data is a buzz word which usually represents enormous data which cannot be processed by a single system due to its bulky size, large variety, and high-speed of generation. Advancement in IT technologies is the primary reason for generation

D. Swami (✉) · B. Sahoo
National Institute of Technology Rourkela, Rourkela 769008, Odisha, India
e-mail: swamx.mi@gmail.com

B. Sahoo
e-mail: bdsahu@nitrkl.ac.in

© Springer Nature Singapore Pte. Ltd. 2018
Y.-C. Hu et al. (eds.), *Intelligent Communication and Computational Technologies*, Lecture Notes in Networks and Systems 19,
https://doi.org/10.1007/978-981-10-5523-2_29

315

Fig. 1 Physical data model
[1]

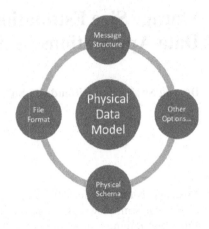

of big data. At any given time period only a fraction of big data is useful for most application domains. Hence, many experts and researchers have recommended the use of cloud for big data to optimally manage and reduce the overall cost of operating such systems. Cloud computing is a model which carters three services of its users, namely dynamisms, abstraction and resource sharing. Generally, a storage structure is defined in the physical data model. A physical data model is a representation of data on the secondary storage device and it also includes other data structures like indexes and others. It also defines the constraints of the database systems, like the data types available to store a data, the number of secondary indexes allowed, and others. As shown in Fig. 1, a physical data model comprises of Message Format, File structure, Physical schema, and other entities. There are two ways in which data in a table may be stored either in row-order or column-order considering options provided by physical schema [1].

The physical schema defines the storage space required to organize the data on secondary storage devices. Also, it defines the number of indexes and limits the data structures which can be used to create the index. A mathematical model can be used to estimate the size of storage space required to store data. We found that storing 1.5GB blogging data with three secondary indexes (including a Text search index) was stored by MongoDB in 2.63GB which was 1.7 times the original size. It is very critical to know storage space requirement because it will impact the decision process of buying a storage space. Also, most cloud service providers limit access to storage by limiting the number of IOPS performed by an application. Hence, it is in the best interest of application developers and designers to have a detailed knowledge of physical schema of a data model or database before deciding to host the data on the Cloud. In Sect. 2, relevant works on physical schema, data models, and past attempts to estimate storage size for different physical schema are discussed. Successively, a mathematical model of storage space requirement for JSON-based databases is proposed. In Sect. 4, a simulation of the derived model would be discussed and the results would be experimental verified. Finally, contributors would conclude the work.

2 Literature Review

A true benchmark in the field of large-scale database management systems was achieved by information retrieval model by E Codd [2]. Only few works discuss and suggest new models for evaluating the pros and cons of big data systems. In Table 1 a list of important trends relating to evaluation of data models is revealed for the period starting from early 1970s to present.

Three of every four companies have found the necessity of using or shifting to Big Data solutions in the next 2 years [3]. These industries would be facing a great challenge of researching and choosing a big data technology as they have a large variety of solutions to choose from. With 10+ Data Models (listed in Table 2) and 45+ DBMS systems (listed in Table 3) are available for various applications. However, a single solution does not fit all purpose of the industry, hence it becomes eventually necessary to combine one or more solutions into a single conglomerated system that solves all the business problems. For instance, Oracle Big Data System, provides both NoSQL and/or Hadoop cluster options to its customer with SQL.

Table 1 Findings and open problems

Research work	Findings and open problems	Year
[2]	The provisions for data description tables in recently developed information systems represent a major advantage toward the goal of data independence	1970
[4]	New metadata information types, such as QoS of service for storage, and algorithms to exploit them, may be needed to meet emerging trends	1996
[6]	Schema-last is probably a niche market	2005
[5]	The high increase of disk usage compared to raw data is due to additional schema as well as version of information that is stored each key-value pair	2012
[7]	Integration of structured and unstructured data and information from distributed, heterogonous virtual clouds need further research	2013
[8]	Data storage and search schemas (or Indexes) are responsible for high latency and overhead	2014
[9]	Applications often drive the design of the underlying storage systems	2014

Table 2 Data models for big data applications

List of data models for big data solutions				
Content stores	Graph	Native XML	RDBMS	Time series
Document stores	Key-value stores	Navigational	RDF stores	Wide-Column
Event stores	Multi-value stores	Object-oriented	Search engines	

Table 3 Database solutions for big data applications

List of databases for big data solutions

Adabas	Db4o	Hypertable	MySQL	Solr
Algebraix	DynamoDB	IDMS	Neo4j	Sphinx
Amaxon CloudSearch	Elasticsearch	IMS	NEventStore	Titan
Azure DocumentDB	Event store	Jack rabbit	ObjectStore	TC-TT
BaseX	Flare	Jena	Oracle BigData SQL	UniData, uniVerse
Cache	Google cloud bigtable	MarkLogic	Oracle SQL	Versant object database
Cassandra	Google cloud datastore	Microsoft Azure search	Redis	Voldemort
Couchbase	Google search applicance	Microsoft SQL server	Scalaris	VoltDB
CouchDB	GraphDB	ModeShape	Sedna	...
D3	HBase	MongoDB	Sesame (or RDF4J)	

A major problem for choosing such technologies is that very few models such as Relational, Object-oriented, and Object-Relational have been built on strong mathematical model. Now, modeling of storage is a nontrivial challenge and in many cases demands evaluation of designs. If resource requirement cannot be justified, it would become increasingly difficult to monitor the growth of the system data and could adversely affect performance considering that scalability issue is not tackled in the right way.

Many prominent tools and technologies have been proposed in past to estimate the size of storage space required. MySQL also provides a perl script to estimate the size of storage space required for storing a database on the cluster-based storage engine named NDB based on size of storage space used by InnoDB storage engine to store the data [10]. InnoDB storage engine uses Barracuda file organization. Neo4j, a graph-based database also provides a calculator to estimate storage space, main memory, and processing power required at a node to store and process the data [11]. Neo4j calculator takes number of nodes, size of a single node, number of edges, and storage size of each edge as input to approximate the storage space required [11].

3 Storage Estimation Model for JSON-Based Databases

JSON has been one of the most influential format in the movement of migration from RDBMS to NoSQL [12]. JSON has found its place among many application domains with semi-structured and unstructured data [13–16]. Many databases and

JSON:

{

"name":Devang"

}

Fig. 2 A simple JSON document

BSON:

\x16\x00\x00\x00	// total document size
\x02	// 0x02 = type String
name\x00	// field name
\x06\x00\x00\x00**Devang**\x00	// field value
\x00	

Fig. 3 Physical schema of MongoDB (BSON)

solutions have extended JSON to suit their needs like BSON. BSON is a communication and storage protocol used by MongoDB, which is derived from JSON.

Figure 2 depicts a JSON document with a single field, "name" and its value "Devang". Figure 3 describes the storage schema of BSON which is a communication and storage protocol used by MongoDB. BSON is a storage structure which is derived from JSON. From the figures, it is also evident that BSON will consume much large storage size than JSON, owing to extra information it keeps for recording the data. Although, this extra information does help in increasing throughput by informing about type and size of data, helping I/O processor makes smart decisions (if relevant technologies are available and programmed to use). Above all, this extra information also helps the I/O processor decide how much bits to skip so as to find next document making read task faster. Nevertheless, one cannot ignore the increment in amount of storage space they require.

We propose to derive a model that can help us to estimate the factor by which storage size of JSON increases in comparison with storage size required by CSV. Although, the model is derived for JSON, it is applicable across all databases and solutions that use JSON or its derivatives (e.g., BSON, MessagePack,[1] etc. [17]).

The storage estimation model is explained by considering the physical schema of CSV and JSON storage schemas. For the purpose of modeling storage space requirement, we proposed comparing storage with flat file databases like CSV as the raw storage size because of all available formats. CSV has been more commonly used by many literatures as a physical schema of choice due to its simplicity and high level of human readability that it offers [18–21].

Consider a source S, which emits data at regular intervals. This data may be stored in Table T with following properties:

---

[1]MessagePack is a JSON-like but comparitively smaller in size [22].

- A Table T consists of N columns and R rows.
- Each column of the table has on average b_i bytes of data for ith column.
- Total number of bytes for each row of the table on average is
  $B = \sum_{i=1}^{N} b_i$.
- Each column header is of size c_i bytes of data for ith column.

For simplicity, we assume that the source releases data at regular intervals. It can be considered that source follows some distribution for generating data. Thus, it can be said that the number of rows for the given Table T can be approximated using the prior attained distribution. Also, generating data is a characteristic of the Source. Hence, the maximum number of bytes required to store data in a file can be estimated. Thus, we can get the value of b_i from the source itself. By getting N, which is the number of data items required to be stored in the table, by using distribution, which predicts when the given source will produce the data. Thus, by knowing, b_i, R and N, we can compute B. Finally, the size of column header c_i can be measured since the developer or DBA decides the column name.

CSV organizes the data in row-order format so that columns are mentioned in the first line and all successive lines store the data. Now amortized size[2] of column stored in CSV file would be $\sum_{i=1}^{N} c_i$ and since B bytes is the average size of a row, data would take $B \times R$. Hence, it can be concluded that for CSV store the size of data would be $CSV\_Size = (B \times R) + \sum_{i=1}^{N} c_i$ bytes.

In JSON-based stores, each row is in the format {column1 name: value, column2 name: value, ...} as shown in Fig. 3. Hence, the size of each row in such a physical schema[3] would be $(B + \sum_{i=1}^{N} c_i)$ bytes. For R number of rows in the table, the size of database would be $MC\_Size = R \times (B + \sum_{i=1}^{N} c_i)$ bytes. Thus, the ratio of storage size for JSON-based store to CSV would be $(R \times (B + \sum_{i=1}^{N} c_i) / ((B \times R) + \sum_{i=1}^{N} c_i)$ bytes.

4 Experiment

Experimental evaluation has been conducted with a simulation for total column field storage size of 136 bytes and Row size of 474 bytes for varying number of Row for NYC Taxi cab database [23] that is used for traffic patterns analysis of Taxi cabs to reduce pollution was utilized. To obtain size of column on an average we created a dummy document with all the values NULL or not set. We used this as a reference since we are only after amortized comparison of the storage size requirement. Figure 4 is a CDF and thus its corresponding PDF is "Exponential." Which suggests

---

[2]We use the term amortize because we donot consider the size of putting other characters like comma, carriage return, space for null values, and other special characters.

[3]We are not including comma, other special characters, and null values since we are only after a rough estimate.

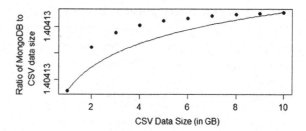

Fig. 4 Simulation: ratio of MongoDB to CSV data size

Table 4 Ratio of MongoDB to CSV data size

Year–month of data generated	No. of records	CSV size (cumulative) (GB)	MongoDB storage size (cumulative) (GB)	Ratio (MongoDB size / CSV size)
2016-01	10906858	1.6	2.3	1.412
2016-02	11382049	4.86	7.05	1.45
2016-03	12210952	9.9	14.75	1.49
2016-04	11934338	16.68	25.18	1.44
2016-05	11836853	24.83	38.95	1.57
2016-06	11165470	34.6	50.17	1.45

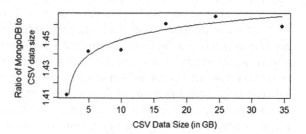

Fig. 5 Experimental evaluation: ratio of MongoDB data size to CSV data size (in GB)

that exponential increase in MongoDB storage size could be noticed when the size of raw data increases linearly. And the results obtained from simulation are produced in Fig. 4.

Results of the simulation were verified by inserting the data of NYC Yellow Taxi dataset in the big data solution, MongoDB (a JSON-based store) using WiredTiger storage engine. MongoDB was used for experiment as it is an open source solution, it uses JSON-like physical schema named BSON and is an extremely popular NoSQL data store [24]. On storing the data in MongoDB the size of stored data increased by 1.4 times the size of storage space used by CSV as shown in Table 4. The results of the experiments are shown in Fig. 5 which confirms the trend suggested by the model. Thus, using the model and simple maths we can devise a storage factor for estimating the size of storage space required by JSON and its derivatives (Table 5).

Table 5   MongoDB throughput (wall clock time)

File	Import start time	Import end time
2016-01	10:33:49	10:46:59
2016-02	10:52:35	11:08:28
2016-03	11:12:48	11:21:39
2016-04	16:13:35	16:25:49
2016-05	16:27:45	16:39:03
2016-06	16:45:50	16:57:00

Above all, from the experiment it is discovered that MongoDB takes on an average 10–13 min to import a csv file of size 1.6 GB on a standard non-commercial grade hard drive with 5400 RPM disk speed on a machine with 8GB RAM and Intel core-i5 6th generation processor.

5   Conclusion

This paper has listed 14 data models and 45+ databases that provide a glimpse of wide range of solutions available in the market for different big data applications. Researchers in the given work had also proposed a model that proved the storage size of determination by using physical schema for JSON-based stores. It has also been proved that the increment in disk utilization is due to the requirement of storing schema and version information into the table so as to allow storing semi-structured or unstructured data. This increased disk usage with respect to raw size shows exponential increment as the size of data increases. In near future, a comprehensive research for uniting structured, semi-structured, and unstructured data from different data inception points needs to be carried out. This research should be from the perspective of storage and QoS achievement using minimum resources so that it assists decision makers to make an optimal choice for their application. Finally, the WiredTiger Storage Engine of MongoDB takes 1.4 times more space than CSV file for NYC Taxi Cab Dataset including a primary index. Also the proposed model varied from the experimental values from 5 to 11%.

References

1. Whitehouse, O.: Fea consolidated reference model document (2005)
2. Codd, E.F.: A relational model of data for large shared data banks. Communications of the ACM 13(6), 377–387 (1970)
3. Gartner.com: Gartner report

4. Gibson, G.A., Vitter, J.S., Wilkes, J.: Strategic directions in storage i/o issues in large-scale computing. ACM Computing Surveys (CSUR) 28(4), 779–793 (1996)
5. Stonebraker, M., Hellerstein, J.: What goes around comes around. Readings in Database Systems 4 (2005)
6. Rabl, T., Gómez-Villamor, S., Sadoghi, M., Muntés-Mulero, V., Jacobsen, H.A., Mankovskii, S.: Solving big data challenges for enterprise application performance management. Proceedings of the VLDB Endowment 5(12), 1724–1735 (2012)
7. Demirkan, H., Delen, D.: Leveraging the capabilities of service-oriented decision support systems: Putting analytics and big data in cloud. Decision Support Systems 55(1), 412–421 (2013)
8. Chen, C.P., Zhang, C.Y.: Data-intensive applications, challenges, techniques and technologies: A survey on big data. Information Sciences 275, 314–347 (2014)
9. Kambatla, K., Kollias, G., Kumar, V., Grama, A.: Trends in big data analytics. Journal of Parallel and Distributed Computing 74(7), 2561–2573 (2014)
10. Ndbcluster size requirement estimator. https://dev.mysql.com/doc/refman/5.7/en/mysql-cluster-programs-ndb-size-pl.html, accessed: 2016-09-30
11. Hardware sizing calculator. https://neo4j.com/hardware-sizing/, accessed: 2016-09-30
12. Padhy, R.P., Patra, M.R., Satapathy, S.C.: Rdbms to nosql: reviewing some next-generation non-relational databases. International Journal of Advanced Engineering Science and Technologies 11(1), 15–30 (2011)
13. Aho, A.V., Sethi, R., Ullman, J.D.: Compilers, Principles, Techniques. Addison wesley Boston (1986)
14. Bollacker, K., Evans, C., Paritosh, P., Sturge, T., Taylor, J.: Freebase: a collaboratively created graph database for structuring human knowledge. In: Proceedings of the 2008 ACM SIGMOD international conference on Management of data. pp. 1247–1250. AcM (2008)
15. Consortium, W.W.W., et al.: Json-ld 1.0: a json-based serialization for linked data (2014)
16. Finn, R.D., Mistry, J., Tate, J., Coggill, P., Heger, A., Pollington, J.E., Gavin, O.L., Gunasekaran, P., Ceric, G., Forslund, K., et al.: The pfam protein families database. Nucleic acids research p. gkp985 (2009)
17. del Alba, L.: Data serialization comparison: Json, yaml, bson, messagepack. https://www.sitepoint.com/data-serialization-comparison-json-yaml-bson-messagepack/, accessed: 2016-09-26
18. Cook, K.B., Kazan, H., Zuberi, K., Morris, Q., Hughes, T.R.: Rbpdb: a database of rna-binding specificities. Nucleic acids research 39(suppl 1), D301–D308 (2011)
19. Cranford, K.: How to excel with sas. In: Proceedings of the 28 th Annual SCSUG Conference, Austin, Texas, September (2007)
20. Shafranovich, Y.: Common format and mime type for comma-separated values (csv) files (2005)
21. Sharma, T.C., Jain, M.: Weka approach for comparative study of classification algorithm. International Journal of Advanced Research in Computer and Communication Engineering 2(4), 1925–1931 (2013)
22. Messagepack. http://msgpack.org/index.html, accessed: 2016-09-26
23. Commission, N.T..L.: Tlc yellow taxi trip record data. http://www.nyc.gov/html/tlc/html/about/trip_record_data.shtml, accessed: 2016-09-30
24. DB-engines.com: Dbms rankings 2017 (2016)

Intelligent Vehicular Monitoring System Integrated with Automated Remote Proctoring

Chetan Arora, Nikhil Arora, Aashish Choudhary and Adwitiya Sinha

Abstract Environmental pollution is one of the crucial challenges confronted by everyone, especially in metropolitan cities. Several authorized governmental agencies and scientists are trying to develop solutions for controlling this dreadful menace. Our research aims to address the challenge by remotely tracking level of hazardous gaseous substance emitted by vehicle within a specified region under certain governmental jurisdiction. Our models assist the concerned authority to proctor, the allowable level of emission, more importantly from vehicles that are older than specified number of years. Further, in this perspective, we have designed a module which will be installed in the exhaust system of a vehicle, so as to measure the accurate level of pollution with an additional feature of displaying the reading on the display panel of the vehicle. This not only would inform the driver of present levels of pollutants, but also warn them if the levels are violated the permitted values.

Keywords Internet of things (IoT) · Intelligent transport system (ITS) Multisensor data · Remote monitoring and proctoring · Air pollution control

C. Arora · N. Arora · A. Choudhary
Department of Electronics, Jaypee Institute of Information Technology,
Noida-62, India
e-mail: arora.chetan23@gmail.com

N. Arora
e-mail: nikhil16.arora@gmail.com

A. Choudhary
e-mail: ashu.sindhu0@gmail.com

A. Sinha (✉)
Department of Computer Science, Jaypee Institute of Information Technology,
Noida-62, India
e-mail: mailtoadwitiya@gmail.com

© Springer Nature Singapore Pte. Ltd. 2018
Y.-C. Hu et al. (eds.), *Intelligent Communication and Computational Technologies*, Lecture Notes in Networks and Systems 19,
https://doi.org/10.1007/978-981-10-5523-2_30

1   Introduction

In today's world, we are experiencing pollution in all possible forms, of which environmental pollution is considered to be the most harmful of all. In a survey conducted in year 2013, more than half of total accumulation of carbon monoxide and nitrogen oxides, and almost quarter of hydrocarbons in atmospheric layer, was contributed by transportation. This data is persistently increasing exponentially every day [1–3]. The vehicles, mainly cars and trucks, produce air pollution at different stages in their functional lifetime. This includes gaseous substances emanated during different stages of vehicular operation, including manufacturing, driving, refueling, and refining, as well as distribution of fuel [4–6].

In order to address this problem of atmospheric pollution, especially caused by vehicles, we have proposed an electronic model that can help individual as well as authorized agencies to keep track of the pollution level of a vehicle. It is also equipped with the provision of alerting the pollution control room directly in case the user has not taken any action to mitigate the high level of emissions from his cars for the last n number of days. This alert information includes car manufacturing details, owner information, as well as the location information via geographical positioning system (GPS). Our model is based on embedded systems with multi-sensor features, extended with Internet of things (IoT) to remotely control, track, and proctor violating cases. IoT is one of the big ideas of the moment and it has grabbed the attention of the researchers rapidly. IoT is employed in our model to send the sensed data to pollution control center (cloud), so that the concerned authority can take legitimate action on the defaulter. The sole aim is to design an electronic device that can be easily installed in every vehicle to have a continuous track on the level of pollutants emitted. Initially, it warns the driver if the level of any of the component is above the threshold and alerts the person to check/service the vehicle. However, if the situation persists for n days, our device is capable to upload the information to authorized governmental portal responsible for pollution control so that appropriate steps could be taken.

2   Related Work in Intelligent Transportation System

There are several works carried out in the domain of intelligent transportation system [7–9], of which few relevant studies are highlighted. In [10] the authors have discussed that modern technical growth observed in electronics and wireless technology has greatly led to the advent of environmental sensor networks (ESN). The ESNs significantly improve the natural environment monitoring, area surveillance, intelligent data gathering, etc. WAPMS is one such system that exemplifies ESN. Another work presented by [11] gives an insight of controlling air pollution and the relative sources emitting harmful gases with the help of sensors. The authors considered that monitoring of hazardous pollutants and harmful emissions is very crucial for preventing loss of life due to environmental accidents. Conventional instruments are bulky and expensive, and require more time to set up as well as maintain. This makes the conventional methods unsuitable for monitoring environment. Therefore, the authors propose the application of robust, miniature, and relatively less-expensive solid-state gas sensors as an effective alternative for performing monitoring tasks [12].

In another research conducted in [13], the researchers have proposed an environmental air pollution monitoring system (EAPMS) for observing the concentrations of pollutants in atmosphere. The developed system complies with IEEE 1451.2 standard and effectively records the presence of CO, SO_2, NO_2, and O_3 by means of semiconductor sensors. The smart transducer interface module (STIM) was instigated using analog device ADuC812 macro-converter. Network capable application processor (NCAP) was established and further linked to STIM through transducer-independent interface to calibrated sensed data using standard calibration methods. In yet another recent study in [14], the authors have illustrated the design, implementation, and evaluation of GasMobile. This is a convenient system for quantification based on off-the-shelf modules and suited to be used for several applications. The accuracy of sampled data was improved by manipulating sensor readings near governmental measurement stations to update sensor calibration and analyze the impact of mobility on the accuracy over the sensed data pattern.

3 Proposed Framework

Our proposed model involves microcontroller board for mounting multiple sensors that could collect data of gaseous substance present in the surrounding (Fig. 1). Certain specifications of the microcontroller are highlighted in Table 1. Detailed specifications of all the electronic devices required for enabling the circuitry to function are also highlighted (Fig. 2).

Arduino Uno: The microcontroller panel which uses ATmega328 has fourteen digital output/input pins. Among total number of pins, six can be utilized as PWM pins and six as analog pins. It operates at 16 MHz oscillator with USB port.

Gas Sensors: This type of sensing device is considered as a subclass of chemical sensors that employ a heater with an electro-chemical device. This category of sensors is sensitive to a wide variety of gases and is normally deployed indoors to monitor the presence of certain gases, under room temperature. Output produced by a gas sensor is an analog signal that is readable with an analog input of Arduino. For instance, the sensitivity of MQ-2 gas sensor can record the presence of LPG, hydrogen, methane, propane, and other combustible gases. Gas sensing module is often used for detecting gas leakage indoor as well as industrial areas.

LCD: LCD module needs to be compatible with Arduino, preferably Nokia 1100 display. This display uses PCD8544 controller chip from Philips.

Wi-Fi Module: Wi-Fi is required to link the proposed model with Internet. ESP8266 is a low-cost Wi-Fi module executing at 3.3 V.

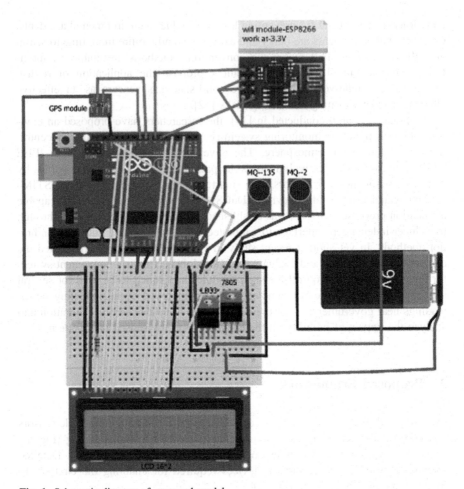

Fig. 1 Schematic diagram of proposed model

Table 1 Microcontroller
specifications

Features	Values
Microcontroller type	ATmega328
Digital input/output pins	14
Analog input pins	6
Functional voltage	5 V
Input range of voltage	7–12 V
Flash memory	32 KB (ATmega328)
SRAM	2 KB (ATmega328)
DC current input/output pin	40 mA
DC current pin	50 mA
EEPROM	1 KB (ATmega328)
Clock rate	16 MHz

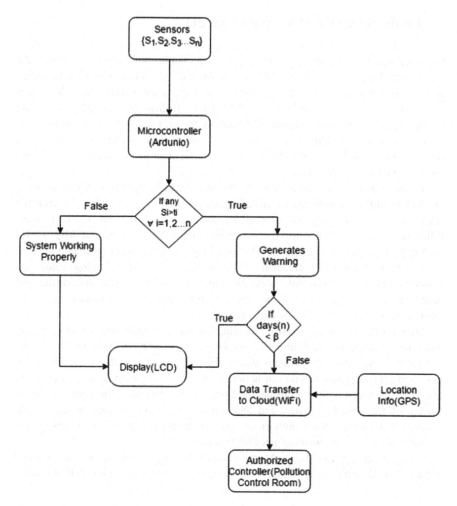

Fig. 2 Control flow diagram of proposed methodology

GPS Module: It provides a current location in form of latitude and longitude of the location that can easily be sent to the base server over cloud via Wi-Fi module.

Voltage Regulator: Voltage regulator maintains a constant voltage upon giving a higher voltage. Our model requires two voltage regulators namely, 7805 and LD33, to regulate 5 V and 3.3 V, respectively.

4   Implementation of Proposed Model

Our proposed model is designed for Nokia 1100 display screen interfaced with the microcontroller. In order to execute ESP8266, v0.9.2.2AT firmware is uploaded and checked with AT commands. Once AT commands are working, esp8266 needs to be get connected with the Wi-Fi module. This is followed by creating an account on the cloud server of the authorized pollution center. For our experimentation, we have deployed over ThingSpeak cloud and created a new channel. Finally, the real-time stream of sensed data is transmitted to the channel and graph begins to get plotted against recorded data.

In this research, we intended to control the dreadful pollution which is being emitted by different vehicles by installing this device which can keep a constant track on each and every gas emitted by them. Gas sensors are used to detect different gases, i.e., MQ2 for carbon monoxide; LPG, smoke, and MQ135 for carbon dioxide and nitrogen gases. These sensing devices report one analog value at each instant of time. To calculate different values from one analog value, the sensitivity curve given in datasheets of both the sensors is used, and carried out equations for each gas and put the single analog value in this equation to get the ppm (part per million) value of each gas.

After getting all the ppm values, the threshold is set, which means that particular gas is being emitted more than the allowed level imposed by the government. If any violation is recorded, the system warns the vehicle driver about the excessive emission and suggests to take an action as soon as possible. However, required action is not taken within a prespecified time frame, and then the complete information of the vehicle, owner, and location that is locally sensed is sent to the concerned authority. Such authorized site can assist pollution controllers to use reported details for performing necessary action.

In Fig. 3, set of graphs highlights the trend of sensed data, recorded by sensors and reported to the authenticated cloud for remotely proctoring vehicles. In addition,

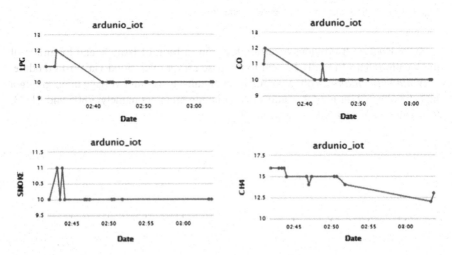

Fig. 3 Sensed information display over cloud (authorized proctor) via IoT

the instantaneous data is displayed on the LCD panel installed in the vehicle, thereby showing the current status or health of the vehicular system.

5 Conclusion

The significant development in sensor technologies along with the urge to control air pollution has resulted in several intelligent mechanisms for monitoring emission of pollutant gases with the help of sensing devices. Our designed system records ppm values of carbon monoxide, smoke, LPG, and CH_4 (methane); and displays them on vehicle's LCD instantaneously. With violation of threshold of any of the emitted pollutant above permissible number of days, warning message is issued to the driver. When warnings are unattended, then in such extreme cases, our proposed model is capable of sending vehicle-specific information to the authorized site for being proctored.

References

1. Hasenfratz, David, Olga Saukh, Silvan Sturzenegger, and Lothar Thiele. "Participatory air pollution monitoring using smartphones." Mobile Sensing (2012): 1–5
2. Jamil, Muhammad Saqib, Muhammad Atif Jamil, Anam Mazhar, Ahsan Ikram, Abdullah Ahmed, and Usman Munawar. "Smart environment monitoring system by employing wireless sensor networks on vehicles for pollution free smart cities." Procedia Engineering 107 (2015): 480–484.
3. Fleming, J. "Overview of automotive sensors." IEEE sensors journal 1, no. 4 (2001): 296–308
4. Al-Ali, A. R., Imran Zualkernan, and Fadi Aloul. "A mobile GPRS-sensors array for air pollution monitoring." IEEE Sensors Journal 10, no. 10 (2010): 1666–1671.
5. Postolache, Octavian A., JM Dias Pereira, and PMB Silva Girao. "Smart sensors network for air quality monitoring applications." IEEE Transactions on Instrumentation and Measurement 58, no. 9 (2009): 3253–3262.
6. Kwon, Jong-Won, Yong-Man Park, Sang-Jun Koo, and Hiesik Kim. "Design of air pollution monitoring system using ZigBee networks for ubiquitous-city." In Convergence Information Technology, 2007. International Conference on, pp. 1024–1031. IEEE, 2007.
7. Monaci, F., F. Moni, E. Lanciotti, D. Grechi, and R. Bargagli. "Biomonitoring of airborne metals in urban environments: new tracers of vehicle emission, in place of lead." Environmental Pollution 107, no. 3 (2000): 321–327.
8. Ma, Yajie, Mark Richards, Moustafa Ghanem, Yike Guo, and John Hassard. "Air pollution monitoring and mining based on sensor grid in London." Sensors 8, no. 6 (2008): 3601–3623.
9. Cordova-Lopez, L. E., A. Mason, J. D. Cullen, A. Shaw, and A. I. Al-Shamma'a. "Online vehicle and atmospheric pollution monitoring using GIS and wireless sensor networks." In Journal of Physics: Conference Series, vol. 76, no. 1, p. 012019. IOP Publishing, 2007.
10. Khedo, Kavi K., Rajiv Perseedoss, and Avinash Mungur. "A wireless sensor network air pollution monitoring system." Doi: 1005.1737(2010).

11. Pummakarnchana, Ornprapa, N. Tripathi, and J. Dutta. "Air pollution monitoring and GIS modeling: a new use of nanotechnology-based solid state gas sensors." Science and Technology of Advanced Materials 6, no. 3 (2005): 251–255.

12. Kramer, Thomas R., H. M. Huang, Elena Messina, Frederick M. Proctor, and Harry Scott. "A feature-based inspection and machining system." Computer-Aided Design 33, no. 9 (2001): 653–669.

13. Kularatna, Nihal, and B. H. Sudantha. "An environmental air pollution monitoring system based on the IEEE 1451 standard for low cost requirements." IEEE Sensors Journal 8, no. 4 (2008): 415–422.

14. Lee, Duk-Dong, and Dae-Sik Lee. "Environmental gas sensors." IEEE Sensors Journal 1, no. 3 (2001): 214–224.

ObfuCloud: An Enhanced Framework for Securing DaaS Services Using Data Obfuscation Mechanism in Cloud Environment

Krunal Suthar and Jayesh Patel

Abstract Cloud computing is now a day's become most attracted phenomena to use for a large-scale organization or for individual who need various network services with least cost. Normally, individual's information is stored on public Cloud which is available to everyone for access. This fundamental raise some issue opposite to flexible services provided by Cloud providers, like Confidentiality, Integrity, Availability, Authorization and many more. To protect the data, lots of options available now a days and most preferable way is to use encryption. Encryption only cannot provide enough protection while considering user's sensitive information, as well as it consumes more time to process encryption and decryption. To remove the burden of Cloud server, as well as to keep adequate security to user's information in Cloud environment, in this paper, we propose a methodology for combining both techniques, viz. obfuscation and encryption. The user data may be encrypted if it requires security for its files or document, and the DaaS service of Cloud must be secured using obfuscation techniques. Using this two-way approach, we can say that the proposed scheme offers enough security towards anonymous access and preserve privacy even of the information available on Cloud Servers. We also aim to provide proper integrity checking mechanism, better access control mechanism which lessens the burden of Client as well as service provider.

Keywords Cloud storage · Data protection · Integrity · Obfuscation
Access control · Privacy preservation

K. Suthar (✉)
Computer Engineering Department, Sankalchand Patel College of Engineering,
Visnagar, Gujarat, India
e-mail: krunal_bece@yahoo.co.in

J. Patel
Computer Science Department, AMPICS, Kherva, Gujarat, India
e-mail: jayeshpatel_mca@yahoo.com

© Springer Nature Singapore Pte. Ltd. 2018
Y.-C. Hu et al. (eds.), *Intelligent Communication and Computational
Technologies*, Lecture Notes in Networks and Systems 19,
https://doi.org/10.1007/978-981-10-5523-2_31

1 Introduction

Cloud computing is the term related to the internet-based high computing whose aim is to provide devices with a shared pool of resources, information or software on demand and pay per-go basis anytime anywhere. The Cloud models come with mainly important characteristics like the pooling of resource, services on demand, access of the broad network, quick elasticity and precise service. A service provider of Cloud provides service models like PaaS, SaaS, IaaS, DaaS that come with a pack of three basics deployment models that are private, public Cloud and hybrid Cloud [1].

1.1 Cloud Services

Services provided by the Clouds are broadly divided into four major categories which are shown in Fig. 1.

1.2 Database and Document Security in Cloud Environment

Users' information security is one of the main concern in Cloud environment. Data protection [2] includes various issues like manage confidentiality, integrity, provides authentication, achieve availability and many more. Data confidentiality means that only authenticated user has access of data. Data integrity means the information must be unchanged while available on a remote system or on the local system. Authentication refers to the method of checking whether the user who tried

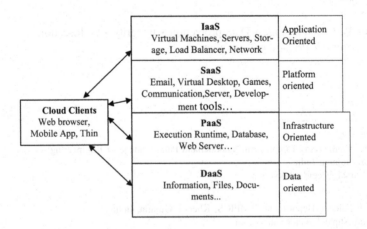

Fig. 1 Services offered [10, 11]

to access information is valid or not. Data availability means to achieve the data anytime whenever required.

Normally, confidentiality achieved by encryption technique, but for the Cloud environment when user data and user personal information are separated from each other. Encryption is only important when we think about the security of user-stored document, but when the personal information is targeted by any attacker then it is required to preserve the privacy of database on Cloud server. Here the encryption scheme can be worked but it consumes lots of time for encryption and decryption process because thousand of processes are done simultaneously. So, to achieve speed and security both required a framework, which deals with users' and Cloud servers' issue using technique mentioned [3].

1.3 Why Data Obfuscation Required in Cloud Computing Environment?

i. Data Confidentiality Protection
   Confidentiality guarantees that the sensitive information has not become available to the person who is not authentic or to other processes, or Devices. The information details should be considered as broken if any of the following conditions are violated: (1) The physical availability of the user data is known to the user (2) Provider able to access the personal information of the Cloud user (3) Meaning of user-uploaded information are disclosed to the user [4].
ii. Some of the major issues in current Cloud system are as follows:
A. Every provider of Cloud work as a provider of different layered services like platform, software layer and infrastructure layer. The user is only able to use the infrastructure service provided as it is while using a Cloud application, and because of this dependency, the service provider knows where the users' data is located and has full access privileges to the data.
B. The providers force a user to only use the interface provided by them, so the users need to submit their data in a fixed format. And because of this, a service provider has an adequate idea about how to access the user information and is able to control this information.

By combining four entities we can achieve protection in Cloud environment: Cloud, Infrastructure Cloud, Encryption, Data Obfuscation and Data De-Obfuscation.

(1) Cloud: A Cloud provides database as a service upon users' requests.
(2) Infrastructure Cloud: An Infrastructure Cloud provides virtualized system resources, such as CPU, memory and Database. An authenticated user can request to retrieve information or files which the user sent to the Cloud.
(3) Encryption: This technique provides an efficient way to covert user information into an unreadable format.

Table 1 Obfuscation techniques [8, 12]

Strength of obfuscation technique [13]: domain	Techniques	Potency	Resilience	Cost
Transform	Code	Medium	One-way	Free
Transform	Data	High	Two-way	Cheap
Transform	Control	Medium	Partial one-way	Costly

(4) Data Obfuscation: its a mechanism that the data converted into a format which makes reverse engineering difficult for the attacker as well as for any automated software [5] (Table 1).

2 Literature Review

Arockiam and Monikandan in [6] proposed a novel technique to achieve confidentiality to address the issue of data security. The author aims to provide confidentiality using two important techniques that are identified as obfuscation and encryption. The key used for encryption purpose is: to keep secret with the user and access of this data is only permitted by passing a corresponding decryption key. Authors not only aim to use encryption but also offer an obfuscation mechanism to increase the security of data.

Atiq ur Rehman, M. Hussain SZABIST in [7] presented a model for managing DaaS confidentiality of data stored in Cloud database. The model consists of two main features. The first focus is on how the user data will be stored in the server. A second feature provided regarding how a user can send query so that even data will be fetched using DaaS service. The database admin has no idea about this process and also about what type of data requested by the user. The model performs query execution on the encrypted and obfuscated data.

Arvind Narayanan and Vitaly Shamatikov in [8], in this, the methodology for achieving privacy are discussed. The owner who want to share the information with different users does not need to hide each data entry separately but to obfuscate the database entry, which provides execution of only particular types of queries. Even if the database details are provided still the database is only accessible with reference to the designed privacy model. Here, the novel concept of database privacy is proposed, other than that, for managing secrecy of individual records only some of the queries are permitted and realized it using provably secure obfuscation technique.

Krunal Suthar [9], in this research has proposed a model to have proper confidentiality, security and integrity of user information. In the proposed scheme, encryption is basically done in client side and obfuscation is done for the Cloud service providers. By using encryption the data which is in transition becomes secure. Data obfuscation helps data which is on rest in the machine of the service provider to get secured. To achieve user as well as server control mechanism authors have also proposed an algorithm which proves that this technique together provides adequate security. From the implementation analysis, authors argue that compared to existing schemes the proposed scheme provides better protection towards stored information on a Cloud which is based on encryption only.

3 Proposed Methodology

The approach mainly focuses on three sections that are uploading, integrity management and secure access control with proper rights management. Figures 2, 3 and 4 give detail description of operations that are carried out.

Upload Data on Cloud

1. CU > CSP (Input Login Detail (Assume that user is already registered))
2. CSP verify > CU (Detail verified at server. If user replies accordingly)

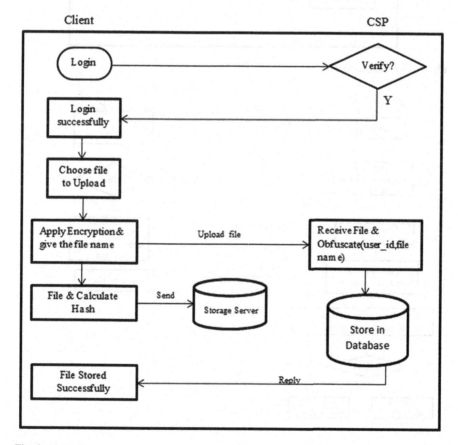

Fig. 2 Uploading

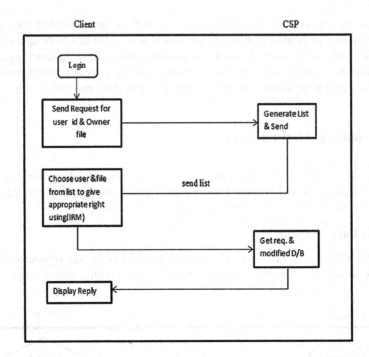

Fig. 3 Right management

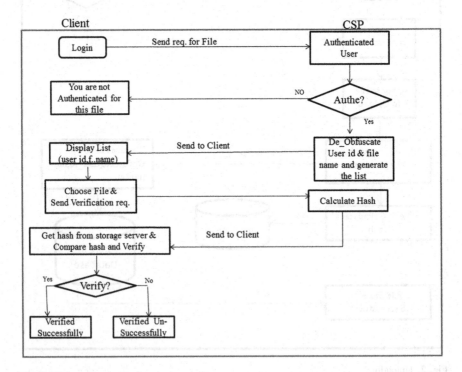

Fig. 4 Integrity management

3. CU Upload > (User uploads the file, this includes encryption mechanism provided by the server to the client interface.)
4. CSP obfuscates and stores (store this information in database after obfuscating it)
5. CU store hash (Hash of that file and store on the storage server)

For Right management

1. CU *Id & Owner file req.* > CSP (Client send request for owner's file list to CSP)
2. CSP *File List* > CU (CSP generate list of owner's file and send to the client)
3. CU *File & IRM* > CSP (Client choose user and file and apply IRM policy then sends to the CSP)
4. CSP un-obfuscate and update (CSP modify database).

For Integrity

1. CU *Req. File* > CSP (Client sends request for files to CSP)
2. CSP *Verify* ≥ CU (Check the user is athenticated or not and reply accordingly)
3. CSP *De-obfuscate files* > CU (CSP De-obfuscates files and sends to the client)
4. CU *File & verify request* > CSP (Client sends verification req. to server)
5. CSP *Hash* > CU (CSP calculates Hash and sends to the client)
6. CU compares and verify (Client compare Hash code store on storage server)

4   Result and Discussion

We run our model on a system with Intel Core I3 processor with 4 GB RAM, system running 64-bit windows and using Cloudsim simulator.

Basic analysis
In the basic analysis, consider a table containing sample information shown in Table 2 and the obfuscated data stored on server shown in Table 3.

Performance analysis
In Performance analysis phase, we analysed developed system using different evaluation parameter shown in Fig. 5 in terms of execution speed. First, demonstrate the result regarding consumption of time for encryption and obfuscation

Table 2 Sample data

User_Id	File name	Upload date	Hdd_name
first@gmail.com	C:\Users\obfu1.txt	10-9-2016	F drive
second@gmail.com	C:\Users\obfu2.txt	10-9-2016	F drive
third@gmail.com	D:\Files\obfu3.txt	17-9-2016	E drive

Table 3 Obfuscated data

User_Id	File name	Upload date	Hdd_name
AnVfdjlkjl88sf8aads=	RTpcQR6haZGVtaWMgRIxkYXRhXNoX0kjQn4lcncopJyL4dA=	MKol3bSxyNg ==	PKk=
Op5nagDjjWnxjk56f=	RTpcQR6lcnNcU09OWS1qA1xEZXNrdG4rtXGNsb3Vkc2lccmNc=	MKol3SaxyNg ==	PKk=
Yu4MashAjh2hdkjsm=	QzpcQZNlcnNcU09OWJ61QQ1xEZXNrdG88c2lFxWRzaEl3Snz=	Kolw7zRSyNg ==	Fzl=

process on different size of the file. We have also shown a comparison between the system that uses obfuscation and the system without obfuscation by varying the file size. This proves obfuscation only increases the little bit time and reduces the burden of Cloud server.

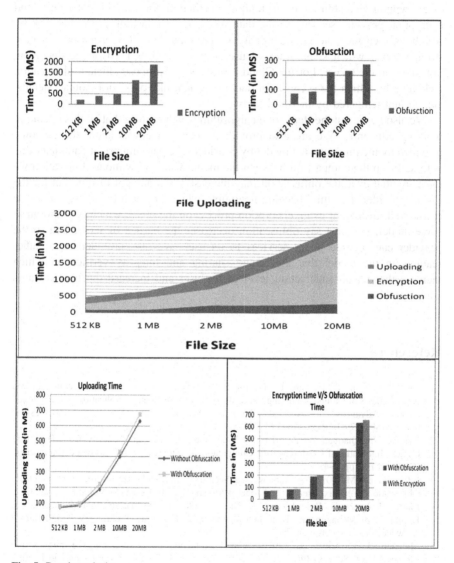

Fig. 5 Result analysis

5 Conclusion

Even Cloud computing provides numerous advantages to the user, but still, due to security issues many users hesitate to adopt it, as well as, the service provider may have an issue about unauthorized access. So, to solve an issue related to both user and service provider, we developed a new framework by proposing a combination of encryption and obfuscation technique together. Before sending data on Cloud encryption provides security to the data which is on transition in the network by which user ensures the confidentiality of his data. We also proposed a secure storage server which keeps track of user keys as well as hash of the document uploaded on the server. For the Cloud providers, we propose efficient obfuscation technique by which the secret information of client like password, contact details, etc. are not tempered by the third party.

We also figure out the steps for an algorithm which ensures that all this operation working efficiently. We are also providing detailed analysis about the outcome provided by the implemented model by considering a very important parameter that is time. From the comparison between the model with and without obfuscation we must say that even the obfuscation may increase small amount of time but for the Cloud provider, this time become negligible when the security of users is to be considered. Instead of using encryption process on the server which is proposed in some model, we must say obfuscation decreases the burden of server and so the provider can provide better services to its user. We also provided some other features in the model that are secure sharing and Integrity verification which increase the overall satisfaction level of the user and increase the trust towards Cloud providers.

References

1. Anca A., Florina P., Geanina U., George S., Gyorgy T. (2013) "Study on advantages and disadvantages of Cloud Computing – the advantages of Telemetry Applications in the Cloud" Recent Advances in Applied Computer Science and Digital Services ISBN: 978-1-61804-179-1- Page 118–123.
2. Xiaojun Yu, Qiaoyan Wen (2010) "A View about Cloud Data Security from Data Life Cycle", International Conference on Computational Intelligence and Software Engineering (CiSE), IEEE, pp 1–4.
3. Juli C., Jayesh M. (2016) "A Framework to Secure Cloud Data Server Information Using Data Obfuscation" Technix International Journal for Engineering Research, Volume 2 Issue 08.
4. Martin M, Agnew G., Bole. J, Page M,, Rhodes W(2012) "Information Right Management & Digital Right Management" Published on the IEEE Emerging Technology Portal. (http://www.ieee.org/go/emergingtech).
5. Muhammad H., Ahmed E. (2012) "Cloud Protection by Obfuscation: Techniques and Metrics," IEEE Seventh International conference on P2P, Parallel, Grid, Cloud and Internet Computing.

6. Dr. L. Arockiam, Monikandan S. (2014) "Efficient Cloud Storage Confidentiality to Ensure Data Security" IEEE International conference on Computer communication and Information, jan. 03–05, coimbatore, India.

7. Atiq R., Hussain M. (2011) "Efficient Cloud Data Confidentiality for DaaS," International Journal of Advanced Science and Technology Vol. 35, October.

8. Arvind N., Vitaly S. (2013) "Obfuscated Database and Group Privacy", The University of Texas at Austin{arvind,shmat}@cs.utexas.edu.

9. Suthar K., Patel J (2015) "EncryScation: A Novel Framework for Cloud IaaS, DaaS security using Encryption and Obfuscation Techniques" In 5th Nirma University International conference on Engineering (NUiCONE) Dec.

10. Rabi P., Manas R., Suresh S. (2011) "Cloud Computing: Security Issues and Research Challenges". (IJCSITS) Vol. 1, No. 2.

11. Shilpashree Srinivasamurthy, David Q. Liu (2010) "Survey on Cloud Computing Security" Department of Computer Science, Indiana University – Purdue University Fort Wayne, Fort Wayne, IN 46805.

12. A Net 2000 Ltd. White Paper, "Data Masking: What You Need to Know" (http://www.Net. 2000Ltd.com & Info@Net2000Ltd.com).

13. Christian C., Clark T., Douglas L. (2012) "A Taxonomy of Obfuscating Transformation", Department of computer science, The University of Auckland, Private Bag 92019 Auckland, New Zealand. {collberg,cthombor,dlow001}@ cs.auckland.ac.nz".

The Interdependent Part of Cloud Computing: Dew Computing

Hiral M. Patel, Rupal R. Chaudhari, Kinjal R. Prajapati and Ami A. Patel

Abstract Consumers educe umpteen advantages by placing private data enclosed by cloud computing utilities, although the hindrance of keeping data in such kind of services is the unavailability of consumers own data in absence of Internet connection. To figure out this enigma in an effective and excellent manner, new computing that is independent as well as collaborative with the cloud computing is emerged denoted as Dew computing. The Dew computing is revealed and realized as a fresh layer in the currently distributed computing hierarchy. Dew computing is placed as the base level for the Fog and Cloud computing archetypes. Hierarchical and interdependent separation from Cloud to Dew Computing satisfies the necessity of low and high-end computing demands in day to day life. These new computing paradigms diminish the expense and enhance the execution especially for ideas like Internet of Everything (IoE) and the Internet of Things (IoT). This paper presents basic concepts as well as cloud–dew architecture with working flow of dew computing, the correlation among Cloud Computing, Fog Computing, and Dew Computing along with comparison among all these paradigms.

Keywords Dew computing · Fog computing · Cloud computing · IoT

H.M. Patel (✉) · R.R. Chaudhari · K.R. Prajapati · A.A. Patel
Sankalchand Patel College of Engineering, Visnagar, India
e-mail: hmpatel.ce@spcevng.ac.in

R.R. Chaudhari
e-mail: rrchaudhari.ce@spcevng.ac.in

K.R. Prajapati
e-mail: kinjal1292@gmail.com

A.A. Patel
e-mail: ameeap90@gmail.com

© Springer Nature Singapore Pte. Ltd. 2018
Y.-C. Hu et al. (eds.), *Intelligent Communication and Computational Technologies*, Lecture Notes in Networks and Systems 19,
https://doi.org/10.1007/978-981-10-5523-2_32

1 Introduction

The way to store and retrieve personal as well as commercial information has been totally refined due to enormous innovations in today's world. With the fast growth of Internet, users can access their data anywhere in the world without carrying the data on physical devices. In early 2000, the new notion was introduced titled as "Cloud Computing," which is the technology of storing and accessing data as well as applications over an Internet network. Cloud computing uses a network of shared large pools of systems, resources, and servers. With the concept of pay on use, cloud computing architecture allows the client to procure services at different level of abstraction such as Platform as a Service, Software as a Service, and Information as a Service, depending upon their requirements [1]. It gives users the freedom from location bounding as users can access services everywhere with the help of availability of Internet and a standard web browser; allow working on a single project from multiple Geo-distributed workplaces.

Today's tech-savvy world is brimful of electronic gadgets, sensors, robots, machines, appliances, equipments, and actuators which all are smart objects programmed to carry out functions mainly as stand apart devices and are interconnected through either wired or wireless links to form a one universal network which is known as into the Internet of Things [2].

Registration to the cloud provides the privilege of accessing data from service providers in any part of the world, but this ease comes up with the hazards of security as well as privacy. This provokes the idea of intercept concentrating on cloud and initiate unraveling how to store and operate the spurt of data that is being yielded by IoT. IoT demands mobility support, broad range of Geo-distribution, location awareness, as well as low latency qualities. Thus, the Internet of Things wields an augmented configuration of Cloud Computing designated as Fog Computing. Fog Computing is an articulation between the powerful Cloud Computing and the network of incalculable smart devices. Whereas the Cloud Computing accredit its client computers by sharing resources for computing and data storage located in the server at a remote data center known as Cloud, Whereas the fog computing provides resources to the networked clients by proximate to the source of the data. The whole system using Cloud and Fog Computing for Internet of People and Internet of Things is known as Internet of Everything.

As we know, Cloud and fog computing require Internet connection. The evident detriment of keeping data using the Cloud utility is forfeiting access in the absence of Internet connection. Because all resources are far from user's premises and out from user's control, if an Internet connection is lost, the user will not be able to access the user's own data. To eliminate this problem one more new concept comes into picture known as "Dew computing" [3].

1.1 What is Dew Computing?

Dew Computing is a prototype whose objective is to wholly grasp the abilities of personal computers as well as cloud services. In such kind of archetype, organization of software on a personal computer is based on the Cloud–dew Architecture which offers lavish utilities independent and collaborates with cloud services. Dew Computing is the future direction of on-premises computer applications [4].

"Dew computing is a method where the on-premises computer provides utilities which are independent and also collaborative with cloud services. The goal of Dew computing is to fully realize the potential of on-premise computer and cloud services." [5, 6]

The independence feature encourages using on-premises resources as far as possible before sending requests to cloud services to fully understand the power of on-premises computers [5]. The meaning of Collaboration is an exchange of information with cloud services automatically during dew computing application's operation. Collaboration may involve synchronization, correlation, or any other type of interoperations. The collaboration feature realizes the potentials of cloud services by promoting the use of cloud services together with on-premises computers. Independence suggests inherently distributed nature of application whereas collaboration suggests inherently connected nature of dew computing application [5].

1.2 The Correlation Amidst Cloud Computing, Fog Computing, and Dew Computing

Like, a cloud is far away from the ground, fog is closer to the ground and dew is on the ground, cloud computing is remote, fog computing is in the neighborhood of

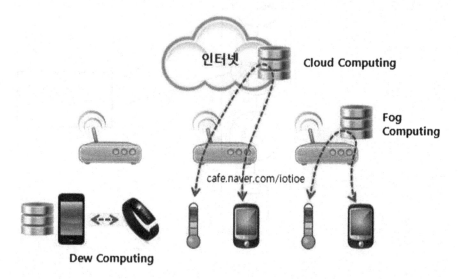

Fig. 1 Pictorial view of cloud, fog, and dew computing [11]

users and dew computing is at the user end itself. Figure 1 shows the pictorial view of cloud, fog, and dew computing.

2   Cloud-Dew Architecture

The task of making synchronization between data on the cloud and local computer is very trivial in the case of complex data. In [7] author's architecture follows the conventions of Cloud architecture, in addition to Cloud servers, there are dew servers which are situated on the native system and act as a buffer between the local user and the Cloud servers, also abstain the enigma of data becoming out of synchronization. This dew server would essentially host scaled-down variants of websites, full of pre-downloaded contents, which the user could access without Internet connection [8]. The major two functions performed by dew server and its related databases are: providing the client services same as services provided by the Cloud server and another is, maintaining synchronization between databases of dew server and cloud server. Cloud-dew architecture is an extension of the client-server architecture [9]. This architecture is represented in Fig. 2. A dew server has the following features [10]:

1. A dew server is a light weight web server, which is able to serve only one user.
2. A dew server can store only user's data because a dew server is very small like a drop of dew while a cloud server is very big like a real cloud.
3. A dew server is as week as a real drop of dew because a dew server's data vanishes easily due to hardware failure, infection of virus, etc.

Fig. 2 Cloud–dew architecture [7]

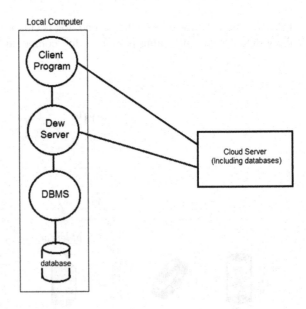

4. Because cloud can provide all the necessities, after disappearance dew will come out again, similarly vanished dew server can be entertained once again because of copy of all dew server data in the cloud servers.
5. A dew server is running on the local computer, so it is available with or without an Internet connection.

This architecture can furthermore be employed to make available websites offline. Suchlike system can diminish the overhead of Internet data for an organization having weak Internet connectivity. Many functions like displaying files or images, playing audio or video would be possible without Internet connection but provided that data had been synchronized to the "dew site" from the web over the last connection interval.

3 Existing Architecture Analysis

We found that only few authors have worked on the concept of Dew computing, so first let us clear what existing scheme is. In existing cloud–dew architecture, two kinds of URLs are considered: regular URL such as https://www.test.com and local URL such as https://mmm.test.com. On user's local computer the website is hosted, which is known as "dew site." Here mmm can be used to indicate dew site whereas www is used to indicate website. All names of the dew sites that user wants on his dew server must be placed in the host file, which is available in almost all operating system and maps host names to IP addresses. A dew server can be accessed with the help of local host. When a user enters URL in the navigation bar of the browser, if a URL corresponds to a website then browser follows the steps to map the domain name to the IP address using DNS server and display website content and if a URL corresponds to a dew site then dew server check the existence of domain name in host file. If host file does not contain requested domain name then dew server will send a request for script and database of requested website to the remote domain (cloud server). When the re requested is approved by remote domain then script and database of the website will be integrated into dew server. Then local URL will be mapped to the local host. Dew server will then find target host name by using an environmental variable. The URL request will be then redirected to the corresponding dew site script. The user will then perform an operation on dew site [7]. Synchronization with cloud server will be started on the availability of the Internet connection. Now, to perform synchronization between the content of dew site and website, user has to be logged into the website. Once the user is logged into the website, after doing internal mapping of user id of dew site and user id of the website, a link is created between a user on dew site with website and synchronization will be started automatically [7]. Through Fig. 3, we speculate the steps describe above of dew computing process.

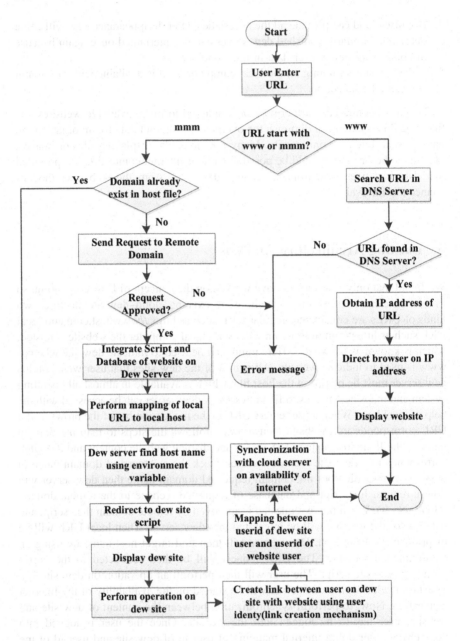

Fig. 3 Dew computing process flow

4 Our Contribution

In Sect. 4.1, we are trying to do basic analysis of currently available distributed computing paradigms based on general parameters in Table 1, based on location-based parameters in Table 2 and networked-based parameters in Table 3. Section 4.2 contains Table 4, which includes comparative exploration of Cloud, Fog, and Dew Computing.

4.1 Basic Analysis

See Tables 1, 2 and 3.

Table 1 Analysis based on general parameters

No	Requirement	Cloud computing	Fog computing	Dew computing
1	Goal	Providing services on demand	Make IoT more efficient	To achieve the potentiality of personal computers and cloud services
2	Peculiar feature	Providing scalable and measured resources on demand using Internet technologies	Proximity to end user, dense geographical distribution, mobility	Productive as well as cloud friendly cooperative services to users
3	Amount of data handle	Medium	Large	Low
4	Computing devices used	High configured computer (server) + Automation devices	Automation devices (sensors, controllers, chips, disks, network devices)	Computer
5	Database used	Huge sized	Large sized	Middle sized
6	Support IoE?	Yes	Yes	Yes
7	Eliminate bottleneck	No	Yes	Yes
8	Fault tolerant	No	Yes	Yes

Table 2 Analysis based on location parameters

No	Requirement	Cloud computing	Fog computing	Dew computing
1	Location of resource	Remote	At proximity of user/on edge of network	Right at user end/on the same machine
2	Location information available during data analytics?	Not available	Available	No need of location information
3	Remoteness between client and server	Numerous hops	One hop	On the same machine
4	Geo-distribution	Centralized	Distributed	Personalized
5	Mobility	Limited	Supported	Limited

Table 3 Analysis based on network parameters

No	Requirement	Cloud computing	Fog computing	Dew computing
1	Data traffic	High	Medium	Low
2	Service latency	High	Medium	Low
3	Satisfy need of IoT	No	Yes	Not proposed for IoT, but used in organizing IoT control
4	Internet or IoT	Internet	IoT	Internet
5	Number of nodes	Few	Large	One
6	Security level	Low	Medium	High
7	Internet connection required?	Yes	Yes	No
8	Delay jitter	High	Very low	Not present
9	Improve QoS	No	Yes	Yes
10	Scalability	No	Yes	Yes
11	Reliability	No	Yes	Yes
12	Barrage on data during routing	High probability	Very low probability	Very low probability
13	No of servers	Few	Large	One

4.2 Comparative Exploration of Cloud, Fog, and Dew Computing

See Table 4.

Table 4 Comparative exploration of cloud, fog and dew computing

No	Cloud computing	Fog computing	Dew computing
1	Processing of data applications is time consuming because of working from centralized cloud	Because of processing of data is done on the network edge, less time is required in operation	Dew works off line, but when Internet is available, data and applications are updated on the cloud. So it is very fast in processing
2	Each bit of data is sent to the centralized cloud which causes bandwidth problems	Relatively less number of bandwidth problem, as every bit of data is collected at some access point, instead of sending over cloud	No bandwidth is required in collection of data but when network is available, synchronization of data and applications done on cloud channel
3	Slow response time and scalability problems as a Result of depending servers that are located at remote places	It is possible to solve response time and scalability issue by placing tiny servers known as edge server in proximity of user	Good response time and no scalability problem
4	It is widely used in the business community to fulfill high computing demands on low cost using utility pricing model	It is preferable for batch processing jobs adopted by business world	It is used in everyday life. For example, integrated traffic control system of a town, which enable auto-adaptive traffic control behavior [9]

4.3 Technical Challenges of Dew Computing

As far as our knowledge is concerned, we are the one among the others to provide various future issues and challenges which may become obstacles for the adoption of dew computing and may also become new research direction. Here we list out the issues as follows:

1. Although dew computing will provide to user some services offline, the time for which will be in an incompatible state with the cloud, it is also one of the important concern. Intricacy of compatible matters will expand integrally with increasing number of users.
2. For synchronization of data on a local computer with cloud services, local machine has to run all the time, which consume too much energy.
3. It is required to design OS which implements and manages the collaborate feature of dew computing.
4. When more than one dew sites are created on a single dew server running on a local computer, there may be chances of conflicts for usage of available ports and other resources between dew sites, so it will raise the requirement for the development of new communication protocol transferring data between dew sites and remote cloud servers.

5. It is not safe to store user credentials in the database of dew server running on the local computer, because there may be chances of accessing a database of dew server by using some malicious software running on the local computer. So, database security is also one of the major concerns.

6. Dew site developer becomes bounded to use set of platforms and databases that are only dew capable. So that it does not allow much freedom in developing dew site. This leads to finding the solution of developing dew site independent of platform and database.

7. As time goes on, dew server will download more and more data which eventually cause available storage out of run, so there must be a requirement of developing a mechanism for replacement of downloaded content.

8. As dew servers must have to synchronize with cloud server periodically, hence there may be chances of out of synchronization related issues.

5 Conclusion

Dew computing is grounded on a micro service idea in vertically distributed computing classification. The ability to provide a web-surfing experience without an Internet connection is the realization of the distributed systems. This paper includes various fundamentals of a new concept of dew computing along with a comparative exploration of cloud computing, fog computing and dew computing, which can be used to understand strengths, limitations as well as applications of these paradigms. We have also discussed various future issues and challenges which may become obstacles for the adoption of dew computing and which may also provide a new direction of research. In summary, we can conclude that dew computing is closely connected to cloud computing. Dew computing which is supportive in perceiving the vigor of cloud computing is not isolated from cloud computing, but it is the interdependent part of cloud computing. Without cloud computing, dew computing would not be possible.

References

1. Evolution of Cloud to Fog Computing. https://blog.rankwatch.com/evolution-of-cloud-to-fogcomputing/.
2. M. Abdelshkour (2015) IoT, from Cloud to Fog Computing. Cisco Blog. http://blogs.cisco.com/perspectives/iot-from-cloud-to-fog-computing.
3. D. Bradley (2016) Dew helps ground cloud services. Science Spot. http://sciencespot.co.uk/dew-helps-ground-cloud-services.html.
4. Andy Rindos, Yingwei Wang (2016) Dew computing: the Complementary Piece of Cloud Computing. In: IEEE International Conferences on Big Data and Cloud Computing (BDCloud), Social Computing and Networking (SocialCom), Sustainable Computing and Communications (SustainCom), pp 15–20.

5. Yingwei Wang (2016) Definition and Categorization of Dew Computing. In: Open Journal of Cloud Computing (OJCC), vol. 3, issue 1.
6. Y. Wang (2015). The Initial Definition of Dew Computing. Dew Computing Research. http://www.dewcomputing.org/index.php/2015/11/10/the-initial-definition-of-dewcomputing/.
7. Yingwei Wang (2015) Cloud-dew architecture. Int. J. Cloud Computing, vol. 4, issue 3, pp 199–210.
8. David Edward Fisher, Shuhui Yang (2016) Doing More with the Dew: A New Approach to Cloud-Dew Architecture. In: Open Journal of Cloud Computing (OJCC), vol. 3, issue 1.
9. Karolj Skala, Davor Davidovi, Enis Afgan, Ivan Sovi, Zorislav Sojat (2015) Scalable Distributed Computing Hierarchy: Cloud, Fog and Dew Computing. In: Open Journal of Cloud Computing (OJCC), vol. 2, issue 1.
10. Yingwei Wang (2015) Cloud-dew architecture: realizing the potential of distributed database systems in unreliable networks. In: International Conference on Parallel and Distributed Processing Techniques and Applications (PDPTA), pp 85–89.
11. Comparison of Cloud and Fog Computing https://media.licdn.com/mpr/mpr/p/8/005/071/c/3e304b5.jpg.

Analysis of Parallel Control Structure for Efficient Servo and Regulatory Actions

Aarti Varshney, Puneet Mishra and Vishal Goyal

Abstract This paper investigates an intriguing issue about the tuning aspects of the parallel control structures. This parallel control structure essentially decouples the servo action from the regulatory action and provides an opportunity to the control engineer for separately deciding the ability of the controllers for servo and regulatory action. This paper provides a thorough comparative study and thereby suggesting an appropriate combination of tuning rules for achieving better efficiency of the control structure. Three different well accepted tuning rules viz. Ziegler Nichols, Direct Synthesis (DS) and Gain Margin Phase Margin formulae have been considered and a critical analysis of the control tuning rules combinations have been performed. The performance of considered tuning rules combinations is assessed on the basis of a transient response criterion, i.e., overshoot, an error-based criterion, i.e., Integral of time-weighted absolute error for both setpoint and disturbance rejection, and a measure of controller output aggression, i.e., Integral of absolute rate of controller output. On the basis of performed studies for a first-order plus dead time system, it may be inferred that DS–DS tuning rule combination provided superior performance among all the considered cases for nominal as well as plant-model mismatch case.

Keywords PID · Parallel control structure · Tuning · Setpoint tracking Disturbance rejection

A. Varshney · P. Mishra · V. Goyal (✉)
Department of Electronics and Communication Engineering, GLA University, Mathura, India
e-mail: vishal.glaitm@gmail.com

A. Varshney
e-mail: aartivarshney11@gmail.com

P. Mishra
e-mail: puneet.mishra@ymail.com

© Springer Nature Singapore Pte. Ltd. 2018
Y.-C. Hu et al. (eds.), *Intelligent Communication and Computational Technologies*, Lecture Notes in Networks and Systems 19,
https://doi.org/10.1007/978-981-10-5523-2_33

357

1   Introduction

Industrial control systems can be thought as interconnection of components forming a system configuration which provide a desired system response [1]. This desired system response can be classified in two major objectives, (1) Setpoint Tracking, (2) Disturbance Rejection. The setpoint tracking problem is also called as servo problem and the disturbance rejection problem is called as load regulation. Most of the time a single controller strategy is normally employed in the control systems to achieve both of these objectives. This strategy is also called as, conventional control scheme (CCS) which mostly uses a Proportional-Integral-Derivative (PID) Controller. The PID control configuration has always proved to be very popular since its inception in the industrial processes. According to Miller and Desborough [2] estimation 98% controllers which are employed in the industry are PID controllers. The reason for their wide acceptance is; their simple structure, simple implementation, low cost and ease in understanding the behavior of the individual controller actions. However, the use of PID controllers in CCS, i.e., in a simple feedback control strategy may not always accomplish different objectives of the control problem. A single PID controller can either provide better servo action or better regulation action. However, in industrial problems, it is normally desired that the controller action be generated in such a way that both of these objectives get accomplished in the best possible manner. Many researchers have reported this issue and in view of this, different notions of control and various control schemes to tackle this issue have been proposed. Ender [3] noted that 30% controllers in the industries are in manual mode and 25% of the controllers which are acting in automatic mode, are poorly tuned. Ogunnaike and Mukati in 2006, proposed a Robust, Tracking, Disturbance rejection-overall Aggressiveness (RTD-A) controller, to provide direct relationship between controller performance attributes and loop features without any deviation [4]. This scheme was well capable of providing a decoupled action between the tracking and regulation action and the tuning of the controller parameters was also addressed. However, the tuning rules in the paper were ambiguous and could be further simplified. To alleviate this compromise, a scheme is further suggested by Karunagaran and Wenjian in [5], known as parallel control structure (PCS) or double controller scheme. This scheme consists of two controllers and could independently and directly tune the three loop features, i.e., servo problem, regulation problem and robustness problem. This directly tuning of three features is also known as Transparent Online Tuning. This scheme solves decoupling problem and provide good tracking as well as avoid disturbance very efficiently.

Further, to deal with the disturbance rejection and robustness of the system, many modifications in PCS have also been proposed but they are mostly based on different process configurations. A modified Smith predictor was proposed in [6] after Smith predictor [7] to get satisfactory disturbance rejection and robustness. In modified Smith predictor, the work was concentrated on plants having larger time-delays. This work addressed decoupling of setpoint tracking response from disturbance rejection under nominal conditions. Further, Astrom Smith Predictor [8] was the first modified Smith predictor which controls the processes with integrator and long dead time. These limitations of Astrom Smith Predictor were taken into consideration in [9]. A two-degree of freedom Smith predictor has also been analyzed in [10] and some other works which are related to parallel control structure also address different types of the processes to be controlled such as, systems with time-delay and double integrators [11–13].

A discussed in literature survey presented above, most of the works related to this PCS control scheme is majorly inclined toward modification of the control structure for achieving better control performance. Another domain of work which has been addressed, deals with different process types which are found in the control industries and each of these works provides a different tuning mechanism and gain values which create a lot of hassle in the mind of the control engineer such as which control strategy should be used with which type of system and which tuning rules will provide the better control performance. In other words, there is a research gap about, how a parallel control structure can be tuned among various available tuning rules, in order to get best from the control structure. Since modifications in the control structure can provide better decoupling but it cannot guarantee better individual tracking response or disturbance rejection response. This paper addresses the same issue and discusses the tuning of the controller parameters used in a parallel control structure. In this work, different tuning rules, which are well established and widely accepted in the industries, have been thoroughly analyzed and a comparative study has been drawn from the studies made in the paper. Three different tuning rules, i.e., Ziegler–Nichols (ZN), Direct Synthesis (DS) and Gain Margin and Phase Margin (GMPM) tuning rules have been considered and their combinations have been investigated to achieve an insight about what gain values should be used in the industrial control loop, if a parallel control structure is employed. The plant on which this study is performed is a first-order plus dead time system, which can model wide variety of the industrial processes. The performance of the tuning rules has been assessed using integral of absolute error (ITAE) for setpoint tracking as well as for disturbance rejection and percentage overshoot in tracking.

Further, this paper is divided into four sections which are as follows. Following, introduction and related literature survey in Sect. 1; Parallel Control Structure (PCS) is described in Sect. 2 along with the problem formulation. The detailed simulation studies have been carried out and are presented in Sect. 3. Finally, the paper is concluded in the Sect. 4 by summarizing the best suitable combination for the proposed parallel control structure.

2 Problem Description

This section discusses the structure of the parallel control structure [5] Fig. 1 shows the structure of the parallel control structure. g_p is the process which is to be controlled, with an output y. The disturbance d is acting directly on the process and is added directly to the process variable. u is the control input applied to the process and is generated through the combination of the two controllers, g_{c1} and g_{c2}. The controller g_{c1} is considered here is responsible for the setpoint tracking and g_{c2} for the disturbance rejection only. However, this decoupled action, i.e., dedicated controllers for individual control objectives can only be achieved if the process dynamics are estimated accurately by an appropriate means. In this work, an estimated transfer function g_m is considered for the implementation of PCS as shown in Fig. 1. Decoupling in processes is the main concern so, the objective is to design a double controller scheme for independent tuning of setpoint tracking response and disturbance rejection response, by which it can be able to fulfill all the three performance aspects of a control loop including tracking response, disturbance rejection response, and robustness.

By normal block diagram reduction method, the process variable y can be written as in (Eq. 1). It may be clearly seen from this equation that if $g_p = g_m$ then the controller g_{c1} can be tuned for the setpoint tracking and g_{c2} can be set for the disturbance rejection.

$$y = \frac{g_p(1 + g_{c2}g_m)}{g_m(1 + g_{c2}g_p)}\left[\frac{g_{c1}g_m}{1 + g_{c1}g_m}\right]r + \left[\frac{1}{1 + g_{c2}g_p}\right]d \tag{1}$$

Where, all the signals and systems are considered to be in Laplace domain.

Under nominal conditions, g_p can be replaced by g_m and the above equation reduces to,

$$y = \left[\frac{g_{c1}g_m}{1 + g_{c1}g_m}\right]r + \left[\frac{1}{1 + g_{c2}g_m}\right]d \tag{2}$$

In this way, this double controller scheme is well capable to provide two main features: first, the process output tracks the setpoint input completely and second, the servo response decouples the regulatory response from tracking response. The tracking is performed by g_{c1} while disturbance is rejected with the help of g_{c2}.

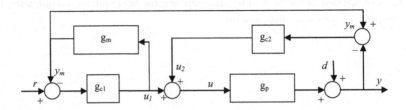

Fig. 1 The parallel control structure [5]

As discussed in Sect. 1, the main problem lies with the tuning of the two controllers and it is very essential to find a good combination for the controller tuning parameters which provide a better control performance. In this work, only simple parallel PID controllers have been considered for both g_{c1} and g_{c2}. The reason for choosing simple PID controllers for the analysis is their wide acceptance in the industrial world and better flexibility associated with them.

$$G_c(s) = k_c(1 + T_I/s + T_D s) \tag{3}$$

Or

$$G_c(s) = k_p + k_I/s + k_D s \tag{4}$$

Where,

k_c, k_p Proportional gain
k_I Integral gain
k_D Derivative gain
T_I Integral time
T_D Derivative time

Further, for the analysis purpose, a first-order system with the dead time (FOPDT) system has been considered for the g_p. Since, the problem associated with PCS is their efficient tuning; this work addresses and assesses different tuning combinations for PID Controllers used in PCS. In this paper three tuning methods viz. ZN method, DS method, and GMPM methods are compared together with the help of different combinations of g_{c1} and g_{c2}. Eight different combinations are considered. For example, DS method tuned PID for g_{c1} and GMPM method for g_{c2} and similarly other combinations have been considered. The next subsection presents a brief overview of the tuning methods considered in this paper for the continuity of the article.

2.1 Open Loop Ziegler–Nichols Method

To determine the parameters of a PID controller, one of the most popular methods is Ziegler–Nichols method [14, 15]. In this work, process reaction curve method for the calculation of the PID controller parameters has been used. This technique is widely used in the industries and therefore is considered in this work.

The FOPTD model taken for the process dynamics is represented as,

$$G_m(s) = \frac{ke^{-ds}}{\tau s + 1} \tag{5}$$

Where k is the system gain, τ is time constant, and d is dead time.

The controller parameters for ZN tuning can be obtained as,

$$k_c = \frac{1.2\,\tau}{k\,d} \qquad (6)$$

$$T_I = 2d \qquad (7)$$

$$T_D = 0.5d \qquad (8)$$

2.2 Direct Synthesis

For a feedback controller, in this case PID, a model-based method can also be used. This DS method is designed to tune the controller settings for a desired closed-loop response which is predetermined. A reference trajectory normally employed is a first-order transfer function having a steady state gain of unity with desired time constant for the optimum setpoint tracking. There is only one parameter to be tuned which is time constant and this is the main advantage of this method. The focus of this method can either be on rejecting setpoint disturbances or load disturbances. However, literature is mostly concentrated on the design of trajectory for setpoint tracking. Since, FOPDT system can represent the behavior of a wide range of processes, the gains of the PID controller for this kind of process as in Eq. (5) are represented as [16, 17],

$$k_p = \frac{\tau}{k(\tau_c + d)} \qquad (9)$$

$$k_I = \frac{1}{k(\tau_c + d)} \qquad (10)$$

$$k_D = 0 \qquad (11)$$

Here, τ_c is the reference trajectory time constant and is considered as 1 in this work.

2.3 GMPM Method

Conventionally, to control a continuous-time system $g_p(s)$ whose nominal model is,

$$g_m(s) = \left(\frac{k}{\bar{\tau}s^2 + \tau s + 1}\right) e^{-ds} \qquad (12)$$

In the Gain Margin and Phase Margin method (GMPM) [18–20], the tuning of the PID controller can be done as,

$$\begin{bmatrix} k_p \\ k_I \\ k_D \end{bmatrix} = \frac{\pi}{2A_m dk} \begin{bmatrix} \tau \\ 1 \\ \tilde{\tau} \end{bmatrix} \tag{13}$$

Where, A_m is the desired open loop gain margin and in this work it is considered as 3, d is the dead time, and k is steady state gain of the system.

2.4 Criteria for Performance Evaluation

For the performance evaluation of the considered tuning methods combinations, a unified framework is not yet established. Overshoot, rise time, decay ratio, etc., are the basic transient characteristics to examine the system performance but these characteristics fail to provide a reasonable information about the error behavior in the process variable. Performance evaluation based upon errors in the process variable can provide a better indication of the process controller efficiency. The considered error-based criteria employed in this work are ITAE for setpoint tracking and disturbance rejection separately. It may be clearly noted that a combination of these performance measures can provide a better indication of the efficiency of the control structure.

2.4.1 Integral of the Time-Weighted Absolute Error (ITAE)

ITAE integrates the absolute error multiplied by the time over time. If a controller is optimized for minimum ITAE then it will strongly suppress the errors which persist for long duration. Therefore, in this paper ITAE is considered as one of the considered performance parameter.

$$ITAE = \int_{t_o}^{t_f} t|e(t)|dt \tag{14}$$

Where, $e(t)$ is the error in process variable, t_o is the time at which setpoint or disturbance is applied. In this work the setpoint is applied at $t = 0$ s and disturbance is applied at $t = 20$ s. For the case of plant-model mismatch ($g_p \neq g_m$), the disturbance is applied at $t = 15$ s. The disturbance magnitude in this work is considered −1 for both cases.

2.4.2 Integral of Absolute Rate of Controller Output (IARCO)

Another interesting performance measure is IARCO. It is desired that the controller output is least oscillatory and the controller action generation is as smooth as possible. If the controller output is more oscillatory in that case the final control element may get damaged and/or its life span may get reduced.

$$IARCO = \int_0^{t_f} \left| \dot{u}(t) \right| dt \tag{15}$$

Where, $\dot{u}(t)$ is first derivative of control input $u(t)$ to plant.

3 Simulation Studies

In this paper, an FOPDT system [5] is considered and the performances of different combinations of $g_{c1}-g_{c2}$ with different tuning rules have been assessed. The simulation studies have been divided into two categories, (1) Nominal case and (2) Plant-model mismatch case. Since it is a frequent case that a plant model gets mismatched from the original one and hence has been considered in this work also.

3.1 Nominal Case

The nominal model of system to be controlled with a PID loop is g_m which is same as g_p.

$$g_p(s) = g_m(s) = \frac{1}{2s+1} e^{-s} \tag{16}$$

The results of controller parameter values obtained for both cases using different tuning techniques are given in Table 1.

The simulation is carried out using MATLAB (version R2013a) software, for setpoint tracking and load regulation. A unit step input is used for both setpoint and load changes for all combinations of different methods. Three methods with eight different combinations are analyzed here. The combinations are formed in the form

Table 1 Controller parameters values for different methods for g_{c1} and g_{c2}

Method	k_p	k_I	k_D
ZN	2.4	1.2	1.2
DS	1.0	0.5	0
GMPM	1.046	0.523	0

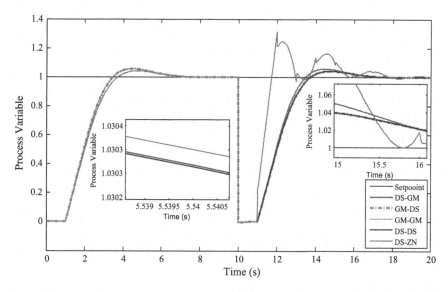

Fig. 2 Process variable (y) variation in PCS for nominal case

Table 2 Performance evaluation and comparison of different tuning methods in nominal case

Method	Overshoot	ITAE$_S$	ITAE$_d$	IARCO	Total
DS-GMPM	4.05	2.1679	2.8438	2.539	11.6007
DS-ZN	4.08	2.1696	2.6794	354.6	363.529
ZN-DS	30.77	2.0161	2.8728	14.15	49.8089
ZN-GMPM	30.77	2.0160	2.8475	14.28	49.9135
DS-DS	**4.05**	**2.1670**	**2.8660**	**2.4095**	**11.4925**
ZN-ZN	30.53	2.0145	2.6857	367	402.2302
GMPM-GMPM	5.60	2.1377	2.8430	2.654	13.2347
GMPM-DS	5.60	2.1377	2.8652	2.5241	13.127

of $g_{c1} - g_{c2}$, i.e., first controller is tuned with the one method and the second controller with other. Here the different combinations are DS-GMPM, DS-ZN, DS-DS, ZN-DS, ZN-GMPM, ZN-ZN, GMPM-DS, and lastly GMPM-GMPM. Figure 2 shows the process variable variation after the application of setpoint at $t = 0$ s, for a unit step setpoint. It can be clearly seen from this figure that the DS method is providing a better setpoint tracking response and better disturbance rejection also. On the other hand, other methods are either providing better tracking or better regulation only. The results are well supported by quantitative assessment as provided in Table 2. It may be noted that the value of IARCO is also very less for DS-DS method which suggests that the controller output is also very smooth. The same is depicted in Fig. 3 also. It is worth mentioning here that some of the gain combination results have been omitted from the figures. For the better representation of the results since it can be easily inferred from the quantitative comparison that these combinations are not suited for PCS at all.

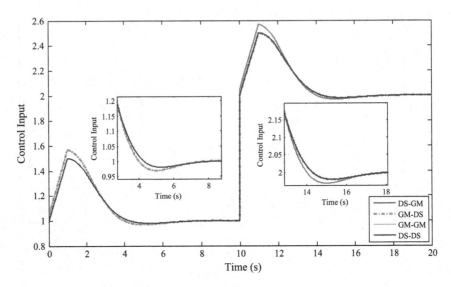

Fig. 3 Control action (u) of PCS in nominal case

3.2 Plant-Model Mismatch Case

The nominal model and plant are not same in mismatched condition.

$$g_m(s) = \frac{1}{2s+1}e^{-s} \neq g_p(s) = \frac{1.2}{1.6s+1}e^{-1.2s} \tag{17}$$

It may be seen from Fig. 4 that the tuning combinations using ZN as one of the controller tuning technique are not able to cope up with the mismatched condition. The result of the mismatched conditions or plant parameter drift has resulted in oscillations in the process as well as the control input variables in the control loop. It is instructive to show here that techniques using DS method is still providing better results and is also coping up with the plant-model mismatched condition. It may be noted here that the overshoot value has been increased but this change is obvious since the controllers were not tuned for this mismatched conditions and due to the uncertainty in the process parameters the tuned controller parameters have been drifted from their normal behavior. The quantitative comparison presented in Table 3 shows that still DS-DS tuning combination performing best from the considered cases.

It may also be noticed from Fig. 4 that the controller output for the mismatched case for the methods using ZN has been significantly oscillatory. This is an undesirable aspect and should be avoided. The reason for the same is that the

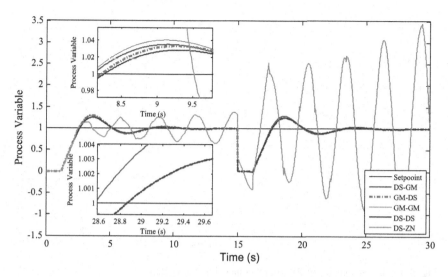

Fig. 4 Process variable (*y*) variation in PCS for plant-model mismatch case

Table 3 Performance evaluation comparison of different tuning methods in mismatched case

Method	Overshoot	ITAE$_S$	ITAE$_d$	IARCO	Total
DS-GMPM	25.47	2.5651	19.6786	4.9093	52.623
DS-ZN	38.38	2.9137	225.338	821.781	1088.41
ZN-DS	84.88	2.9089	18.5893	14.8658	121.244
ZN-GMPM	85.08	2.9332	19.7767	15.1831	122.973
DS-DS	**25.63**	**2.5552**	**18.527**	**4.5106**	**51.2228**
ZN-ZN	66.15	7.0596	507.277	1093.30	1673.78
GMPM-GMPM	29.47	2.6393	19.6868	5.1052	56.9013
GMPM-DS	29.61	2.6284	18.532	4.6995	55.4699

control elements may get damaged from this excessive overshoot in the controller output which may decrease the plant life and productivity also. The controller output for only some cases has been shown in Fig. 5 for better understanding and representation of the performances of the considered tuning methods. From the considered case, it may be easily seen that the controller performance using DS-DS approach is most appropriate for the industrial use since this reduces overshoot, ITAE for setpoint tracking, ITAE for disturbance rejection even with the generation of a smooth controller output.

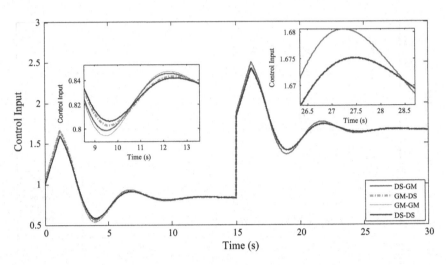

Fig. 5 Control actions (u) of PCS in plant-model mismatch case

4   Conclusion

This paper presented a thorough comparative study between different tuning rules for a parallel control structure. The parallel control structure is usually used to obtain decoupled servo and regulatory actions. Since it is quite hard to find a unified set of tuning rules which can be easily implementable to the Parallel control structure, this work strives to find a best suitable method for the parallel control structure which can give most satisfactory performance parameters of a system, i.e., tracking performance, disturbance rejection, and robustness. To achieve and analyze this, eight different combinations have been tried out using some widely accepted tuning rules for simple conventional control structures and their extension to PCS has been assessed. The assessment has been done on the basis of overshoot in setpoint tracking, ITAE for setpoint tracking and ITAE for disturbance rejection and the algebraic summation of all these. The analysis has been done for two different cases, i.e., nominal and mismatched case. For both the cases, it was found that direct synthesis method outperformed all the other considered tuning rule combinations and gives good and reasonable result for both tracking performance as well as disturbance rejection.

References

1. Dorf R. C., Bishop R. H.: Modern Control System, Eleventh Edition, Pearson Prentice Hall.
2. Desborough L., Miller R.: Increasing customer value of industrial control performance monitoring-Honeywell's experience, Proceedings of CPC VI Tucson, AZ, (2001).
3. Ender D.: Process control performance: not as good as you think. Control Eng. 180, (1993).

4. Ogunnaike B. A., Mukati K.: An alternative structure for next generation regulatory controllers: Part 1: basic theory for design, development and implementation. Journal of Process Control. 499–509, (2006).
5. Karunagaran G., Wenjian C.: The Parallel Control Structure for Transparent Online Tuning. Journal of Process Control. 21, 1072–1079 (2011).
6. Tian Y. C., Gao F.: Double-controller scheme for separating load rejection from set-point tracking, Chemical Engineering Research and Design. 76, 445–450 (1998).
7. Smith O. J. M.: Closer control of loops with dead time. Chemical Engineering Progress. 53, 217–219 (1957).
8. Astrom K. J., Hang C. C., Lim B. C.: New smith predictor for controlling a process with an integrator and long dead-time. IEEE Transactions on Automatic Control 39, 343–345 (1994).
9. Zhang W. D., Sun Y. X.: Modified smith predictor for controlling integrator/time delay processes. Industrial and Engineering Chemistry Research 35, 2769–2772 (1996).
10. Zhang W., Sun Y., Xu.X.: Two degree-of-freedom smith predictor for processes with time delay. Automatica 34, 1279–1282 (1998).
11. Liu.T., He X., Gu D. Y., Zhang W. D.: Analytical decoupling control design for dynamic plants with time delay and double integrators. Control Theory and Applications, IEEE Proceedings 151, 745–753 (2004).
12. Tang W., Meng D., Wang X., Zhang K., Wang M.: Controller design of a two-degree-of-freedom smith predictor scheme for improved performance. IEEE International Conference on Information and Automation, ICIA, 242–247 (2009).
13. Wei T., Songjiao S., Mengxiao W.: A novel smith predictor with double-controller structure. Proceedings of the American Control Conference. 6, 5074–5079 (2002).
14. Korsane D. T., Yadav Vivek, Raut K. H.: PID Tuning Rules for First Order plus Time Delay System. International Journal Of Innovative Research In Electrical, Electronics, Instrumentation and Control Engineering, Vol. 2, (2014).
15. Shahrokhi M., Zomorrodi A.: Comparison of PID Controller Tuning Methods. Chemical and Petroleum Engineering.
16. Dighe Y. N, Kadu C. B., Parvat B. J.: Direct Synthesis Approach For Design of PID Controller. International Journal of Application or Innovation in Engineering and Management. 3 (2014).
17. Jung C. S., Song H. K., Hyunc J. C.: A direct synthesis tuning method of unstable First-order-plus-time-delay processes. Journal of Process Control. 9, 265–269 (1999).
18. Xiong Q., Cai W. J., He M. J., M. He: Decentralized control system design for multivariable processes - a novel method based on effective relative gain array. Industrial and Engineering Chemistry Research 45, 2769–2776 (2006).
19. Xiong.Q., Cai W. J.: Effective transfer function method for decentralized control system design of multi-input multi-output processes. Journal of Process Control 16, 773–784 (2006).
20. Xiong.Q., Cai W. J., He M. J.: Equivalent transfer function method for PI/PID controller design of MIMO processes. Journal of Process Control 17, 665–673 (2007).

Fine Grained Privacy Measuring of User's Profile Over Online Social Network

Shikha Jain and Sandeep K. Raghuwanshi

Abstract From last few years, social networking has become part of everyone's life. A person known to computer has its online social accounts like Facebook, Twitter, or MySpace. Considerable amount of work has been done to address privacy issues in online social networks. Many authors have discussed about information breach since users share their personal information which reveals their identity. A serious attention is required to pay to reduce privacy risk of users posed by their daily information sharing activities. There is no classified measure to measure privacy. This paper tries to quantify privacy and evaluate it. This paper also includes privacy issues raised from the individual user's viewpoint then propose a framework to compute the privacy index for a user on OSN [1] and shows the applicability and requirement of privacy index. This score indicates that the user is aware of their privacy profile or not. The index can be used for the recommendation to enhance the privacy settings of the users in the group. The framework proposes a mathematical model of basic commodity index to calculate privacy index of any user in OSN. Our definition of index takes an index number which is a number indicating sensitivity, more sensitive information a user discloses, the higher his or her privacy risk and so is its index number. The framework considers both sensitivity and visibility of information of user's profile and computes index value on the basis of them. The sensitivity of profile items over survey data is calculated using naïve formula. Based on privacy measurement function values, the users on OSN are classified in three categories as Secure, Mediocre, and Vulnerable to privacy attack. We further compare the normal indexing technique to our privacy measurement function. This along with propose algorithm shows better efficacy and reduces the possibility false classification of user's categories.

S. Jain (✉)
CSE Department, SATI Vidisha, Vidisha, India
e-mail: shikha03jain@gmail.com

S.K. Raghuwanshi
CSE Department, MANIT, Bhopal, India
e-mail: sraghuwanshi@gmail.com

© Springer Nature Singapore Pte. Ltd. 2018
Y.-C. Hu et al. (eds.), *Intelligent Communication and Computational Technologies*, Lecture Notes in Networks and Systems 19,
https://doi.org/10.1007/978-981-10-5523-2_34

371

Keywords Privacy score · Online social network (OSN) · Naïve theory
Personalization · Customization

1 Introduction

The process of flow of information through the communication channel is called
information diffusion. Online Social Networks (OSN) is the most used means for
this nowadays. Online social networks (OSNs) [2] like Facebook, MySpace, Orkut,
and Twitter help user to create community with other users of their interest and also
help them to create new contact. Every user over social network creates his/her
profile that contains information out of which some information is public and some
is private. The base for users to create group is their common interest. People share
information in this way to their contacts. The most striking thing about social
network is that it allows users to search for people they know or to they want to
know [3].

Privacy is the major concern raised in this place, since a user footprints are very
clear to understand and they are very easily identifiable by the attackers. This gives
rise to one of the major attacks called identity theft and Spam identities. Location
tracking system will help them to reach easily since users share their location
information more easily (check-in) [1].

The main contributions of this paper is to introduce a privacy measurement
function by which a user can assess his/her privacy setting over OSN, with the same
it can also be used in organization to know and classify group of users based on
their privacy awareness and profile item settings. We work by creating clumps of
dimensions that define the users in three levels high, medium, and low sensitive.
Each clump is distilled against separate threshold and classify entire crowd source
in three categories namely secure, mediocre, and vulnerable to attack.

The organization of the paper is as follows: Sect. 2 includes background of this
research work followed by Sect. 3 which introduces our experimental model and
our proposed privacy measurement functions, followed by result and evaluation in
Sect. 4, including comparison and analysis. Section 5 summarizes the paper and
future work to be done in this field.

2 Background

Social Media not only provides communication features but also affect users by
making them more vulnerable, by making them more exposed to the network, it
raises privacy concern.

Out of which, identity theft and digital stalking are important issues. To avoid it,
so many researches have been done which address privacy policy conflicts and
privacy preservation [4–6]. It includes measurement of privacy because of raising

privacy issues [3, 7–9]. The basis they choose for measurement is online information sharing behavior. In research work [10], the author used public information to measure privacy. However, it did not provide any measurement functions.

Several works have been done in this field which calculate privacy and measure privacy using sensitivity and visibility [2], which has been further modified in later research [11]. Commonly, Item Response Theory (IRT) is used to evaluate sensitivity and visibility of attributes in order to evaluate privacy score [2, 12]. The authors in [3] develop a tool, Privometer, to measure leakage of information. The leakage of information is measured in numeric values derived from the combination of probability of inference. The tool suggests most vulnerable entity based on these scores [2, 11].

The author in [11] also uses online information sharing behavior as a theme of measurement to provide an ultimate method to evaluate privacy by using the container independent networks that fit the real-world data. These methods use user's profile item information and their privacy setting information in mathematical item response theory underpinning [7].

The work in [1] uses the similar concept of sensitivity and visibility. The author has extended the work by proposing following privacy measurement (ranking) also called privacy index for social network actor model [10], i.e., the weighted privacy index (w-PIDX), where privacy index is maximum, i.e., (m-PIDX), and composite privacy index (w-PIDX). They further provide a test which evaluates the effectiveness of three PIDXes [10, 1].

Research also provides support to content references and social network concepts. It also includes the cause that does not directly harm users but indirectly makes them uniquely identifiable [9]. Thus, it triggers an urge to have a system which measures privacy and helps to govern users who are more vulnerable.

3 Preliminaries

We briefly introduce the experimental model that we have used for our calculation and generated formula. Following components define our experimental model:

Social Entities—Social entity refers to anything which has its unique virtual presence like a person, a company, or anything. Social entities on the web mean persons and organization those who are part of social web network [1].

Attribute—Attributes may be defined as the profile item of a user like his/her name, gender, address, phone number, etc. Attributes play a very crucial role on privacy; they have either direct or indirect effect or impact on privacy of an online user [1].

Some attributes directly disclose privacy (like phone no, residential address, bank details, photos) and considered as more sensitive attribute. While some attributes collectively identify a user in a network. They are called as quasi-identifiers. In simple words, attribute describes a group of behavior or data which affects privacy [1].

Sensitivity—Sensitivity is defined as the degree of openness of an attribute. It ranges from 0 to 1. This property depends on the item itself [7]. The sensitivity directly impacts on privacy of the user. The item which reveals more information about a user is more sensitive than the other. Thus, that particular item has high sensitivity value.

Visibility—We can simply say that V (i, j) = 1 if the attribute value is visible in network otherwise 0 [11]. The visibility of item i by user j is represented by V (i, j).

Consider a social network represented by G having N nodes and every node $i \in$ *{1. . . N}* represents a user or any real time social entity. Each user is characterized by their user profile created at the time of joining of network represented as p_j, $j \in$ *{1, 2, 3, ...n}*, With each profile item a privacy level is defined by the user which determines their willingness to disclose item associated information [7]. The response matrix (N × n) represents the level of privacy of N users corresponding to n profile items (i, j) defines the privacy setting of user I*i corresponding to user j*. Response matrix R can only take values either 0 or 1 (i.e., 1 for visible and 0 for not visible) [11].

Each profile item is assigned a privacy impact factor known as sensitivity level of it. Attributes sensitivity is represented by $\beta_j = \{\beta_1, \beta_2, \beta_3.....\beta_n \}$

Thus in any online social network to calculate over all privacy score of users a simple index formula may be used based on the sensitivity and visibility of profile items [11]:

$$PIDX = \sum \beta_j \times R(i,j) \tag{1}$$

Privacy Measurement Function

Let (P, S) represent complete attributes list and the attribute's sensitivity.

$P = \{p_1, p_2, p_3..............p_n\}$

$\beta = \{\beta_1, \beta_2, \beta_3.....\beta_n \}$

Let (P_H, P_M, P_L) represent these of attributes, as highly sensitive attributes, Medium Sensitive Attributes, and Low Sensitive attributes, respectively, as per algorithm1.1.

Privacy_Measurement_Function

Input: R and (P_h, P_m, P_l) //Response Matrix of N × n and profile item categories.

Output: *FGPIDX*

Initialize: X_h, X_m, and X_l to zero.

α_h, α_m, and α_l, to zero.

for i = 1 to n

if $p_i \in P_h$ then

$X_h = X_h + R_i * \beta_i$

$\alpha_h = \alpha_h + \beta_i$

else if $p_i \in P_m$ Then

$X_m = X_m + R_i * \beta_i$
$\alpha_m = \alpha_m + \beta_i$

Else if $p_i \in P_l$ Then

$X_l = X_l + R_i * \beta_i$
$\alpha_l = \alpha_l + \beta_i$

$$FGPIDX = \frac{(\alpha_h \times X_h + \alpha_l \times X_l + \alpha_l \times X_l)}{\Sigma} \qquad (2)$$

4 Experiment and Result Analysis

This section presents the experiment and result data.

Dataset: For this experiment, we have collected dataset from Facebook user's profile.

Here is a list of profile items that covers personal and professional features with respect to a user like age, birth date, marital status, political views, etc.

The privacy levels a user is either 0 (means user did not disclose the information) and 1 (means information is available for its friend list). This is also defined as the visibility of any item corresponding to a user.

Profile items selection and their sensitivity are important for privacy measurement. We used naïve theory for calculating the sensitivity of user's data based on their visibility level.

Naive theory makes use of natural language to define sets and various operations to be performed on sets [13].

Naïve computation of sensitivity: If |Ri| denotes the number of users who set R $(i, j) = 1$, then the sensitivity βi can be computed as the proportion of users that are reluctant to disclose item i [7]. That is,

$$\beta_i = N - Ri \vee \frac{}{N} \qquad (3)$$

The higher value of β_i more sensitivity of item i.

Sensitivity of profile items is visualized and represented in Fig. 1, with a word cloud using the sensitivity of the profile items assigned to profile items in our dataset which we evaluated using naïve Eq. (3). The larger font size represents large sensitive value (Table 1).

Postal Address

Contact Number

Current City

School

Relationship Status

Gender

Photos Name

Hometown Date of Birth

Friend List College

Work/ Employer

Email Id

Fig. 1 Profile item sensitivity calculated using naïve approach on collected dataset

Table 1 Sensitivity of profile items calculated using naïve equation

Sno.	Profile Item		Value
1	Low sensitive items	Name	0.00
2		Gender	0.00
3		DOB	0.13
4		Marital status	0.15
5		School	0.21
6		College	0.13
7		Home town	0.06
8	Medium sensitive items	Photos	0.03
9		Friend list	0.30
10		Work/employer	0.34
11		Current city	0.06
12	High sensitive items	Contact no	0.79
13		Email id	0.77
14		Postal address	0.83

As per the proposed scheme, we first categorized the profile items into three categories based on their sensitivity as highly sensitive attributes, Medium Sensitive attributes, and Low Sensitive attributes. Based on this there can be eight cases of three classes of to be known or not known completely. We first evaluate and see the change in the overall participation of profile items in the index value calculation. As per the proposed scheme, we first categorized the profile items into three categories as highly sensitive attributes, Medium Sensitive attributes and Low Sensitive attributes [17]. Based on this there can be eight cases of three classes of to be known or not known completely [17]. We first evaluate and see the change in the overall participation of profile items in the index value calculation (Fig. 2).

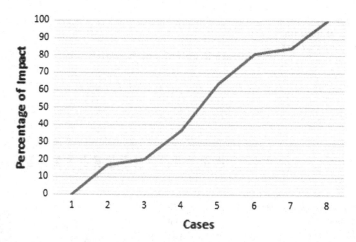

Fig. 2 Profile items participation in index calculation

Fig. 3 Index values over
complete set of users (in
sequence) using *PIDX*

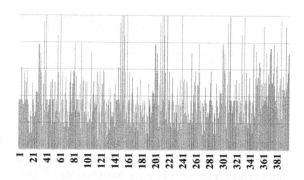

We further tested simple privacy index with our fine grained privacy indexes on the user dataset. With Figs. 3 and 4 it is clearly visible that finely grained privacy index calculates index value with more accuracy and has better user classification as compared to basic privacy index.

Collective value α_h, α_m, and α_l will help us to distribute user in secure, mediocre and vulnerable categories based on separately specified threshold. Results reveal that our methods work better in reducing false entries and is doing more accurate user classification. We tested the dataset with both indexes and distribute users in three categories as specified on basis of threshold. Figure 5 is showing both methods have a different set of user classification.

Dataset comprises collection 400 users' profile of Facebook. Using the privacy indexing [11], 19 out of 400 users fall into the category of the secure list. However, using our Fine Grained index distils 10 users as secure. Results show considerable change in filtering of users and reduction in false alarming.

Fig. 4 Index values over complete set of users (in sequence) using *FGPIDX*

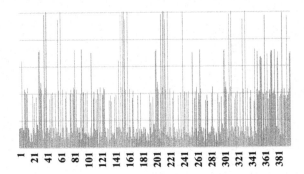

Fig. 5 User classification in secure, mediocre, and vulnerable categories

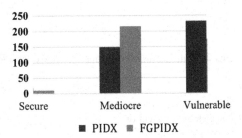

5 Conclusion and Future Work

Security and privacy are the major issues in social Medias. Since users relay in much of information obtained from such Medias, this become important to study and evaluate privacy. Here we have measured the privacy score by assigning privacy indexes [14]. Finally, we obtained the users separated in three different categories called secure, mediocre and vulnerable from input response matrix. We evaluate and demonstrate our suggested algorithm and formula on an experimental data obtained from Facebook. Thus, we have presented a practice approach to measure privacy for a user and suggest correcting his/her privacy settings in case if they fall into the vulnerable category.

Future work involves using more refined analysis of privacy settings of users. Analysis can be carried out on the basis of different profile items and new conclusions can be derived.

References

1. Y. Wang and R. N. Kumar, "Privacy Measurement for Social Network Actor Model," in The 5th ASE/IEEE International Conference on Information Privacy, Security, Risk and Trust, 2013.

2. D. M. Boyd and N. B. Ellison, "Social Network Sites: Definition, History, and Scholarship," J. Comp.-Mediated Communication., vol. 13, no. 1, Oct. 2007, pp. 210–230.

3. E. M. Maximilien, T. Grandison, T. Sun, D. Richardson, S. Guo, and K. Liu, "Privacy-as-a-Service? Models, algorithms, and results on the facebook platform," in Web 2.0 Security and privacy workshop, 2009.

4. L. Sweeney, "K-anonymity: a model for protecting privacy," International Journal on uncertainty, Fuzziness and knowledge-based system, vol. 10, no. 5, pp. 557–570, 2002.

5. A. Machanavajjhala, J. Gehrke, D. Kifler, and M. Venkitasubramaniam, "L-diversity: Privacy beyond k-anonymity," in Proceedings of the 22nd IEEE International Conference on Data Engineering, 2006.

6. F. Buccafurri, G. Lax, and V. Graziella, "Privacy-Preserving Resource Evaluation in Social Networks," in *Tenth Annual International Conference on Privacy, Security and Trust*, 2012, pp. 51–58.

7. J. Becker and H. Chen, "Measuring Privacy Risk in Online Social Networks," in Web 2.0 security and privacy Workshop, 2009.

8. N. Talukder, M. Ouzzani, A. K. Elmagarmid, H. Elmeleegy, and M. Yakout, "Privometer: Privacy protection in social networks," 2010 IEEE 26th International Conference on Data Engineering Workshops (ICDEW 2010), pp. 266–269, 2010.

9. J. Anderson, C. Diaz, F. Stajano, K. U. Leuven, and J. Bonneau, "Privacy-Enabling Social Networking over untrusted networks," in WONS, 2009, pp. 2–7.

10. Y. Liu, K. P. Gummadi, and A. Mislove, "Analyzing Facebook Privacy Settings: User Expectations vs. Reality," in *IMC' 11*, 2011.

11. K. U. N. Liu, "A Framework for Computing the Privacy Scores of Users in Online Social Networks," Knowledge Discovery. Data, vol. 5, no. 1, pp. 1–30, 2010.

12. F. B. Baker and S. -H. Kim. *Item Response Theory: Parameter Estimation Techniques.* Marcel Dekkerm, Inc., 2004.

13. Naïve set theory definition https://en.wikipedia.org/wiki/Naive_set_theory.

14. Michel Jakob, Zbynek Moler, Michal Pechoucek, Roman Vaculin "Content-based Privacy Management on the Social Web" ACM International Conferences on Web Intelligence and Intelligent Agent Technology-2011 pp 277–280.

Analyzing the Behavior of Symmetric Algorithms Usage in Securing and Storing Data in Cloud

Ashok Sharma, Ramjeevan Singh Thakur and Shailesh Jaloree

Abstract It has been found that the Cloud computing facilities have been used by users of various categories to meet their flexible computing needs which include Processing Powers, Storages, Servers, etc., and among these needs, Public Cloud storage has been used for storing huge data files of different formats and different sizes in Cloud. Because of data security concerns in Cloud, symmetric algorithms have been preferred for encryption and decryption of these data files. Because of limited Internet speed, we must have provision to identify fast and efficient symmetric algorithms for Cloud storages. In the literature survey, we have not found any framework to investigate the efficiency of symmetric algorithms in securing and storing data in Cloud based on various parameters like Encryption Time, Decryption Time, and Throughput, etc.

Keywords Encryption time · Decryption time · SLA · EDaS

1 Introduction

Presently, Cloud storages like Google drive, Wuala, Rackspace, Onedrive, etc., have attracted various IT users by various offerings. These storages are part of Public Cloud platform where different IT users can share the storages with credentials. No doubt, adoption of Public Cloud will definitely help for efficient use of

A. Sharma (✉)
Department of Computer Science, BU, Bhopal, India
e-mail: ashoksharmamca@yahoo.co.in; myofficialid@yahoo.com

R.S. Thakur
MANIT, Bhopal, India
e-mail: ramthakur2000@yahoo.com

S. Jaloree
SATI, Vidisha, India
e-mail: rsjaloree@gmail.com

© Springer Nature Singapore Pte. Ltd. 2018
Y.-C. Hu et al. (eds.), *Intelligent Communication and Computational Technologies*, Lecture Notes in Networks and Systems 19,
https://doi.org/10.1007/978-981-10-5523-2_35

resources. However, in Public Cloud major concern of data security has refrained the Potential IT users from adopting the Public Cloud storages [1–4].

In case of data security, various symmetric algorithms are available in Standard Literature which includes DES, 3DES, AES, RSA, RC6, Blowfish, Twofish, etc., and it has been seen most of the cloud storage providers are using AES-256. The findings from Literature survey are very astonishing that most of the Cloud Storage providers are using AES-256 for securing data in Cloud [5–7]. In this paper, we have presented a framework to investigate the efficiency of five symmetric algorithms in terms of Encryption Time, Decryption Time, Memory Utilization, and Throughput [8–10].

2 Framework for Investing the Optimized Symmetric Algorithms in Cloud

The framework has mainly six components namely key store, set of integrated algorithm, data files storage, EDaS, customers.

Before moving any type of data files in Cloud Storages, all files are forwarded to crypter tool deployed in Amazon Cloud and for encryption of files, one by one; each algorithm is selected by user along with the key of fixed size or variable size depending upon the case. The encrypted text files of different sizes/fixed sizes are stored in Cloud storage, and the encryption time Et and log details get stored in log storage and in decryption process, decrypted time D_t is recorded in case of each algorithm (Fig. 1).

In first phase, for encryption process, each algorithm is tested for the input files of different sizes and different formats with key of fixed size and corresponding log details get stored in log storage.

In second phase, for decryption process, each algorithm is tested for the input files of different sizes and different formats with key of fixed size and corresponding log details get stored in log storage.

In third phase, for encryption process, each algorithm is tested for the input files of fixed size with variable keys and corresponding log details get stored in log storage.

In the last phase, for decryption process, each algorithm is tested for the input files of fixed size with variable keys and corresponding log details get stored in log storage.

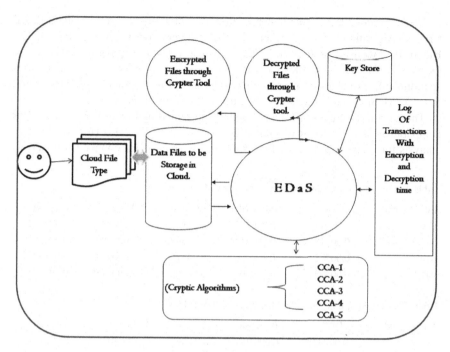

Fig. 1 Framework for investigating the behavior of symmetric algorithms

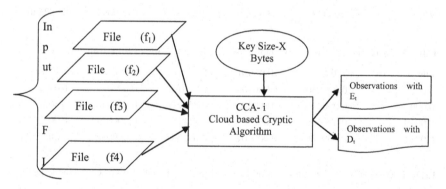

Fig. 2 Schematic flow of analysis work done for investigation of file size variation with fixed key size

3 Experimentation Process and Results

For experimentation process, we have conducted the investigation using a machine with the specifications: Processor—Intel i3-3230 M (3rd generation) with 2.6 GHz clock speed and RAM-2 GB.

Figure 2, represents the experimentation flow performed with respect to variation in parameter for a particular cryptic algorithm deployed in Cloud.

CCA-i represents one particular cryptic algorithm from a set of Cloud-Based Cryptic Algorithms (CCA), where in the present case CCA = {AES, 3DES, Blowfish, Twofish, RC6}. The CCA-i is chosen w.r.t. a particular key size (X-bytes). The variety of files (of different sizes) are supplied to CCA and corresponding effect of algorithm is recorded (in terms of Encryption time E_t, Decryption time D_t, Memory utilization, Throughput) in the form observations.

The outcomes of experimentation done is presented in Tables 1, 2, 3, and 4 (for comparative analysis for text input files); Tables 5, 6, 7, and 8 (for comparative analysis for image input files); Tables 9, 10, 11, and 12 (for comparative analysis

Table 1 Encryption time taken by algorithms with fixed key

File size (kb)	3DES	Blowfish	RC6	AES	Twofish
9	0.000441	0.000107	0.051372	0.000087	0.002332
38	0.0001819	0.000417	0.215909	0.000300	0.009452
61	0.002827	0.000650	0.345756	0.000453	0.001501
94	0.004384	0.000997	0.929664	0.000715	0.002296

Table 2 Decryption time taken by algorithms with fixed key

File size (kb)	3DES	Blowfish	RC6	AES	Twofish
9	0.000429	0.000102	0.048837	0.000078	0.002139
38	0.001772	0.000388	0.206466	0.000293	0.008904
61	0.002834	0.000612	0.33329	0.000449	0.014308
94	0.004392	0.000936	0.511467	0.000694	0.022266

Table 3 Encryption time taken by algorithms with fixed text file of 9 kb with variable key

File size (kb)	3DES	Blowfish	RC6	AES	Twofish
0.008	0.000430	0.000106	0.050767	0.000083	0.002271
0.009	0.000426	0.000105	0.050712	0.000077	0.002227
0.0010	0.000426	0.000104	0.051185	0.000077	0.002213
0.0011	0.000429	0.000107	0.051587	0.000076	0.002258

Table 4 Decryption time taken by algorithms with fixed text file of 9 kb with variable key

Key size (kb)	3DES	Blowfish	RC6	AES	Twofish
0.008	0.004600	0.000105	0.047965	0.00090	0.047965
0.009	0.00443	0.000102	0.048753	0.00077	0.048753
0.0010	0.00428	0.000103	0.048098	0.00078	0.048098
0.0011	0.00429	0.000105	0.048797	0.00077	0.048797

Table 5 Encryption time taken by algorithms with fixed key size

File size (kb)	3DES	Blowfish	RC6	AES	Twofish
20	0.001987	0.00456	0.240671	0.000435	0.010307
19	0.002731	0.00634	0.331026	0.000327	0.014319
225	0.030305	0.006738	5.170204	0.004665	0.158268
242	0.032781	0.007266	3.970895	0.005039	0.169557

Table 6 Decryption time taken by algorithms with fixed key size

File size (kb)	3DES	Blowfish	RC6	AES	Twofish
20	0.002017	0.000432	0.225164	0.000318	0.009802
19	0.002810	0.000598	0.314295	0.000455	0.013941
225	0.030604	0.006432	3.493699	0.004699	0.153527
242	0.033052	0.006976	3.778684	0.005122	0.166428

Table 7 Encryption time taken by algorithms with variable key size

File size (kb)	3DES	Blowfish	RC6	AES	Twofish
0.008	0.002011	0.000463	0.240892	0.000321	0.010253
0.009	0.001947	0.000472	0.239001	0.000319	0.010224
0.0010	0.002004	0.000458	0.238568	0.000318	0.010179
0.0011	0.002014	0.000459	0.239194	0.000330	0.010305

Table 8 Decryption time taken by algorithms with fixed image file with variable key files

Key size (kb)	3DES	Blowfish	RC6	AES	Twofish
0.008	0.001981	0.000429	0.230787	0.000324	0.010129
0.009	0.002007	0.00043	0.228417	0.000318	0.010145
0.0010	0.001982	0.000428	0.228192	0.000315	0.009952
0.0011	0.001984	0.000425	0.229594	0.000319	0.009908

Table 9 Encryption time taken by algorithms with fixed key

File size (kb)	AES	Blowfish	3DES	RC6	Twofish
9	0.000223	0.000317	0.001437	0.166067	0.007242
10	0.000237	0.000348	0.001466	0.175565	0.007594
13	0.000309	0.000427	0.001815	0.218299	0.009609
19	0.000416	0.000576	0.002583	0.310257	0.013468

Table 10 Decryption time taken by algorithms with fixed key

File size (kb)	3DES	Blowfish	3DES	RC6	Twofish
9	0.000209	0.000298	0.001416	0.157980	0.007013
10	0.000228	0.000306	0.001488	0.166228	0.007462
13	0.000285	0.000390	0.001819	0.205997	0.009284
19	0.000412	0.000553	0.002577	0.294764	0.01314

Table 11 Encryption time taken by algorithms with fixed audio files of 9 kb with variable key

File size (kb)	3DES	Blowfish	3DES	RC6	Twofish
8	0.00295	0.000404	0.001838	0.217727	0.009557
9	0.000272	0.000388	0.001819	0.217961	0.009381
10	0.000289	0.000390	0.001841	0.217961	0.009395
11	0.000275	0.000411	0.001827	0.217126	0.009454

Table 12 Decryption time taken by algorithms with fixed audio files of 9 kb with variable key files

Key size (kb)	3DES	Blowfish	3DES	RC6	Twofish
8	0.000283	0.000399	0.001842	0.206077	0.008804
9	0.000280	0.000381	0.001803	0.203170	0.009055
10	0.000277	0.000383	0.001816	0.204957	0.009042
11	0.000305	0.000390	0.001859	0.205468	0.008887

for audio input files); and Tables 13, 14, 15, and 16 (for comparative analysis for video input files)

In this section, we are investigating the performance of AES, 3DES, Blowfish, Twofish, RC6 ciphers on text files of different sizes with fixed/variable key.

Table 13 Encryption time taken by algorithms with fixed key

File size (kb)	3DES	Blowfish	3DES	RC6	Twofish
35.35	0.000742	0.001014	0.004787	0.550354	0.023882
63.28	0.0001365	0.001857	0.008754	0.997955	0.043841
75.06	0.0001617	0.002171	0.010467	1.196142	0.052912
121.86	0.0002621	0.003620	0.017150	1.964087	0.086838

Table 14 Decryption time taken by algorithms with fixed key

File size (kb)	AES	Blowfish	3DES	RC6	Twofish
35.35	0.000785	0.001059	0.004780	0.576682	0.024642
63.28	0.001309	0.001902	0.008714	1.052934	0.044923
75.06	0.001618	0.002316	0.010430	1.261896	0.053677
121.86	0.002646	0.003647	0.017116	2.064917	0.088245

Table 15 Encryption time taken by algorithms with fixed video files

File size	3DES	Blowfish	3DES	RC6	Twofish
8 byte	0.000733	0.0001078	0.00478	0.578464	0.024826
9 byte	0.000745	0.0001066	0.004771	0.576448	0.024833
10 byte	0.000756	0.0001074	0.00473	0.577740	0.024897
11 byte	0.000745	0.0001068	0.004795	0.577162	0.025026

Table 16 Decryption time taken by algorithms with fixed video files

Key size	3DES	Blowfish	3DES	RC6	Twofish
8 byte	0.00758	0.001006	0.004747	0.542694	0.024327
9 byte	0.00746	0.001013	0.004785	0.546983	0.024198
10 byte	0.00738	0.001017	0.004815	0.544347	0.023991
11 byte	0.00741	0.001013	0.004785	0.546527	0.024025

Initially, data files with sizes 9, 38, 61, 94 kb with key size of 8 bytes are given as input.

For the supplied input of different text files with fixed key and fixed text files with different key files to crypter tool, result obtained is tabulated for AES, 3DES, Blowfish, RC6, and Twofish, and is depicted in Tables 1, 2, 3, and 4.

3.1 Comparative Analysis of Image Files for Usage Optimizations

In this section, we are investigating the performance of AES, 3DES, Blowfish, Twofish, RC6 ciphers on image files of different sizes with fixed/variable key.

Input: data files with sizes 19, 20, 224, 242 kb and Key size: with fixed key of 0.008 kb.

For the supplied input of different image files with fixed key and different key files with fixed image file to crypter tool, result obtained is tabulated for AES, 3DES, Blowfish, RC6, and Twofish, and is depicted in Tables 5, 6, 7, and 8.

3.2 Comparative Analysis of Audio Files for Usage Optimization

In this section, we are investigating the performance of AES, 3DES, Blowfish, Twofish, RC6 ciphers on audio files of different sizes with fixed/variable key.

Input: data files with sizes 9, 10, 13, 19 kb with fixed key of 8 kb

For the supplied input of different audio files with fixed key and different key files with fixed audio file to crypter tool, result obtained is tabulated for AES, 3DES, Blowfish, RC6, and Twofish, and is depicted in Tables 9, 10, 11, and 12.

3.3 Comparative Analysis of Video Files for Usage Optimization

In this section, we are investigating the performance of AES, 3DES, Blowfish, Twofish, RC6 ciphers on video files of different sizes with fixed/variable key.

Input: data files with sizes 35.35, 63.28, 75.06, 121.86 kb with fixed key of 8 kb

For the supplied input of different video files with fixed key and different key files with fixed video file to crypter tool, result obtained is tabulated for AES, 3DES, Blowfish, RC6, and Twofish, and is depicted in Tables 13, 14, 15, and 16.

4   Conclusion

i. In case of encryption time consumption, AES and Blowfish must be preferred among all.
ii. Decryption time is more in comparison to encryption time taken by ciphers.
iii. In case of encryption time consumption, AES and Blowfish must be preferred among all in case of fixed size text document.
iv. In case of encryption and decryption time consumption, AES and Blowfish must be preferred among all in case of image and audio files.
v. In case of encryption and decryption time consumption, AES and Blowfish must be preferred among all in case of video files.

5   Future Work

This paper discusses the scope of improvement for using optimized symmetric key algorithm for Cloud environment. After careful investigation of candidate Algorithms, we have optimized the implementation of algorithm for deciding the use of efficient algorithm in Cloud depending upon the outcomes of investigations. Since AES is secure among all symmetric algorithms but in terms of some parameters like memory utilization and throughput is not better than Blowfish. However, we can look into improving AES by replacing conventional key expansion routine with the genetic concept routine in the AES.

References

1. R. Rajan, "Evolution of Cloud Storage as Cloud Computing Infrastructure Service", IOSR Journal of Computer Engineering, vol. 1, no. 1, pp. 38–45, 2012.
2. D.J. Abadi, "Data management in the Cloud: limitations and opportunities", Bulletin of the IEEE Computer Society Technical Committee on Data Engineering, vol. 32, pp. 3–12, 2009.
3. S. Kumar and R.H. Gouda, "Cloud Computing: Research Issues, Challenges, Architecture, Platforms and Applications: A Survey", International Journal of Future Computer and Communication, vol. 1, no. 4, pp. 356–361, 2012.
4. P. Singh and A. Jain, "Survey paper on Cloud Computing," International Journal of innovations in Engineering and Technology (IJIET), vol. 3, no. 4, pp. 84–89, 2014.
5. D. Rani and R. Ranjan, "A comparative study of SaaS, PaaS and IaaS in Cloud Computing", International Journal of Advanced Research in Computer science and software Engineering, vol. 4, no. 6, pp. 458–462, 2014.
6. J. Ding, L. Sha and X. Chen, "Modeling and evaluating IaaS Cloud using performance evaluation process Algebra", in 22nd Asia-Pacific Conference in Communications (APCC), 2016.
7. T. Karnwal, T. Sivakumar and G. Aghila, "Cloud Services in Different Cloud Deployment Models: An Overview", International Journal of Computer Applications, vol. 34, no. 8, pp. 30–36, 2011.

8. A. Sharma, R.S. Thakur and S. Jaloree, "Investigation of Efficient Cryptic Algorithm for Image Files Encryption in Cloud", International Journal of Scientific Research in Computer Science and Engineering, vol. 4, no. 5, pp. 5–11, 2016.
9. A. Sharma, R.S. Thakur and S. Jaloree, "Investigation of Efficient Cryptic Algorithm for Cloud Storage", in Fourth International Conference on Recent Trends in Communication and Computer Networks ComNet Narosa Publication, pp. 58–63, Nov 2016.
10. A. Sharma, R.S. Thakur and S. Jaloree, "Investigation of Efficient Cryptic Algorithm for Video Files Encryption in Cloud", International Journal of Scientific Research in Computer Science and Engineering, vol. 4, no. 6, pp. 8–14, 2016.

Author Index

A
Adhao, Asmita Sarangdhar, 157
Ahmad, Tauseef, 237
Arefin, Sayed E., 43
Arora, Chetan, 325
Arora, Nikhil, 325
Aruna, N.S., 119
Asawa, Krishna, 3
Ashrafi, Tasnia H., 43

B
Bansal, Gaurav, 247
Baranwal, Gaurav, 287
Bhesdadiya, Rajnikant, 191
Bhimani, Purvi, 15

C
Chakrabarty, Amitabha, 43
Chaturvedi, Abhay, 247
Chaudhari, Rupal R., 345
Chopra, Deepti, 259
Choudhary, Aashish, 325

D
Das, Asit Kumar, 97
Das, Kowshik D.J., 43
Das, Priyanka, 97
Das, Ujjal Kumar, 141
Dilip, Ladumor, 191

G
Gadag, Mahesh, 181
Gadag, Nikit, 181
Goyal, Puneet Kumar, 77
Goyal, Vishal, 357

H
Haque, Misbahul, 237
Hariharan, S., 107, 119

Hossain, Md. Arshad, 43

I
Imran, Mohd., 237

J
Jain, Shikha, 371
Jaloree, Shailesh, 381
Jangir, Pradeep, 191
Jhaveri, Rutvij H., 213
Johri, Era, 129
Joshi, Shreedhar, 181

K
Kamalesh, V.N., 203
Kaur, Arvinder, 259
Keserwani, Pankaj Kumar, 141
Khare, Ashish, 85
Kolhe, Mohan L., 77
Kumar, Sachin, 23
Kuwor, Priyanka, 129

M
Maiti, Prasenjit, 57
Makawana, Pooja R., 213
Malhotra, Sachin, 171
Mavani, Monali, 3
Mishra, Puneet, 357
Mulla, C., 67

N
Naik, Aditya Kumar, 287
Nethravathi, B., 203

P
Panchal, Gaurang, 15, 35
Pant, Bhasker, 295
Patel, Ami A., 345
Patel, Hiral M., 345

© Springer Nature Singapore Pte. Ltd. 2018
Y.-C. Hu et al. (eds.), *Intelligent Communication and Computational
Technologies*, Lecture Notes in Networks and Systems 19,
https://doi.org/10.1007/978-981-10-5523-2

Patel, Jaimin, 35
Patel, Jayesh, 333
Pathak, Shantanu, 305
Pawar, Vijaya Rahul, 67, 157
Popat, Dhaval, 129
Prajapati, Kinjal R., 345

R
Raghuwanshi, Sandeep K., 371
Rakshit, Soubhik, 97
Rao, D. Rajeshwar, 305
Raut, R., 67

S
Sahoo, Bibhudatta, 57, 315
Samaddar, Shefalika Ghosh, 141
Satpathy, Suchismita, 57
Sawant, Tejashri, 227
Sen, Arnaja, 129
Shah, Hardik, 129
Sharma, Ashok, 381
Shrimali, Manish, 77
Singh, Devesh Pratap, 295

Singh, Narinder, 269
Singh, S.B., 269
Singh, Vijay, 295
Singh, Vikas Kumar, 77
Sinha, Adwitiya, 325
Sirsikar, Sumedha, 227
Sreeja, P., 107
Srivastava, Prashant, 85
Suthar, Krunal, 333
Swami, Devang, 315

T
Tanwar, Sudeep, 23
Thakur, Ramjeevan Singh, 381
Trivedi, Indrajit, 191
Trivedi, Munesh C., 77, 171
Turuk, Ashok Kumar, 57
Tyagi, Sudhanshu, 23

V
Varma, H., 67
Varshney, Aarti, 357

Printed in the United States
by Bookmasters

Printed in the United States
By Bookmasters